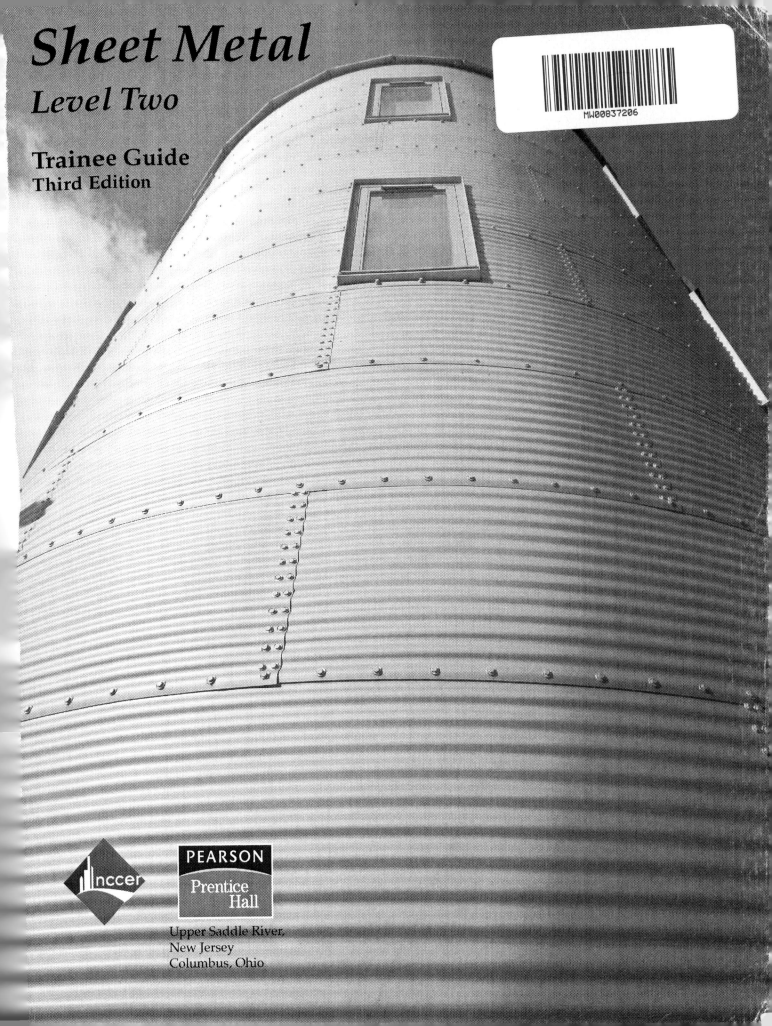

Sheet Metal
Level Two

Trainee Guide
Third Edition

nccer

PEARSON
Prentice Hall

Upper Saddle River,
New Jersey
Columbus, Ohio

National Center for Construction Education and Research

President: Don Whyte
Director of Curriculum Revision and Development: Daniele Stacey
Sheet Metal Project Managers: Carla Sly and Angela Jonas
Production Manager: Tim Davis
Quality Assurance Coordinator: Debie Ness
Editors: Rob Richardson and Matt Tischler
Desktop Publishing Coordinator: James McKay
Production Assistant: Brittany Ferguson

NCCER would like to acknowledge the contract service provider for this curriculum: MetaMedia Training International, Germantown, Maryland.

This information is general in nature and intended for training purposes only. Actual performance of activities described in this manual requires compliance with all applicable operating, service, maintenance, and safety procedures under the direction of qualified personnel. References in this manual to patented or proprietary devices do not constitute a recommendation of their use.

10 9 8 7 6 5 4 3 2
ISBN 0-13-604484-0
ISBN 13 978-0-13-604484-0

PREFACE

TO THE TRAINEE

By choosing sheet metal training, you are taking a step toward a satisfying and rewarding career. Continuing your craft education is important, as technology and the materials with which you'll work are changing all the time. Sheet metal workers mainly work in heating, ventilation, air conditioning, and refrigeration (HVACR), but there are many other opportunities. You might help build airplanes, automobiles, or even billboards. You might install hoods and vents for restaurants, or build grain silos for farmers. You could also express an artistic talent by constructing striking ceilings and roofs for commercial buildings.

There were approximately 200,000 sheet metal workers in the United States in 2004, according to the Bureau of Labor and Statistics. By 2014, the demand is estimated to grow by more than 10%. Careers can progress from ductwork installer, to fabrication foreman, to architectural sheet metal designer.

You'll discover that a career in sheet metal work can be financially rewarding. Average wages start around $37,000, and a sheet metal foreman often earns twice what foremen in other construction trades earn.

This is the second installment of a four-level curriculum that meets the requirements of a standard sheet metal apprenticeship program (4 years and 8,000 hours of on-the-job training). If you complete all four levels, you will earn NCCER credentials as a Sheet Metal craft worker. NCCER credentials are increasingly being required for construction projects across the country, and credentialed craft workers often earn more than other qualified persons. We wish you the best as you begin your construction career.

We invite you to visit the NCCER website at www.nccer.org for the latest releases, training information, newsletter, and much more. You can also reference the Contren® product catalog online at www.nccer.org.

Your feedback is welcome. You may email your comments to curriculum@nccer.org or send general comments and inquiries to info@nccer.org.

CONTREN® LEARNING SERIES

The National Center for Construction Education and Research (NCCER) is a not-for-profit 501(c)(3) education foundation established in 1995 by the world's largest and most progressive construction companies and national construction associations. It was founded to address the severe workforce shortage facing the industry and to develop a standardized training process and curricula. Today, NCCER is supported by hundreds of leading construction and maintenance companies, manufacturers, and national associations. The Contren® Learning Series was developed by NCCER in partnership with Pearson Education, Inc., the world's largest educational publisher.

Some features of NCCER's Contren® Learning Series are as follows:

- An industry-proven record of success
- Curricula developed by the industry for the industry
- National standardization, providing portability of learned job skills and educational credits
- Compliance with the Office of Apprenticeship requirements for related classroom training (*CFR 29:29*)
- Well-illustrated, up-to-date, and practical information

NCCER also maintains a National Registry that provides transcripts, certificates, and wallet cards to individuals who have successfully completed modules of NCCER's Contren® Learning Series. *Training programs must be delivered by an NCCER Accredited Training Sponsor in order to receive these credentials.*

Contents

Contren® Curricula

NCCER's training programs comprise over 50 construction, maintenance, and pipeline areas and include skills assessments, safety training, and management education.

Boilermaking
Cabinetmaking
Carpentry
Concrete Finishing
Construction Craft Laborer
Construction Technology
Core Curriculum:
   Introductory Craft Skills
Drywall
Electrical
Electronic Systems Technician
Heating, Ventilating, and
   Air Conditioning
Heavy Equipment Operations
Highway/Heavy Construction
Hydroblasting
Industrial Maintenance
   Electrical and Instrumentation
   Technician
Industrial Maintenance Mechanic
Instrumentation
Insulating
Ironworking
Masonry
Millwright
Mobile Crane Operations
Painting
Painting, Industrial
Pipefitting
Pipelayer
Plumbing
Reinforcing Ironwork
Rigging
Scaffolding
Sheet Metal
Site Layout
Sprinkler Fitting
Welding

Pipeline
Control Center Operations,
   Liquid
Corrosion Control
Electrical and
   Instrumentation
Field Operations, Liquid
Field Operations, Gas
Maintenance
Mechanical

Safety
Field Safety
Safety Orientation
Safety Technology

Management
Introductory Skills for the
   Crew Leader
Project Management
Project Supervision

Spanish Translations
Andamios
Currículo Básico
   Habilidades
   Introductorias
   del Oficio
Instalación de Rociadores
   Nivel Uno
Orientación de Seguridad
Seguridad de Campo

Supplemental Titles
Applied Construction Math
Careers in Construction

Acknowledgments

This curriculum was revised as a result of the farsightedness
and leadership of the following sponsors:

Ivey Mechanical
Ohio Valley CEF
Cinfab LLC
New England Mechanical Services
Metropolitan Community Colleges
Associated Builders and Contractors – Heart of America Chapter
Construction Education Center

This curriculum would not exist were it not for the dedication and
unselfish energy of those volunteers who served on the Authoring Team.
A sincere thanks is extended to the following:

Douglas Grey
John Hasselbeck
Donald Poulin
James Wright

NCCER PARTNERING ASSOCIATIONS

American Fire Sprinkler Association
Associated Builders and Contractors, Inc.
Associated General Contractors of America
Association for Career and Technical Education
Association for Skilled and Technical Sciences
Carolinas AGC, Inc.
Carolinas Electrical Contractors Association
Center for the Improvement of Construction
    Management and Processes
Construction Industry Institute
Construction Users Roundtable
Design-Build Institute of America
Electronic Systems Industry Consortium
Merit Contractors Association of Canada
Metal Building Manufacturers Association
NACE International
National Association of Minority Contractors

National Association of Women in Construction
National Insulation Association
National Ready Mixed Concrete Association
National Systems Contractors Association
National Technical Honor Society
National Utility Contractors Association
NAWIC Education Foundation
North American Crane Bureau
North American Technician Excellence
Painting and Decorating Contractors of America
Portland Cement Association
SkillsUSA
Steel Erectors Association of America
Texas Gulf Coast Chapter ABC
U.S. Army Corps of Engineers
University of Florida
Women Construction Owners and Executives, USA

04201-08

Trade Math Two

04201-08
Trade Math Two

Topics to be presented in this module include:

Overview

When you first began doing sheet metal work, it may not have seemed obvious how you would use math skills on the job. But at this point in your career you probably realize how often you apply mathematics in your trade. Math is used in layout and duct design. It is also used to calculate angles and distances between points. The math you learn and work with as a sheet metal worker is not abstract and theoretical, but applies to solving work-related problems. When you see this connection, you may find it easier to learn than you thought.

Objectives

When you have completed this module, you will be able to do the following:

1. Apply mathematical formulas to solve problems.
2. Solve problems sequentially with simple equations.
3. Solve linear, circular, and angle measurement problems.
4. Solve percentage problems.
5. Define and solve ratio and proportion problems.
6. Use a protractor, a vernier caliper, and a micrometer.
7. Calculate selected seam allowances.
8. Demonstrate competence in solving selected field measuring problems.
9. Apply standard rules and practice for solving selected field measurement problems.

Trade Terms

Coefficient	Hypotenuse
Complement	Ratio
Cosine	Sine
Fractional percentage	Tangent
Function	

Required Trainee Materials

1. Pencil and paper
2. Protractor
3. Calculator
4. Micrometer
5. Vernier caliper

Prerequisites

Before you begin this module, it is recommended that you successfully complete *Core Curriculum* and *Sheet Metal Level One*.

This course map shows all of the modules in the second level of the Sheet Metal curriculum. The suggested training order begins at the bottom and proceeds up. Skill levels increase as you advance on the course map. The local Training Program Sponsor may adjust the training order.

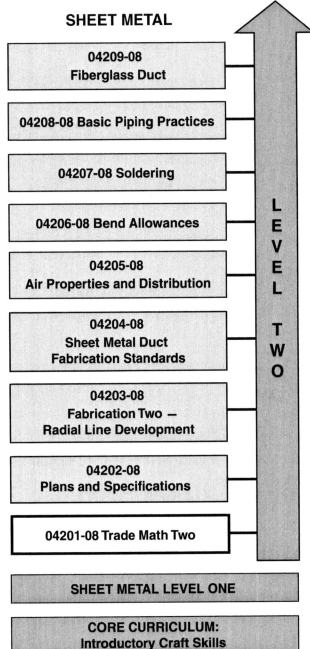

SHEET METAL

04209-08
Fiberglass Duct

04208-08 Basic Piping Practices

04207-08 Soldering

04206-08 Bend Allowances

04205-08
Air Properties and Distribution

04204-08
Sheet Metal Duct
Fabrication Standards

04203-08
Fabrication Two —
Radial Line Development

04202-08
Plans and Specifications

04201-08 Trade Math Two

LEVEL TWO

SHEET METAL LEVEL ONE

CORE CURRICULUM:
Introductory Craft Skills

201CMAP.EPS

1.0.0 ◈ INTRODUCTION

As a professional in the sheet metal trade, you will use applied mathematics on the job every day. Once you learn the basic principles, applied math is not very difficult and is useful to everything that you do. The exercises and practice drills in this module are designed to help you practice solving the types of math problems you will find at work. To be successful at trade math, learn the basic principles, ask for help when you need it, and practice, practice, practice.

2.0.0 ◈ FORMULAS AND SYMBOLS

You will solve many construction specification problems by using formulas. Formulas are algebraic equations made up of letters that represent values and symbols (mathematical signs, such as + and ×) that tell you what to do. Study the following formula for calculating area:

$$a = \pi r^2$$

To read this formula, you would say, "Area is equal to pi times radius squared." When you see an expression such as πr^2 in a formula, it means that you must multiply the values. In this case, you would multiply the radius squared by π.

Some values in formulas are variables. Any number can replace a variable. In this sample formula, r is a variable. No matter what size a circle is, and thus no matter what its radius is, we can determine its area with this formula. Some values, however, are constant. Pi is a constant. Its value is always 3.14. Whenever you see the word pi or its symbol (π) in a formula, you know to replace it with 3.14.

For this example, let's say that the radius (r) is 2 inches. You can plug this number into the equation in place of the letter r. Look at the formula with the numbers that represent the variable and the constant plugged into the equation:

$$a = (3.14)(2^2)$$

Notice that 3.14 and 2^2 have been placed inside parentheses. When numbers are enclosed in parentheses, you must finish any calculations inside each set of parentheses before completing any other calculations. In this case, you will first multiply 2×2:

$$a = (3.14)(4)$$

$$a = 12.56 \text{ square inches}$$

The parentheses in this equation are a type of grouping symbol. Grouping symbols are just like punctuation in writing. They help you make sense

DID YOU KNOW?
Variables and Constants

It can seem confusing at first to look at a formula with letters, numbers, and symbols in it, but formulas are really just a shorthand way of writing mathematical statements. You can create your own simple formula. For example, if you pay Sam $5.00 per hour, the $5.00 is a constant. Unless you raise Sam's wages, it will not change from week to week. However, Sam works a different number of hours each week. The number of hours is the variable. You can construct a simple formula to calculate Sam's weekly salary:

$$S = 5H$$

S stands for salary, $5.00 is the hourly pay rate, and H is the number of hours worked. Once you understand what the letters and symbols mean, a formula becomes easier to read. All formulas, even the complex ones, are logical statements. Take each formula a step at a time until you understand how it works.

of an equation, just like punctuation helps you make sense of a sentence. Grouping symbols tell you which numbers belong together and which **functions** to perform first. It's important to pay attention to how terms are grouped in a formula and to do the calculations in the right order. In more complex problems additional types of grouping symbols, such as square brackets [ ] and braces { }, are used.

2.1.0 Fractions Refresher

A fraction divides whole units into parts. Common fractions are written as two numbers separated by a slash or by a horizontal line. The top number is the numerator. The bottom number is the denominator. Equivalent fractions have the same value or are equal. For example, ½, ¾, ⅝, and ⁹⁄₁₆ are all equivalent. When you are working with fractions, it is often best to reduce them to lowest terms. If there is no number (other than 1) that will divide evenly into the numerator and the denominator, the fraction is already expressed in lowest terms.

Add and subtract fractions by finding the lowest common denominator and then adding or subtracting the numerators. Then reduce the resulting fraction to the lowest terms. Multiply fractions by multiplying one numerator by the other and one denominator by the other, and reduce the resulting fraction to lowest terms. To divide fractions, invert one fraction and multiply the top numbers and then multiply the bottom numbers. Then reduce the resulting fraction to lowest terms.

Like fractions, decimals represent values that are less than one whole unit. You are already familiar with decimal points in the form of money:

$$25 \text{ cents} = 0.25 \text{ or } {}^{25}\!/_{100}$$

$$10 \text{ cents} = 0.10 \text{ or } {}^{10}\!/_{100}$$

$$50 \text{ cents} = 0.50 \text{ or } {}^{50}\!/_{100}$$

To convert a fraction to a decimal, divide the numerator by the denominator: $\frac{3}{4} = 0.75$. To convert decimals into fractions, think of the numbers to the right of the decimal point as being numerators above an invisible denominator. Take 2.75, for example. Rewrite it as ${}^{275}\!/_{100}$. Now reduce the fraction to its lowest terms, and you end up with $2\frac{3}{4}$. Remember to use the same number of zeros in the denominator as there are numbers to the right of the decimal point. In some cases, you will have to round up or down.

Notice the relationship between decimals and fractions in the previous examples. Think of a whole number divided into 100 parts. You can express any part of the whole as a percentage. In the above examples, 0.25 is 25 percent of 100, 0.10 is 10 percent of 100, and 0.50 is 50 percent of 100.

To find the percentage of a given number, think of the percentage as a decimal. For example, 4 percent is 0.04, 44 percent is 0.44, and 444 percent is 4.44. Simply multiply the given number by the decimal equivalent of the percentage.

2.1.1 Study Problems: Adding, Subtracting, Multiplying, and Dividing Fractions

Add the following fractions.

1. $\frac{1}{8} + \frac{3}{8} = $ _____

2. $\frac{1}{4} + \frac{3}{8} + \frac{9}{16} = $ _____

3. $\frac{2}{3} + \frac{7}{8} + {}^{19}\!/_{16} = $ _____

4. $\frac{7}{12} + \frac{4}{9} + \frac{1}{32} = $ _____

5. $\frac{1}{15} + \frac{1}{10} + \frac{1}{64} = $ _____

Multiply the following fractions.

6. $\frac{3}{8} \times {}^{15}\!/_{16} = $ _____

7. $\frac{2}{32} \times \frac{4}{16} = $ _____

8. $\frac{3}{16} \times {}^{24}\!/_{16} = $ _____

9. $1{}^{13}\!/_{16} \times \frac{5}{8} = $ _____

10. $\frac{9}{16} \times \frac{7}{16} = $ _____

DID YOU KNOW?

Common Denominators

When you are adding and subtracting fractions, they must have the same denominator. This is known as finding the lowest common denominator, or LCD. Let's say you want to add $\frac{1}{3} + \frac{3}{5} + \frac{5}{12}$. The three denominators of these fractions are 3, 5, and 12. To find the LCD, write down the first several multiples of each denominator; in other words, multiply each by 1, then 2, then 3, then 4, etc. Here's what that would look like for these fractions:

$\frac{1}{3}$: 3, 6, 9, 12, 15, 18, 21, 24, 27, 30, 33, 36, 39, 42, 45, 48, 51, 54, 57, 60

$\frac{3}{5}$: 5, 10, 15, 20, 25, 30, 35, 40, 45, 50, 55, 60

$\frac{5}{12}$: 12, 24, 36, 48, 60

Continue this process until you have a number that is common in each list. In this case, 60 is the smallest number that is common to all three lists, so 60 is the LCD for these three fractions.

Now convert each of the fractions to 60ths by following these steps. First look at the denominator in the fraction $\frac{1}{3}$. Multiply it by the number that will give you a product of 60: $3 \times 20 = 60$. Then, since any fraction with the same numerator and denominator is equal to 1, multiply ${}^{20}\!/_{20}$ by $\frac{1}{3}$. You wil get the fraction ${}^{20}\!/_{60}$.

Follow the same steps with $\frac{3}{5}$ an $\frac{5}{12}$. Here's what you should come up with:

$\frac{1}{3} \times {}^{20}\!/_{20} = {}^{20}\!/_{60}$

$\frac{3}{5} \times {}^{12}\!/_{12} = {}^{24}\!/_{60}$

$\frac{5}{12} \times \frac{5}{5} = {}^{25}\!/_{60}$

Now add the fractions together:

${}^{20}\!/_{60} + {}^{24}\!/_{60} + {}^{25}\!/_{60} = {}^{69}\!/_{60}$

Finally, simplify and reduce:

${}^{69}\!/_{60} = {}^{60}\!/_{60} + {}^{9}\!/_{60} = 1{}^{9}\!/_{60}$, or $1{}^{3}\!/_{20}$

NOTE:

Another way to find a common denominator—although not the *lowest* common denominator—would be simply to find the product of all three denominators ($3 \times 5 \times 12 = 180$) and convert the fractions to that denominator. Then simplify and reduce. The answer will be the same either way.

Divide the following fractions.

11. $\frac{7}{8} \div \frac{3}{4} = $ _____

12. $\frac{2}{32} \div \frac{7}{32} = $ _____

13. ${}^{17}\!/_{32} \div 1\frac{1}{8} = $ _____

14. $\frac{1}{64} \div \frac{8}{8} = $ _____

15. ${}^{20}\!/_{16} \div \frac{7}{16} = $ _____

2.1.2 Study Problems: Decimals and Percentages

Convert the following numbers into fractions to the nearest sixteenth. Reduce the fractions in your answers.

1. 2.875 = _____

2. 11.333 = _____

3. 1.562 = _____

4. 0.375 = _____

5. 1.187 = _____

Convert the following numbers into decimals or percentages as required.

6. 0.55 = _____

7. 7% = _____

8. 3.5% = _____

9. 1.77 = _____

10. 15% = _____

Find the following percentages.

11. 7% of $35.00 = _____

12. 3% of $900.00 = _____

13. 1% of $85.00 = _____

14. 4% of $16.00 = _____

15. 125% of $940.00 = _____

2.2.0 Sequence of Operations

You must perform mathematical operations in the right order to get the right answer. What would you do first if you had to solve a problem such as $3 + 3 \times 2 - 6 \div 3 + 6$? Would you move from left to right, taking each calculation in turn? In *Trade Math One* you learned the phrase My Dear Aunt Sally to remind you of the right sequence to follow for problems like this one. To get the correct answer, you multiply, divide, add, and subtract (MDAS), in that order. Whether you solve a problem by hand or with a calculator, you must follow the correct sequence of operations. Study the following example. Calculate the numbers shown in bold type in each step:

$3 + 3 \times 2 - 6 \div 3 + 6$

Multiply:

$3 + \mathbf{3 \times 2} - 6 \div 3 + 6$

To get:

$3 + 6 - 6 \div 3 + 6$

Divide:

$3 + 6 - \mathbf{6 \div 3} + 6$

To get:

$3 + 6 - 2 + 6$

Add:

$\mathbf{3 + 6} + \mathbf{6} - 2$

To get:

$15 - 2$

Subtract:

$\mathbf{15 - 2} = 13$

The MDAS sequence also applies when an algebraic expression appears in an equation. Look at the following problem:

$x = 5a[2(a + b)(b)]$

The equation $(a + b)$ with the **coefficient** 2b is multiplied first, resulting in the following:

$x = 5a[2ab + 2b^2]$

Then the terms within the bracket are multiplied by a, resulting in the following:

$x = 5[2a^2b + 2ab^2]$

Finally the terms in the brackets are multiplied by the coefficient 5, resulting in the following:

$x = 10a^2b + 10ab^2$

2.3.0 Coefficients

A coefficient is a number directly in front of a variable, such as the 3 in 3x or the 5 in 5b. Writing 5b is the same thing as writing $5 \times b$. A coefficient may also be a plus or minus sign directly in front of a grouping symbol, such as the plus sign in this equation:

$+(2b - c).$

You will often find algebraic expressions inside grouping symbols. If there is no numerical coefficient outside of the grouping symbol, it means that + l or –1 is the coefficient. The following rules apply when removing grouping symbols preceded by coefficients:

Rule 1 – When the coefficient is +1

When the grouping symbol is preceded by a plus sign and no other coefficient, drop the grouping symbol. Doing this is the same as multiplying each term within the grouping symbol by +1.

$3x + (2b - d) = 3x + 2b - d$

Rule 2 – When the coefficient is –1

When the grouping symbol is preceded by a minus sign and no other coefficient, multiply each term within the symbol by (–1), then drop the grouping symbol.

$$3x - (2b - d) = 3x - 2b + d$$

Rule 3 – When the coefficient is a plus number or letter

When the grouping symbol is preceded by a plus numerical or literal (letter) coefficient, multiply each term within the symbol by the coefficient and drop the grouping symbol.

$$3x + 2a(2b - d) = 3x + 4ab - 2ad$$

Rule 4 – When the coefficient is a minus number or letter

When the grouping symbol is preceded by a minus numerical or literal coefficient, the grouping symbol may be removed by one of two methods:

Method 1

Multiply each term within the symbol by the coefficient and its sign and drop the grouping symbol.

$$3x - 6a(2b - d) = 3x - 12ab + 6ad$$

Method 2

Multiply each term within the symbol by only the coefficient and then proceed as in Rule 2.

$$3x - 6a(2b - d) = 3x - (12ab - 6ad)$$
$$3x - (12ab - 6ad) = 3x - 12ab + 6ad$$

Study the following example in which the algebraic expression is simplified by removing parentheses following these rules:

$$3a + 2(3b) - 3(4c - 5d) = 3a + 6b - 12c + 15d$$

2.4.0 Solving Simple Equations

To solve simple equations, follow these steps:

Step 1 Remove the grouping symbols.

Step 2 Eliminate the fractions.

Step 3 Group like terms.

Step 4 Add like terms.

Step 5 Transpose terms. Place unknown terms to the left of the equal sign. Place known terms to the right of the equal sign.

Step 6 Convert negative unknowns to positive by multiplying the entire equation by –1.

Step 7 Divide both sides by the coefficient and sign of the unknown term.

Study the following example to solve for x.

$$\frac{3(2x - 4)}{2} - \frac{6(3x + 5)}{3} = 0$$

Remove the grouping symbols by multiplying.

$$\frac{6x - 12}{2} - \frac{18x + 30}{3} = 0$$

Eliminate the fractions by dividing each term by its denominator.

$$3x - 6 - (6x + 10) = 0$$
$$3x - 6 - 6x - 10 = 0$$

Group like terms.

$$3x - 6x - 6 - 10 = 0$$

Add like terms.

$$-3x - 16 = 0$$

Transpose terms (unknowns to the left, knowns to the right.)

$$-3x - 16 + 16 = 0 + 16$$
$$-3x = 16$$

Convert negative unknowns to positive by multiplying the entire equation by –1.

$$-1(-3x = 16)$$
$$3x = -16$$

Divide both sides by the coefficient of the unknown:

$$x = -1\tfrac{6}{3}$$
$$x = -5\tfrac{1}{3}$$

2.4.1 Study Problems: Simple Equations

Solve for x:

1. $(3x - 7) - (x - 4) = 0$

   x = _____

2. $2(5x - 5) - 5(3x + 10) = 0$

   x = _____

3. $x - 3(x + 2) + 2(x + 1) = x - 4$

   x = _____

4. $\dfrac{4(x - 2)}{2} = \dfrac{3(6 - x)}{2}$

   x = _____

5. $3x(4 + 3) + 9(x - 2) = 0$

   x = _____

6. $\dfrac{4(3x - 4)}{2} - \dfrac{7(4x + 4)}{4} = 0$

   x = _____

7. $5(2x + 4) - 3(5x + 8) = 0$

   x = _____

8. $\dfrac{4x(6 + 3)}{12} + 6(x + 3) = 0$

   x = _____

9. $2x + 4(x - 3) + 2(x - 4) = x + 6$

   x = _____

10. $\dfrac{(8x - 4)}{4} + \dfrac{(10x + 5)}{5} = 0$

    x = _____

3.0.0 ◆ LINEAR MEASURE

In the English system of measurement, linear measure is expressed in inches and feet, which are further divided into fractional or decimal parts. You know that there are 12 inches in one foot and that there are 36 inches, or 3 feet, in one yard. If you want to convert inches to yards, you must first divide the number of inches by 12 and then divide that number by 3, or you can divide the number of inches by 36. Study the following example:

How many yards are in 72 inches?

   72 inches ÷ 12 inches = 6 feet ÷ 3 feet = 2 yards

or

   72 inches ÷ 36 inches = 2 yards

In the metric system, linear measure is expressed in meters, which are further divided into centimeters and millimeters. The metric system uses the number 10 as the basis for calculations. Although the United States has not completely converted to the metric system, many products and tools used in the U.S. are measured using the metric system. Some job specifications also require the use of this system, so using the metric system and learning how to convert from one system to the other will help you in your career.

Because the metric system is based on 10, calculations are fairly easy. There are 100 centimeters in a meter and 1,000 millimeters in a meter. One meter is equal to 39.37 inches. How many inches are there in one centimeter? To answer this question, divide 39.37 by 100. One centimeter equals 0.3937-inch (or 0.394 inch rounded off). To find how many inches are in one millimeter, divide 39.37 by 1,000. One millimeter equals 0.03937 inch (or 0.0394 inch rounded off). Notice that to solve each problem, you can simply move the decimal point to the left two places when you divide by 100 and three places when you divide by 1,000.

Riveting Ideas

Thinking Metric

Conversion tables, special calculators, and tools that are marked in both the English and the metric systems are available to help you measure and make conversions accurately. However, it's helpful to know how measurements in one system compare to measurements in the other. Study the following to help you learn to think metric.

- One millimeter is slightly more than ¹⁄₃₂-inch.
- One centimeter is approximately ⅜-inch.
- One meter is about 1¹⁄₁₀ yards.
- One kilometer is about ⅝ of a mile.

How many centimeters are in one inch? You know that one centimeter equals 0.3937 inch, so divide as follows:

1 centimeter ÷ 0.3937 inch = 2.54 centimeters

To find how many millimeters are in one inch, divide by 0.03937:

1 millimeter ÷ 0.03937 inch = 25.4 millimeters

The answers to these problems—2.54 centimeters and 25.4 millimeters—are called conversion factors. To convert an English measure to a metric measure, multiply the English measure by the appropriate metric conversion factor. Notice that when the problem involves a fraction, you must convert the fraction to a decimal and then multiply. Study the following examples.

5 inches = 12.7 centimeters (5 × 2.54 centimeters)

5 inches = 127 millimeters (5 × 25.4 millimeters)

¾-inch = 1.905 centimeters (0.75 × 2.54 centimeters)

¾-inch = 19.05 millimeters (0.75 × 25.4 millimeters)

3.1.0 Study Problems: Linear Conversions

Apply what you have learned to the following problems. Note that the conversion factor you will need to solve the problems is given at the beginning of each set of problems. Additional conversion factors are included in *Appendix B*.

Convert each measurement to millimeters.
(1 inch = 25.4 millimeters)

1. 0.25-inch = _____ mm

2. ⅝-inch = _____ mm

3. ²⁵⁄₃₂-inch = _____ mm

4. 7.375 inches = _____ mm

5. 3¾ inches = _____ mm

Convert each measurement to centimeters.
(1 inch = 2.54 centimeters)

6. 1⅝ inches = _____ cm

7. 39.37 inches = _____ cm

8. 12 inches = _____ cm

9. 30 inches = _____ cm

10. 18 inches = _____ cm

Convert each measurement to centimeters.
(1 foot = 30.48 centimeters)

11. 1½ feet = _____ cm

12. 8½ feet = _____ cm

13. 6⅔-foot = _____ cm

14. 5½ feet = _____ cm

15. 2 feet 6 inches = _____ cm

Convert each measurement to meters.
(1 foot = 0.3048 meter)

16. 2.5 feet = _____ m

17. 100 yards = _____ m

18. 16½ feet = _____ m

19. 660 feet = _____ m

20. 5,280 feet = _____ m

4.0.0 ◆ CIRCLES AND ARCS

You will work often with circles and arcs when developing layouts for fittings. The three measurements you will work with the most are the circumference, diameter, and radius. The circumference is the distance around a circle. The diameter is the length of a straight line that crosses from one side of the circle to the other through its center point. The radius is the length of a straight line from the center point of the circle to any point on the circumference.

To find the circumference of a circle when you know its diameter, use the following formula in which c = circumference, π = 3.14, and d = diameter.

$$c = \pi d$$

An arc is a segment or a fraction of a circle. You know that a complete circle has 360 degrees. So an arc that is 10 degrees is equal to ¹⁰⁄₃₆₀ of the circle. To determine the length of an arc, use the following formula where n is a variable representing the degrees of arc:

$$1 = \frac{n}{360}(\pi d)$$

4.1.0 Study Problems: Circles and Arcs

Answer the following questions (show all your work on a separate sheet).

1. Calculate the circumference of a pipe that is 7 inches in diameter (*Figure 1*).

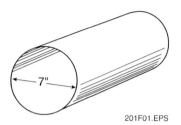

201F01.EPS

Figure 1 ◈ Pipe with 7-inch diameter.

Use *Figure 2* for Questions 2 and 3.

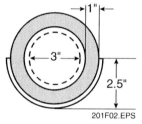

201F02.EPS

Figure 2 ◈ Cross section of a length of insulated pipe with a pipe shield.

2. What is the area of the insulation cross section? Given: Area of a circle = πr^2

3. What is the circumference of the inner surface of the pipe shield?

4. A 90-degree angle drawn from the center of a circle creates an arc with a length of 12 inches. What is the circumference of the circle?

5. A circle has a radius of 6 inches. What is the circumference?

6. What is the length of a 45-degree arc in a circle with a diameter of 8 inches?

7. What is the length of a 22.5-degree arc in a circle with a radius of 3 inches?

8. What is the combined length of two 45-degree arcs in a circle with a diameter of 15 inches?

9. What is the length of a 90-degree arc in a circle with a circumference of 72 inches?

10. What is the length of a 22.5-degree arc in a circle with a circumference of 400 inches?

5.0.0 ◈ ANGLE MEASURE

Two straight lines that meet at a point, or vertex, form an angle. The right triangle is one of the most important shapes in design and construction work. It is the basis for rectangular coordinates (*Figure 3*), which are part of a system for precision hole location in fabrication work.

A right triangle (*Figure 4*) contains one 90-degree angle. The side opposite the right angle is called the **hypotenuse** and is the longest side. Side BC is the altitude or height and side AC is the base. (If BC is the base, then AC becomes the altitude.)

Recall that an acute angle measures between 0 and 90 degrees. The sum of two acute angles of a right triangle is 90 degrees. Therefore, if one acute angle of a right triangle equals 50 degrees, the other angle must equal 40 degrees. Each angle, therefore, is the **complement** of the other angle.

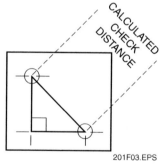

201F03.EPS

Figure 3 ◈ Rectangular coordinates.

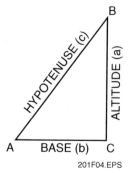

201F04.EPS

Figure 4 ◈ Right triangle.

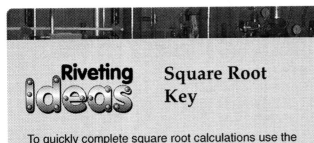

Riveting Ideas

Square Root Key

To quickly complete square root calculations use the square root key (marked with the symbol $\sqrt{\ }$) on your calculator.

The Pythagorean Theorem states the relationship between the three sides of a right triangle: the square of the hypotenuse is equal to the sum of the squares of the other two sides. As a formula, this theorem can be stated as follows:

$$c^2 = a^2 + b^2$$

Algebraic variations of the basic formula are as follows:

$$a^2 = c^2 - b^2$$
$$b^2 = c^2 - a^2$$

In the following example, a = 3 inches and b = 4 inches:

$$c^2 = a^2 + b^2$$
$$c^2 = 9 + 16$$
$$c^2 = 25$$

In the following example, c = 5 inches and b = 4 inches:

$$a^2 = c^2 - b^2$$
$$a^2 = 25 - 16$$
$$a^2 = 9$$

In the following example, c = 5 inches and a = 3 inches:

$$b^2 = c^2 - a^2$$
$$b^2 = 25 - 9$$
$$b^2 = 16$$

These formulas give the square of the sides of a triangle. Construction workers, however, are more interested in the side itself. In order to find the measurement of the side, you must take the square root of both sides of each formula as follows:

$$c^2 = a^2 + b^2 \text{ becomes } c = \sqrt{a^2 + b^2}$$
$$a^2 = c^2 - b^2 \text{ becomes } a = \sqrt{c^2 - b^2}$$
$$b^2 = c^2 - a^2 \text{ becomes } b = \sqrt{c^2 - a^2}$$

5.1.0 Study Problems: Angle Measure

Round your answers to the nearest tenth (0.1).

1. In *Figure 5*, solve for c.

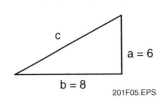

201F05.EPS

Figure 5 ◆ Study problem: solve for c.

2. Find the hypotenuse of a right triangle in which the base is 15 inches and the altitude is 7 inches.

3. In *Figure 6*, solve for b.

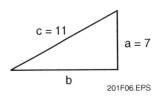

201F06.EPS

Figure 6 ◆ Study problem: solve for b.

4. Find the base of a right triangle in which the altitude is 9 feet and the hypotenuse is 10.3 feet.

5. In *Figure 7*, solve for a.

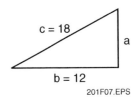

201F07.EPS

Figure 7 ◆ Study problem: solve for a.

6. Find the altitude of a right triangle in which the base is 12 feet and the hypotenuse is 17 feet.

7. Find the hypotenuse of a right triangle in which the base and altitude are each 9 inches.

8. Find the base of a right triangle in which the altitude is 15 inches and the hypotenuse is 25 inches.

_____

9. In *Figure 8*, solve for x.

_____

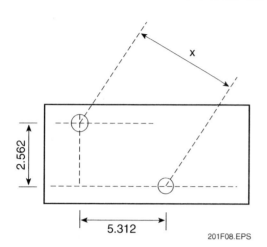

Figure 8 ◈ Study problem: solve for x.

10. In *Figure 9*, how long is the ladder? Take into consideration that it extends 3 feet beyond its top resting point, in accordance with OSHA standards.

_____

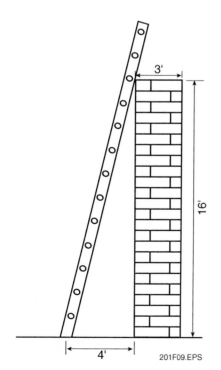

Figure 9 ◈ Study problem: how long is the ladder?

6.0.0 ◈ PERCENTAGE

Percentages are used to express a fractional part of a whole. The term *percentage* means per hundred. For example, you know that there are 100 pennies in one dollar. One hundred pennies is 100 percent of a dollar, so fifty pennies is 50 percent (or one-half) of a dollar.

Sometimes it is more convenient to express percentages as decimals, such as when you do calculations. To express any percentage as a decimal, divide by 100. Study the following examples. Because the % sign represents two decimal places, you can simply remove the % sign and move the decimal point two places to the left. Moving the decimal point is the same as dividing by 100.

10% = 0.10

25% = 0.25

7% = 0.07

75% = 0.75

1% = 0.01

Fractional percentages are less than 1 percent. You must take an extra step when expressing fractional percentages. For example, how would you express ¼ percent as a decimal? First, change the percentage to a fraction by dividing it by 100:

¼% = ¼ ÷ 100 = ¼ × ¹⁄₁₀₀ = ¹⁄₄₀₀

Then, divide the numerator by the denominator:

¹⁄₄₀₀ = 0.0025

Study the following examples:

10% = 0.10, but ¹⁄₁₀ of 1% = 0.001

25% = 0.25, but ¼ of 1% = 0.0025

50% = 0.50, but ½ of 1% = 0.005

75% = 0.75, but ¾ of 1% = 0.0075

Whole-number percentages also can be expressed as fractions. To express whole-number percentages as fractions, divide the number by 100, and then reduce the fraction to its lowest terms. Study the following examples:

1% = ¹⁄₁₀₀

20% = ⅕

25% = ¼

33⅓% = ⅓

50% = ½

66⅔% = ⅔

75% = ¾

To calculate the percentage of any number, convert the given percentage to a decimal and multiply by the base number. Study the following examples:

What is 18% of 600?

(600)(0.18) = 108

What is 60% of 600?

(600)(0.60) = 360

To figure out what percent one quantity is of another, divide the given quantity by the base number. Then convert the decimal answer into a percentage by moving the decimal point two places to the right and adding a % sign. Study the following examples:

What percent of 1,200 is 240?

$^{240}\!/_{1200} = {}^{20}\!/_{100} = 20\%$

What percent of 480 is 120?

$^{120}\!/_{480} = \frac{1}{4} = 25\%$

6.1.0 Study Problems: Percentage of a Number

1. 38% of 490 = _____

2. 7% of 425 = _____

3. 33⅓% of 425 = _____

4. 8% of 96 = _____

5. 30% of 5 = _____

6. 79% of 153 = _____

7. 12% of 19 = _____

8. 10% of 10 = _____

9. 42% of 905 = _____

10. 37% of 139 = _____

6.2.0 Study Problems: Solving for Percent

1. What percent of 40 is 36? _____

2. What percent of 500 is 96? _____

3. What percent of 150 is 95? _____

4. What percent of 750 is 30? _____

5. What percent of 300 is 30? _____

6. What percent of 12 inches is ⅛-inch? _____

7. What is 15 percent of 125 times $14.00? _____

8. What is 3½ percent of 2,080? _____

9. What is 45 percent of 150? _____

10. What is 50 percent of 75? _____

7.0.0 ◈ RATIO AND PROPORTION

A **ratio** is a comparison of two quantities. The two quantities in a ratio are called its terms. If your classroom has 8 tables and 12 chairs, the ratio of tables to chairs is 8 to 12. A ratio is usually written with a colon between the two numbers (8:12). However, because a ratio is a fraction, this ratio may also be written as $^{8}\!/_{12}$. This is the same as saying "8 divided by 12." It is important to write the ratio in the right order; 8:12 is not the same as 12:8.

You must be careful when using ratios with units of measure. The ratio of the length of a one-foot rule to the length of a one-yard stick is not 1:1 but 1:3, because one yard is equal to three feet.

A ratio is always written in the same sequence as stated. For example, the ratio in the sentence "Engine A is running at twice the revolutions per minute of Engine B" is written as 2:1 or $^{2}\!/_{1}$.

A proportion is an equation involving two ratios. Proportion problems allow you to compare two ratios. For example, the ratio a:b = c:d is read "a is to b as c is to d." The second and third terms (b and c) are called the means, and the first and fourth terms (a and d) are called the extremes. The rule of proportions is that the product of the means is equal to the product of the extremes. Study the following example:

2:3 = 4:6

Because ratios are fractions, you can write these terms as follows:

$^{2}\!/_{3} = {}^{4}\!/_{6}$

To get the product of the means and the extremes, cross-multiply:

$2 \times 6 = 12$

$3 \times 4 = 12$

If you are given three of the terms of a ratio, you can find the fourth term by cross-multiplying. Study the following example:

The diameter of the first pulley in a belt transmission system is three times the diameter of the second pulley, or 3:1 (*Figure 10*). The diameter of the second pulley is 5 inches. What is the diameter of the first pulley?

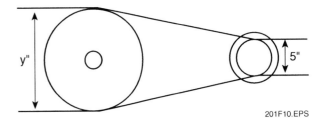

201F10.EPS

Figure 10 ◈ Ratio of pulleys.

Determine the proportion:

$$3:1 = \frac{\text{Diameter of first pulley}}{\text{Diameter of second pulley}}$$

You know that the diameter of the second pulley is 5 inches. Because the diameter of the first pulley is unknown, call it y:

$$\tfrac{3}{1} = \tfrac{y}{5}$$

Cross-multiply the fractions:

$$3 \times 5 = 1 \times y$$
$$y = 15 \text{ inches}$$

You will also use ratios to calculate slope angles used in figuring sheet metal fittings. Some common slope angles are shown in *Figure 11*.

Ratios are directly proportional when an increase in one ratio causes an increase in the other ratio. Study the following example:

One foot equals 12 inches. How many inches are there in 6 feet?

State the ratio of feet to inches:

$$1 \text{ ft}{:}12 \text{ in.} = \tfrac{1}{12}$$
$$6 \text{ ft}{:}y \text{ in.} = \tfrac{6}{y}$$

Equate the ratios:

$$\tfrac{1}{12} = \tfrac{6}{y}$$

Cross-multiply:

$$12 \times 6 = 1y$$
$$y = 72$$

An indirect proportion, also called an inverse ratio, occurs when an increase in one ratio causes a decrease in the other. When solving speed problems and problems pertaining to pulleys and gears, you will use inverse ratios. Study the following example:

An electric motor running at 1,750 revolutions per minute (rpm) has a 2.5-inch diameter pulley. What size pulley is required on the drive shaft to turn it at 875 rpm?

State the speed ratio:

$$\frac{1{,}750 \text{ (motor speed)}}{875 \text{ (shaft speed)}}$$

State the pulley diameter ratio:

$$\frac{2.5 \text{ (motor shaft pulley)}}{x \text{ (shaft pulley)}}$$

Cross-multiply:

$$(1{,}750)(x) = (875)(2.5)$$
$$1{,}750x = 2{,}187.5$$
$$x = 1.25$$

7.1.0 Study Problems: Ratio, Direct Proportion, and Indirect Proportion

1. What is the ratio of 40 to 10? What is the inverse ratio?

   _____

2. What is x in the following problem? _____

   $$\frac{3}{12} = \frac{18}{x}$$

3. The ratio of teeth in gears A and B is 2.5:1. If gear A has 15 teeth, how many teeth does gear B have?

   _____

4. Grade is measured as the number of feet of rise per 100 feet horizontally. A 10-foot rise in a distance of 100 feet is designated as a 10 percent grade. On a hill, a road rises 8 feet for every 100 feet. What is the percentage of grade?

   _____

5. The volume of a quantity of gas is inversely proportional to the pressure exerted on it. If the volume of a quantity of gas is 720 cubic feet when under a pressure of 16 pounds per square inch absolute (psia), how many cubic feet will there be when the pressure rises to 28 psia?

   _____

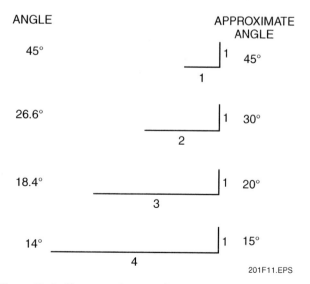

ANGLE   APPROXIMATE ANGLE

45°   1 45°

26.6°   1 30°

18.4°   1 20°

14°   1 15°

201F11.EPS

Figure 11 ◆ Common slope angles.

7.2.0 Mean Proportional

As you have learned, when you know three of the four terms in a proportion problem, you can calculate the fourth term by cross-multiplying. A special type of problem arises when the second and third terms in a proportion (the means) are not known. When this occurs, the two unknown terms are referred to as the mean proportional. Study the following example in which the means are unknown:

What is the mean proportional between 32 and 2?

Write the proportions with the unknown terms designated by x (the mean proportional):

$$\frac{32}{x} = \frac{x}{2}$$

Cross-multiply:

$$(x)(x) = (32)(2)$$
$$x^2 = 64$$

Extract the square root from each side:

$$x = 8$$

7.2.1 Study Problems: Mean Proportional

1. What is the mean proportional between 98 and 2?

_____

2. Two engines have equal ratios of horsepower to cubic inch displacement. One has a displacement of 424 cubic inches. The other has a rated horsepower of 200. If the horsepower of the first engine equals the displacement of the second engine, what is the displacement of the second engine?

_____

7.3.0 Compound Ratios and Proportions

Compound ratios are the product of two or more simple ratios. In many calculations, more than one condition exists that can affect their solutions. Two conditions can change the value of a number: When a number is multiplied by a ratio that is less than one, the value of that number is decreased. For example: $(25)(\frac{3}{5}) = 15$ (a decrease). When a number is multiplied by a ratio that is greater than one, the value of the number is increased. For example: $(25)(\frac{5}{3}) = 41.67$ (an increase).

In solving proportions with a series of ratios (compounding), you must put the individual ratios in the proper order stated in the problem. Study the following example:

A load of six steel bars, 8 feet long by 4 inches wide and 1 inch thick, weighs 652.8 pounds. How much will a load of eight steel bars, 7 feet long by 3 inches wide and 2 inches thick, weigh?

Write the ratio of the unknown quantity to the known quantity (weight of second load to weight of first load):

$$\frac{x \text{ pounds}}{652.8 \text{ pounds}}$$

Write the ratio of the number of bars in Load Two to the number of bars in Load One.

$$\frac{8 \text{ bars}}{6 \text{ bars}}$$

Write the ratio of the length of the bars in Load Two to the length of the bars in Load One.

$$\frac{7 \text{ feet}}{8 \text{ feet}}$$

Write the ratio of the width of the bars in Load Two to the width of the bars in Load One.

$$\frac{3 \text{ inches}}{4 \text{ inches}}$$

Write the ratio of the thickness of the bars in Load Two to the thickness of the bars in Load One.

$$\frac{2 \text{ inches}}{1 \text{ inch}}$$

Write the proportion as the ratio of the weights equated to the bar description ratios. Note that the Load Two terms are the numerators of each ratio and the Load One terms are the denominators.

$$\frac{x}{652.8} = \frac{8 \times 7 \times 3 \times 2}{6 \times 8 \times 4 \times 1}$$

Cancel terms:

$$\frac{x}{652.8} = \frac{\cancel{8}\ 7\ \overset{1}{\cancel{3}}\ \overset{1}{\cancel{2}}}{\underset{2}{\cancel{6}}\ \cancel{8}\ \underset{2}{\cancel{4}}\ 1}$$

Multiply factors:

$$\frac{x}{652.8} = \frac{7}{4}$$

Cross-multiply:

$$4x = (7)(652.8)$$

Solve for x:

$$4x = 4{,}569.6 \text{ pounds}$$

$$x = 1{,}142.4 \text{ pounds}$$

7.3.1 Study Problems:
Compound Ratios and Proportions

1. Two workers using forklifts can unload 10 boxcars in 8 hours. How many boxcars can 4 workers unload in 6 hours?

2. Seven workers can complete an excavation project in five 8-hour days. How many days would it take five workers working 10-hour days to complete 60 percent of the work?

3. A crew of three sheet metal workers is installing an HVAC duct system having 11,280 pounds of duct. Twenty-five percent of the job must be completed per week. How many pounds must be installed per week?

4. What is a contractor's price per hour to hire a first-year employee, if the employee's base rate is $8.16 per hour, plus $2.78 fringe per hour, with combined overhead and profit an additional 22 percent? Calculate to the nearest one cent.

5. A sheet metal shop has the storage capacity to stock 80,000 pounds of 24-gauge metal. Company policy is to re-order metal whenever the stock is reduced to 15 percent of capacity. How many pounds of metal are on hand when it is time to re-order?

6. Find the cost of gutter and downspouts marked at $750.00 with a 12.5 percent discount and an additional 1.5% off the discounted price for cash. The bill was paid with cash at the time of purchase.

7. A shop owner bid $10,000.00 on a job. After paying the labor and material costs, he had $500.00 left. What rate of profit did he realize on this job?

8. The engineer requires that the HVAC system air flow be increased from 3,800 CFM to 4,180 CFM. What is the percent increase?

9. If a sheet metal worker receives 65 percent of his weekly wage as disability insurance, and his average weekly wage is $465.00, how much will he receive per week if he is disabled?

10. Unable to get any calculation from the engineer, the TAB technician decides to divide the air flow from a 6,000 CFM air handler based on floor area. Area A is 1500 square feet while area B is 3,000 square feet. How much air will he distribute to area A? How much air will he distribute to area B?

7.4.0 Mixture Proportions

Mixtures present a special type of problem involving ratio and proportion. Alloys, for example, are composed of two or more metals, the proportions of which determine the physical characteristics of the alloy. Therefore, care must be used to ensure that an exact amount of each alloy metal is added to the mixture. Alloy quantities are expressed either as percentages of the whole or as parts. When percentages are used, the separate percentages must add up to the whole (100 percent). Study the following example:

An alloy is composed of five parts lead, three parts copper, and one part tin. How many pounds of each base metal will be needed to cast a 250-pound bar?

Add the parts to obtain the whole:

$$5 + 3 + 1 = 9$$

Form the proportions:

$\frac{5}{9} = \frac{x}{250}$ (lead)

$\frac{3}{9} = \frac{x}{250}$ (copper)

$\frac{1}{9} = \frac{x}{250}$ (tin)

Cross-multiply and solve for each proportion:

$$(5)(250) = 9x$$

$$x = 138.889$$

$$(3)(250) = 9x$$

$$x = 83.333$$

$$(1)(250) = 9x$$

$$x = 27.778$$

Add the weights of each metal proportion:

138.889 lead

83.333 copper

+ 27.778 tin

250.000 pounds of alloy

7.4.1 Study Problems: Mixture Proportions

1. A piece of solder is 40 percent tin and 60 percent lead. How many pounds of solder can be made from 40 pounds of tin and how many pounds of lead will be used?

_____

2. A casting has six parts copper, four parts lead, and one part tin. How many pounds of each are required to form the 185-pound casting?

_____

3. You are mixing a soldering dip to clean your soldering irons. Your shop's dip solution uses 1 pound of sal ammoniac in 10 gallons of water. How much sal ammoniac would you add for a 1 quart container?

_____

8.0.0 ◈ TRIGONOMETRY

Trigonometry is the study of the properties of triangles and trigonometric functions and of their applications. By using trigonometric formulas, one can calculate the sizes of angles or the lengths of the sides of right triangles. As the acute angles (angles less than 90°) of a right triangle change in size, the sides opposite these angles increase or decrease at a predictable rate or ratio. This ratio, as the acute angles of a right triangle change, is a function of any given angle. In other words, if an angle of a right triangle is changed, so are the sides that depend upon that angle. If an angle increases, so does the side opposite that angle; if the angle decreases, the side opposite that angle also decreases.

8.1.0 Complementary Angles

The sum of two acute angles of a right triangle (*Figure 12*) equals 90°. Therefore, if one acute angle of a given right triangle equals 40°, the other acute angle must equal 50°. Each angle, therefore, is the complement of the other.

When a given acute angle changes in value, the ratio of the side to the hypotenuse changes. This ratio is the quotient resulting from division of one side by the other side, or division of the side by the hypotenuse (review the Pythagorean Theorem).

8.2.0 Functions

Every angle forms its own ratio of sides and hypotenuse. This ratio is called a function of the angle. There are three trigonometric functions that are commonly used by sheet metal workers. These include **sine** (sin), **cosine** (cos), and **tangent** (tan).

These functions can be defined by observing the following conventions (*Figure 13*). Abbreviations will be used to refer to the various sides of the triangle: H for hypotenuse, SA for side adjacent, and SO for side opposite.

The sine of angle A equals:

$$\frac{SO}{H} \text{ or } \frac{a}{c}$$

The cosine of angle A equals:

$$\frac{SA}{H} \text{ or } \frac{b}{c}$$

The tangent of angle A equals:

$$\frac{SO}{SA} \text{ or } \frac{a}{b}$$

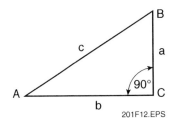

Figure 12 ◈ A right triangle.

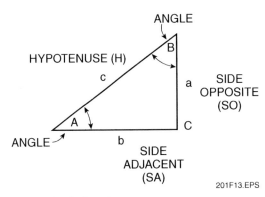

Figure 13 ◈ Angle functions.

The functions of the other acute angle (B) can be defined as follows:

The sine of angle B equals:

$$\frac{b}{c}$$

The cosine of angle B equals:

$$\frac{a}{c}$$

The tangent of angle B equals:

$$\frac{b}{a}$$

Referring to the above conventions, notice the following:

- The sine of one acute angle equals the cosine of the other acute angle
- The cosine of one acute angle equals the sine of the other acute angle
- The tangent of one acute angle equals the cotangent (cot) of the other acute angle

For example, if angle A in *Figure 13* is equal to 25°, its complementary angle B is equal to 65°. Thus:

sin 25° = cos 65°

cos 25° = sin 65°

tan 25° = cot 65°

Refer to the following trigonometric functions to identify any function of an angle from a drawing of the triangle of which the angle is a part. Note: These three formulas are very useful in calculating angles. Keep them close at hand.

$$\sin = \frac{SO}{H}$$

$$\cos = \frac{SA}{H}$$

$$\tan = \frac{SO}{SA}$$

To find the sides and hypotenuse, use the following formulas:

SO = (H)(sin A)
or
a = (c)(sin A)

SA = (H)(cos A)
or
b = (c)(cos A)

SO = (SA)(tan A)
or
a = (b)(tan A)

$$SA = \frac{SO}{\tan A}$$
or
$$b = \frac{a}{\tan A}$$

$$H = \frac{SA}{\cos A}$$
or
$$c = \frac{b}{\cos A}$$

$$H = \frac{SO}{\sin A}$$
or
$$c = \frac{a}{\sin A}$$

The number of degrees in the acute angles of a right triangle can vary from 0° to 90°. Consequently, the values of the trigonometric functions expressed in ratios will change as the size of the acute angle changes. Tables are available that give the values of the three trigonometric functions most used in the sheet metal trade (*Table 1*). The values of the ratios are numbers that are usually expressed in five decimal places.

For practical purposes, this portion of the module will use just the three ratios pertaining to trigonometric functions: sine, cosine, and tangent. In addition, for purposes of consistency and accuracy, solve only for angle A for any given field problem in this module.

To use a table of trigonometric functions, substitute the following values for the sides and hypotenuse of the triangle in *Figure 13*:

a equals 4 inches

b equals 3 inches

c equals 5 inches

Step 1 Find the sine of angle A (sin equals side opposite divided by hypotenuse).

sin A equals ⅘

A equals 0.80000

Use the table to locate the nearest angle in the angle column that is equal to the value nearest to 0.80000. Find 0.7986 in the sine column as being the value nearest 0.80000. Read left to the degree column and find the angle, 53°.

Step 2 Find the cosine of angle A (cos equals side adjacent divided by the hypotenuse).

cos equals ⅗

cos equals 0.60000

Find the value nearest to 0.6000 (0.6018) under the cosine column. Read left to the degree column and find that cos A equals 53°.

Table 1 Table of Trigonometric Functions

Angle	Sine	Cosine	Tangent	Angle	Sine	Cosine	Tangent
1°	.0175	.9998	.0175	46°	.7193	.6947	1.0355
2°	.0349	.9994	.0349	47°	.7314	.6820	1.0724
3°	.0523	.9986	.0524	48°	.7431	.6691	1.1106
4°	.0698	.9976	.0699	49°	.7547	.6561	1.1504
5°	.0872	.9962	.0875	50°	.7660	.6428	1.1918
6°	.1045	.9945	.1051	51°	.7771	.6293	1.2349
7°	.1219	.9925	.1228	52°	.7880	.6157	1.2799
8°	.1392	.9903	.1405	53°	.7986	.6018	1.3270
9°	.1564	.9877	.1584	54°	.8090	.5878	1.3764
10°	.1736	.9848	.1763	55°	.8192	.5736	1.4281
11°	.1908	.9816	.1944	56°	.8290	.5592	1.4826
12°	.2079	.9781	.2126	57°	.8387	.5446	1.5399
13°	.2250	.9744	.2309	58°	.8480	.5299	1.6003
14°	.2419	.9703	.2493	59°	.8572	.5150	1.6643
15°	.2588	.9659	.2679	60°	.8660	.5000	1.7321
16°	.2756	.9613	.2867	61°	.8746	.4848	1.8040
17°	.2924	.9563	.3057	62°	.8829	.4695	1.8807
18°	.3090	.9511	.3249	63°	.8910	.4540	1.9626
19°	.3256	.9455	.3443	64°	.8988	.4384	2.0503
20°	.3420	.9397	.3640	65°	.9063	.4226	2.1445
21°	.3584	.9336	.3839	66°	.9135	.4067	2.2460
22°	.3746	.9272	.4040	67°	.9205	.3907	2.3559
23°	.3907	.9205	.4245	68°	.9272	.3746	2.4751
24°	.4067	.9135	.4452	69°	.9336	.3584	2.6051
25°	.4226	.9063	.4663	70°	.9397	.3420	2.7475
26°	.4384	.8988	.4877	71°	.9455	.3256	2.9042
27°	.4540	.8910	.5095	72°	.9511	.3090	3.0777
28°	.4695	.8829	.5317	73°	.9563	.2924	3.2709
29°	.4848	.8746	.5543	74°	.9613	.2756	3.4874
30°	.5000	.8660	.5774	75°	.9659	.2588	3.7321
31°	.5150	.8572	.6009	76°	.9703	.2419	4.0108
32°	.5299	.8480	.6249	77°	.9744	.2250	4.3315
33°	.5446	.8387	.6494	78°	.9781	.2079	4.7046
34°	.5592	.8290	.6745	79°	.9816	.1908	5.1446
35°	.5736	.8192	.7002	80°	.9848	.1736	5.6713
36°	.5878	.8090	.7265	81°	.9877	.1564	6.3138
37°	.6018	.7986	.7536	82°	.9903	.1392	7.1154
38°	.6157	.7880	.7813	83°	.9925	.1219	8.1443
39°	.6293	.7771	.8098	84°	.9945	.1045	9.5144
40°	.6428	.7660	.8391	85°	.9962	.0872	11.4301
41°	.6561	.7547	.8693	86°	.9976	.0698	14.3007
42°	.6691	.7431	.9004	87°	.9986	.0523	19.0811
43°	.6820	.7314	.9325	88°	.9994	.0349	28.6363
44°	.6947	.7193	.9657	89°	.9998	.0175	57.2900
45°	.7071	.7071	1.0000	90°	1.0000	.0000	

Step 3 Find the tangent of angle A (tan equals side opposite divided by the side adjacent).

tan equals ⅘

tan equals 1.33333

Step 4 Find the value nearest to 1.33333 under the tangent column. Locate 1.3270. Read left to the degree column and find that tan A equals 53°.

Note: The value of the sine or cosine can never be greater than 1.00000, whereas the value of the tangent extends almost to infinity.

In the following example, the values of one side and the hypotenuse are given. If one acute angle is found, the missing value for the hypotenuse or side can also be found.

Find angle A and the opposite side (a) of *Figure 14*.

Step 1 Calculate angle A. Refer to the formulas to calculate cosine A (divide the side adjacent by the hypotenuse: thus, 18 divided by 24 is equal to 0.75000). Use the tables to find that cosine A, 0.75000, equals approximately 41°.

Step 2 Using the recommended formulas, notice that the side opposite (a) can be calculated by multiplying the adjacent value (b equals 18) times tan A. Tan A (41°) equals 0.8693, thus:

(18)(0.8693) = 15.65

8.2.1 Study Problems: Complementary Angles

Use the table of trigonometric functions to locate the ratios for these angles:

1. Sine

   10° = _____

   15° = _____

   20° = _____

   25° = _____

   31° = _____

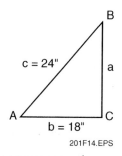

Figure 14 ◆ Trigonometry example.

2. Cosine

   10° = _____

   15° = _____

   20° = _____

   25° = _____

   59° = _____

3. Tangent

   10° = _____

   15° = _____

   20° = _____

   25° = _____

   89° = _____

Use the table of trigonometric functions to locate the angles corresponding to the following ratios:

4. Sine

   0.2419 = _____

   0.4067 = _____

   0.6947 = _____

   0.9455 = _____

   1.0000 = _____

5. Cosine

   0.0349 = _____

   0.3746 = _____

   0.7771 = _____

   0.9816 = _____

   0.9998 = _____

6. Tangent

   57.2900 = _____

   1.9943 = _____

   0.6009 = _____

   0.2493 = _____

   0.0875 = _____

Using the recommended formulas, calculate the following values as indicated.

7. Calculate the angle A for *Figure 15*.

   _____

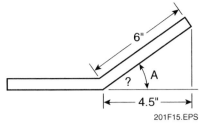

Figure 15 ◈ Study problem: angle A.

8. Calculate the side adjacent for *Figure 16*.

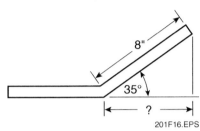

Figure 16 ◈ Study problem: side adjacent.

9. Calculate the value of angle A (*Figure 17*) in the following problems:

   a. c = 8" and b = 6" _____

   b. c = 9" and b = 7.5" _____

   c. c = 14.5' and b = 12' _____

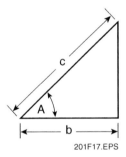

Figure 17 ◈ Study problem: angle A values.

10. Calculate angle A (*Figure 18*).

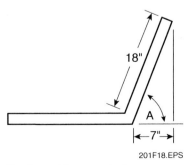

Figure 18 ◈ Study problem: angle A.

11. Calculate the length of adjacent side (b) (*Figure 19*).

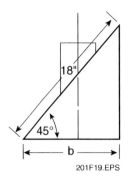

Figure 19 ◈ Study problem: adjacent side.

12. Calculate the length of opposite side (a) (*Figure 20*).

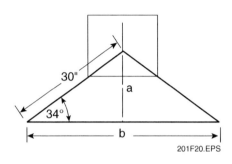

Figure 20 ◈ Study problem: opposite side.

13. Find angle A and the hypotenuse of the triangle depicted in *Figure 21* for the following values:

   a. adjacent (b) equals 3 feet
       opposite (a) equals 1.9 feet

   b. adjacent (b) equals 6.28125 inches
       opposite (a) equals 4.125 inches

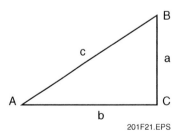

Figure 21 ◈ Study problem: angle A and hypotenuse.

9.0.0 ◆ THE PROTRACTOR

A protractor is used to measure and construct angles. The lines that make up angles are called the sides, and the point where the lines touch is called the vertex. The five basic types of angles are shown in *Figure 22*:

- The straight angle is 180 degrees.
- The acute angle is more than 0 degrees and less than 90 degrees.
- The obtuse angle is more than 90 degrees but less than 180 degrees.
- The right angle is exactly 90 degrees. There are two right angles in a straight angle and four right angles in a circle.
- The reflex angle is more than 180 degrees but less than 360 degrees.

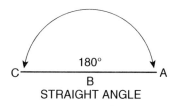
STRAIGHT ANGLE

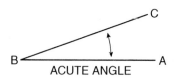
ACUTE ANGLE

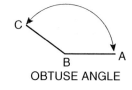
OBTUSE ANGLE

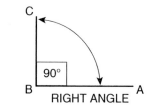
RIGHT ANGLE

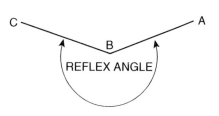
REFLEX ANGLE

201F22.EPS

Figure 22 ◆ Basic angles.

9.1.0 Measuring Angles

To measure an angle with a protractor, place the vertex of the angle at the bottom center of the protractor so that one side of the angle passes through the 0-degree mark. (If the sides of the angle are too short to pass through this mark, carefully extend the sides so that they are long enough.) Read the measurement of the angle at the point where the other side of the angle intersects the curved edge of the protractor.

9.1.1 Study Problems: Measuring Angles with a Protractor

With a protractor, measure the angles in *Figure 23*.

1. _____

2. _____

3. _____

4. _____

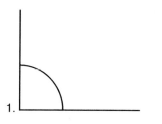
1.

2.

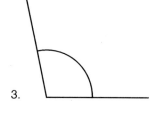
3.

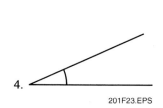
4.

201F23.EPS

Figure 23 ◆ Study problems: measuring angles.

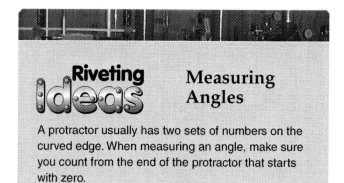

Riveting Ideas — **Measuring Angles**

A protractor usually has two sets of numbers on the curved edge. When measuring an angle, make sure you count from the end of the protractor that starts with zero.

The Micrometer

The micrometer, or mike, is a precision measuring tool. The best way to understand how to use this tool is to practice with test pieces of fixed sizes. You do not have to be a mathematician to read the micrometer. You simply have to understand the divisions on the spindle and the thimble. The spindle has 40 threads per inch. One turn of the thimble advances the spindle one-fortieth inch, or 0.025-inch. To read the micrometer, you note the total number of 0.025-inch graduations uncovered by the thimble. To this total you will add the number on the thimble closest to the index line on the barrel.

To measure an item, hold the micrometer in your hand with your little finger curled around the frame. Use your thumb and forefinger to turn the thimble. The spindle revolves with the thimble. Place the item to be measured between the anvil and the spindle. Carefully tighten the spindle using light pressure between your thumb and forefinger.

The measurement of the opening between the spindle and the anvil is indicated by lines and figures on the barrel and the thimble. One complete turn of the thimble changes the opening between the spindle and the anvil 0.025-inch. Each line on the barrel also represents 0.025-inch. Each number, therefore, represents 0.100-inch.

Micrometers are available in various sizes. The size of the opening between the anvil and the spindle must be added to the barrel and thimble readings.

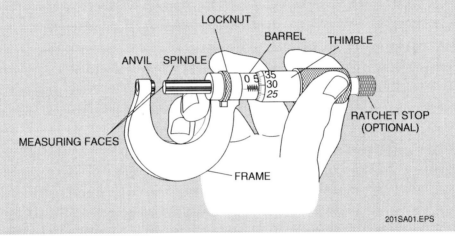

201SA01.EPS

9.2.0 Constructing Angles

To construct an angle using a protractor, follow these steps:

Step 1 Draw a straight line to form one of the sides of the angle.

Step 2 Place the center hole of the protractor over one end of the line.

Step 3 Line up the zero on the edge of the protractor with the line.

Step 4 Make a mark at a point on the curved edge of the protractor for the angle you want. For example, make a mark at 45 for a 45-degree angle.

Step 5 Use a ruler or the straight edge of the protractor to connect the mark with the end of the line.

Practice using the protractor by constructing the following angles:

- 90 degrees
- 45 degrees
- 37 degrees
- 115 degrees
- 135 degrees

10.0.0 ◆ SEAM ALLOWANCE

Seams and hems eliminate raw edges on sheet metal jobs and add strength to the fabricated workpiece. When you lay out a pattern, you must make allowance for additional metal for the seams and hems. The allowance will vary with the gauge of the metal and with the type of seam or hem. The allowances quoted in this section of the module pertain to 24-gauge or lighter metal unless otherwise indicated.

10.1.0 Single Hem

The allowance for a single hem is equal to the width of the seam (*Figure 24*). The formula a = w is used for this calculation. Study the following example:

Calculate the allowance for a single hem that is ³⁄₁₆ of an inch wide.

$a = w$

$a = ³⁄₁₆\text{-inch}$

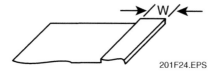

Figure 24 ◆ Single hem.

10.1.1 Study Problems: Single Hem Allowance

1. Hem ³⁄₁₆-inch, one edge

_____

2. Hem ¼-inch, one edge

_____

3. Hem ⁵⁄₃₂-inch, two edges

_____

10.2.0 Double Hem

The allowance for a double hem is twice the width of the intended seam (*Figure 25*).

The formula a = 2w is used for this calculation. Study the following example:

Calculate the allowance for a double-edge hem that is ⁵⁄₃₂ of an inch wide.

$a = 2w$

$a = (2)(⁵⁄₃₂\text{-inch})$

$a = ⁵⁄₁₆\text{-inch}$

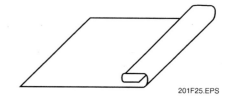

Figure 25 ◆ Double hem.

Tolerances

Tolerance is the total permissible variation of a size. It is the difference between the maximum and minimum limits of a size. For example, the opening of an open-end wrench must be a minimum size or it will not fit over the bolt heads for which it was designed. If the opening is too large, the wrench will not fit snugly, but between the maximum size and the minimum size of the opening is an area of size in which the wrench will still work as intended. Tolerances may be expressed as + a number, – a number, or ± a number. Tolerances on a plan or shop drawing may be general (specified with a note) or they may be specific (specified with a dimension).

Every tolerance has a nominal size and an upper and lower limit. Sometimes the nominal size is also the upper limit or lower limit of the tolerance. For example, the nominal size for the opening of an open-end wrench may be 0.562-inch and the tolerance listed as +0.005-inch or –0.000-inch. This means the lower limit is the same as the nominal size (0.562-inch) and the upper limit is 0.567-inch (0.562 + 0.005). This is an example of unilateral tolerance. This type of tolerance runs in only one direction—in this case, a plus or positive direction.

Tolerance also can run in two directions. This is called bilateral tolerance. For example, if you must bore a hole that is specified as 1.625 inches ±0.005-inch, the nominal size is 1.625 inches. The upper limit is plus 0.005-inch and the lower limit is minus 0.005-inch. The permissible tolerance is the total of the plus and minus limits, or 0.010-inch. Therefore, the hole could vary in size from 1.630 inches to 1.620 inches and still be acceptable. Whether the tolerance is unilateral or bilateral, the total clearance is the difference between the upper limit and the lower limit.

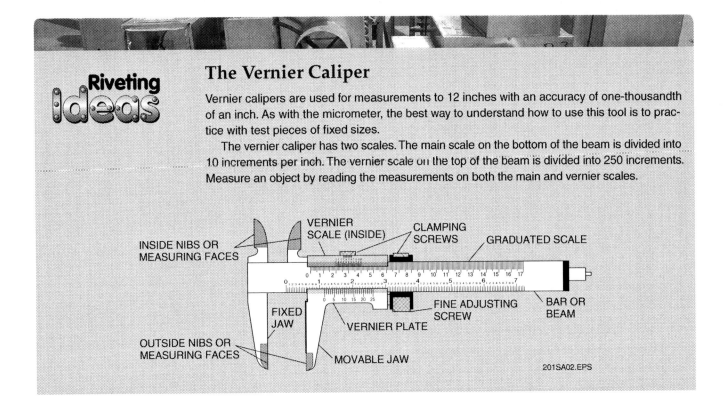

Riveting Ideas

The Vernier Caliper

Vernier calipers are used for measurements to 12 inches with an accuracy of one-thousandth of an inch. As with the micrometer, the best way to understand how to use this tool is to practice with test pieces of fixed sizes.

The vernier caliper has two scales. The main scale on the bottom of the beam is divided into 10 increments per inch. The vernier scale on the top of the beam is divided into 250 increments. Measure an object by reading the measurements on both the main and vernier scales.

201SA02.EPS

10.2.1 Study Problems: Double-Hem Allowance

1. Hem ³⁄₃₂-inch, one edge

_____

2. Hem ⁵⁄₃₂-inch, two edges

_____

10.3.0 Lap Seam

The allowance for lap seams is equal to the seam width (*Figure 26*). The simple formula a = w can be used for this calculation. Study the following example:

Calculate the allowance for a lap seam ³⁄₁₆ of an inch wide.

a = w

a = ³⁄₁₆-inch

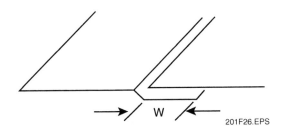

201F26.EPS

Figure 26 ◆ Lap seam.

10.4.0 Grooved Lock Seam

The allowance for a grooved lock seam is usually three times the width of the seam, one-half of which is added to each edge of the pattern (*Figure 27*). If the metal is 22-gauge or heavier, the allowance is three times the width of the seam plus five times the thickness of the metal. Assuming the metal is 24-gauge or lighter, the formula a = 3w is used for this calculation. Study the following example:

Calculate the allowance for a grooved lock seam ³⁄₁₆-inch wide.

a = 3w

a = (3)(³⁄₁₆)

a = ⁹⁄₁₆-inch (allowance for total)

The allowance for each edge is ½ a or ⁹⁄₃₂-inch.

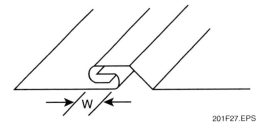

201F27.EPS

Figure 27 ◆ Grooved lock seam.

1. Lock ⅛-inch, each edge

_____

2. Lock ⁵⁄₁₆-inch, both edges (total)

_____

3. Lock ³⁄₁₆-inch, each edge, 22-gauge (0.0336-inch)

_____

10.5.0 Riveted Lap Seam

The allowance for a riveted lap seam is two times the diameter of the rivet for each edge of the seam (*Figure 28*). The formula $a = 2d$ (where d = rivet diameter) is used for this calculation. Study the following example:

Calculate the allowance for one edge of a riveted seam with rivets having a diameter of ⁷⁄₆₄-inch (0.109-inch).

$a = 2d$

$a = (2)(0.109\text{-inch})$

$a = 0.218 \text{ inch } (⁷⁄₃₂\text{-inch})$

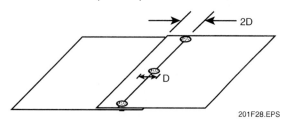

Figure 28 ◆ Riveted lap seam.

10.5.1 Study Problems:
Riveted Lap Seam Allowance

1. Rivet diameter 0.120-inch, one edge

_____

2. Rivet diameter 0.160-inch, two edges

_____

10.6.0 Dovetailed Seam

The allowance for a dovetailed seam is equal to the seam width (*Figure 29*). The simple formula $a = w$ is used for this calculation. Study the following example:

Calculate the allowance for a 1-inch dovetailed seam.

$a = w$

$a = 1 \text{ inch}$

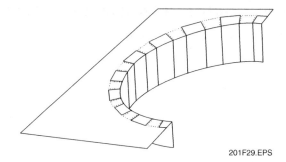

Figure 29 ◆ Dovetailed seam.

10.7.0 Burred Bottom Seam

The allowance for one side of a burred bottom seam is equal to the width of the burr (*Figure 30*). The simple formula $a = w$ is used for this calculation. Study the following example:

Calculate the allowance for a ³⁄₁₆-inch burred bottom seam.

$a = w$

$a = ³⁄₁₆\text{-inch}$

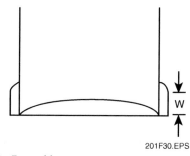

Figure 30 ◆ Burred bottom seam.

10.8.0 Set-In Bottom Seam

The allowance for the body of a set-in bottom seam is equal to two times the width of the seam (*Figure 31*). The formula $a = 2w$ is used for this calculation. Study the following example:

Calculate the allowance for a ³⁄₁₆-inch set-in bottom seam on the body portion of the pattern.

$a = 2w$

$a = (2)(³⁄₁₆\text{-inch})$

$a = ⅜\text{-inch}$

Figure 31 ◆ Set-in bottom seam.

10.9.0 Body and Bottom Allowances

The allowance on one side of the bottom of a fitting with either a single or double seam is equal to two times the width of the flange on the body. However, you must distribute the allowance so that the outer allowance is equal to the width minus 1/32-inch, and the inner allowance is equal to the width plus 1/32-inch.

To calculate this type of allowance you must make the following calculations: the allowance for one side of the body (a); the allowance for one side of the bottom (b); and the two bend allowances for the bottom (c and d) (*Figure 32*). The formulas for these calculations follow:

$a = w$

$b = 2w$

$c = 2w + \frac{1}{32}\text{-inch}$

$d = 2w - \frac{1}{32}\text{-inch}$

Study the following example:
Calculate the allowances for one side of the body, one side of the bottom, and the two bend allowances for a fitting with a single seam with a flange width of 3/16-inch.

$a = \frac{3}{16}\text{-inch}$

$b = (2)(\frac{3}{16}) \text{ or } \frac{3}{8}\text{-inch}$

$c = \frac{3}{8}\text{-inch} + \frac{1}{32}\text{-inch} = \frac{13}{32}\text{-inch}$

$d = \frac{3}{8}\text{-inch} - \frac{1}{32}\text{-inch} = \frac{11}{32}\text{-inch}$

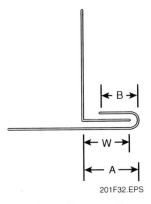

Figure 32 ◆ Body and bottom allowances.

10.10.0 Pittsburgh Lock Seam

The allowance for the pocket part of one side of a Pittsburgh lock seam (*Figure 33*) can generally be determined by multiplying the width of the lower bending leaf of the brake by 2½. Machine-formed Pittsburgh lock seam allowances will vary with metal gauge and type of machine, usu-

ally ranging from 1 inch to 1½ inches. Machines designed for lighter gauge metals will require smaller allowances, while those designed for heavier gauge metals require larger allowances. Study the following example:
Calculate the allowance for the pocket of one side of a Pittsburgh lock seam with a leaf width of ½-inch (*Figure 34*).

$a = (2\frac{1}{2})(w)$

$a = (2\frac{1}{2})(\frac{1}{2})$

$a = 1\frac{1}{4}\text{-inch}$

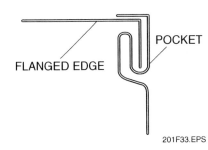

Figure 33 ◆ Pittsburgh lock.

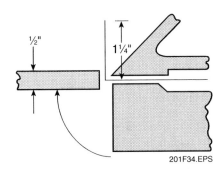

Figure 34 ◆ Allowance for a machine-formed Pittsburgh lock.

10.10.1 Pittsburgh Lock Seam

Calculate the allowance for a Pittsburgh lock seam with a bending leaf equal to the measurements shown in each problem:

1. 5/8-inch bending leaf

   _____

2. 3/4-inch bending leaf

   _____

3. 1/4-inch bending leaf

   _____

10.11.0 Bend Allowances

Bend allowances are related to the thickness of the metal. For inside bend dimensions, you must add 20 percent of the thickness of the metal on each side of the bend lines. For outside dimensions you must subtract 80 percent of the thickness of the metal on each side of the bend lines. Study the following example (*Figure 35*):

Calculate the stretchout (inside dimensions) for a channel 6 inches deep, by 7 inches wide, made with ⅛-inch thick metal.

Add 20 percent of the thickness of the metal to each side of the bend lines. Note that you have two bends on the width.

6 inches + (0.20 × 0.125) = 6.025 inches

7 inches + (0.20 × 0.250) = 7.050 inches

6 inches + (0.20 × 0.125) = 6.025 inches

10.11.1 Study Problem: Bend Allowance

1. Calculate the stretchout (outside dimensions) for a channel 6 inches deep by 7 inches wide made with ¼-inch thick metal.

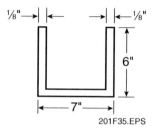

Figure 35 ◈ Bend allowances.

Remember to show all your work when solving mathematical problems.

1. In a formula, a value that can be any number is called a _____.
   a. constant
   b. variable
   c. radical
   d. coefficient

2. Parentheses, brackets, and braces are called _____.
   a. operational symbols
   b. directional symbols
   c. grouping symbols
   d. sequence symbols

3. 175% of $325.00 is _____.
   a. $243.75
   b. $406.25
   c. $500.00
   d. $568.75

4. A developer of a shopping center is willing to pay a 3 percent bonus if the project is completed 30 days ahead of schedule. The HVAC portion is $968,000. The company is willing to share 50 percent of the bonus with the three subcontractors on the project. How much bonus will each of the subcontractors get?
   a. $1,452
   b. $4,840
   c. $14,520
   d. $29,040

5. A shop owner bid $10,000 on a job. After paying the labor and material costs, the shop owner had $500 left. What is the rate of profit that the shop owner realized on this job?
   a. 0.05%
   b. 0.25%
   c. 2.50%
   d. 5.00%

6. A number that is directly in front of a variable is called a _____.
   a. literal expression
   b. coefficient
   c. constant
   d. grouping symbol

7. In the following equation, which step is performed first?

   $3x + 8(2x) + 4 \div 5 - 5$

   a. Remove the grouping symbols by multiplying.
   b. Add all the variables and all the numbers together.
   c. Divide 4 by 5.
   d. Subtract 5.

8. Solve for x in the following equation:

   $$\frac{4(2x+6)}{8} - \frac{2(5x-10)}{5} = 0$$

   a. 1
   b. 7
   c. −1
   d. −7

9. One meter is equal to _____.
   a. 39.37 inches
   b. 33.37 inches
   c. 0.3937 inches
   d. 0.3337 inches

10. When you work with circles and arcs, the measurements used most include _____.
    a. segment, tangent, and radius
    b. angle, diameter, and radius
    c. circumference, diameter, and radius
    d. circumference, segment, and diameter

11. The formula for the circumference of a circle is _____.
    a. $c = \pi d$
    b. $c = \pi d^2$
    c. $c = \pi r$
    d. $c = \pi r^2$

12. The hypotenuse of a right triangle is _____.
    a. the side opposite the right angle
    b. the shortest side
    c. the square root of the longest side of the triangle
    d. equal to the height or altitude

13. To express any percentage as a decimal, divide by _____.
    a. 0.10
    b. 10
    c. 0.100
    d. 100

14. To calculate the percentage of any number, convert the given percentage to a decimal and _____.
    a. multiply by the base number
    b. divide by the base number
    c. multiply by 100 then divide by the denominator
    d. multiply by 100 then divide by the numerator

15. To figure out what percent one quantity is of another, _____.
    a. multiply the given quantity by the base number
    b. divide the given quantity by the base number
    c. convert the percentage to a fraction and cross-multiply
    d. convert the percentage to a decimal and divide

16. The ratios 2:3 and 4:6 _____.
    a. are inversely proportional
    b. represent imbalanced terms
    c. are equal
    d. represent a mean proportional

17. Ratios are directly proportional when an increase in one ratio causes a(n) _____.
    a. decrease in the other
    b. increase in the other
    c. inverse quantity in the other
    d. proportional doubling

18. An angle that is more than 0 degrees and less than 90 degrees is a(n) _____ angle.
    a. obtuse
    b. right
    c. acute
    d. reflex

19. Seam allowance varies depending on _____.
    a. the gauge of the metal and the type of seam
    b. the gauge of the metal only
    c. the type and number of seams
    d. the type of seam only

20. For a bend allowance using ⅛-inch thick metal, you must add _____ to the outside dimension for a single bend.
    a. 0.015625 inches
    b. 0.1 inches
    c. 0.125 inches
    d. 0.2 inch

Summary

One of the most important tools in your toolbox cannot be purchased in a store—the skill you develop in applied math. You will use a calculator on the job, but not even the best calculator can help you if you do not understand the basic principles behind the measurements used every day.

Taking the time now to learn the principles of measuring, percentages, ratios, and proportions will pay off on the job. You will use math to read plans and detail drawings, to calculate stretchout material needs and fitting tolerances, to construct or fabricate fittings for special applications, and to solve problems you may encounter on the job.

Learning the basic principles and practicing problems will enable you to become skilled in math so that you can do your job accurately and well.

Notes

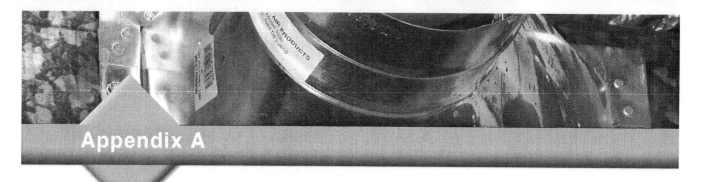

English System Weights and Measures

Linear Measures
1 foot	=	12 inches
1 yard	=	36 inches
	=	3 feet
1 mile	=	5,280 feet
	=	1,760 yards

Area Measures
1 square foot	=	144 square inches
1 square yard	=	9 square feet
1 acre	=	43,560 square feet
1 square mile	=	640 acres

Volume Measures
1 cubic foot	=	1,728 inches
1 cubic yard	=	27 cubic feet
1 freight ton	=	40 cubic feet
1 register ton	=	100 cubic feet

Liquid Volume Measures
1 gallon	=	128 fluid ounces
4 quarts	=	1 gallon
1 barrel	=	31.5 gallons
1 petroleum barrel	=	42 gallons

Weights
1 pound	=	16 ounces
1 hundredweight	=	100 pounds
1 ton	=	2,000 pounds

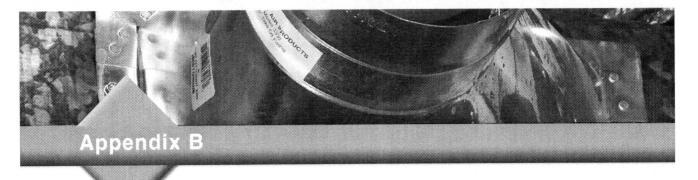

English-Metric Equivalent Measurements

(1 of 2)

Length Measures

1 inch (in.)	= 25.4 millimeters (mm)
1 inch (in.)	= 2.54 centimeters (cm)
1 foot (ft.)	= 0.3048 meters (m)
1 yard (yd.)	= 0.9144 meters (m)
1 mile (mi.)	= 1.609 kilometers (km)
1 millimeter (mm)	= 0.03937 inch (in.)
1 centimeter (cm)	= 0.39370 inch (in.)
1 meter (m)	= 3.28084 feet (ft.)
1 meter (m)	= 1.09361 yards (yd.)
1 kilometer (km)	= 0.62137 miles (mi.)

Area Measures

1 square inch (sq. in.)	= 645.16 square millimeters (mm^2)
1 square inch (sq. in.)	= 6.4516 square centimeters (cm^2)
1 square foot (sq. ft.)	= 0.092903 square meters (m^2)
1 square yard (sq. yd.)	= 0.836127 square meters (m^2)
1 square millimeter (mm^2)	= 0.001550 square inch (sq. in.)
1 square centimeter (cm^2)	= 0.15500 square inch (sq. in.)
1 square meter (m^2)	= 10.763910 square feet (sq. ft.)
1 square meter (m^2)	= 1.19599 square yards (sq. yd.)

Angle Measures

1 degree (°)	= 60 minutes (')
1 minute (')	= 60 seconds (")

Volume Measures for Solids

1 cubic inch (cu. in.)	= 16.387064 cubic centimeters (cm^3)
1 cubic foot (cu. ft.)	= 0.028317 cubic meter (m^3)
1 cubic yard (cu. yd.)	= 0.764555 cubic meter (m^3)
1 cubic centimeter (cm^3)	= 0.061024 cubic inch (cu. in.)
1 cubic meter (m^3)	= 35.314667 cubic feet (cu. ft.)
1 cubic meter (m^3)	= 1.307951 cubic yards (cu. yd.)

English-Metric
Equivalent Measurements

Volume Measures for Fluids

1 gallon (gal.)	= 3,785.411 cubic centimeters (cm^3)
1 gallon (gal.)	= 3.785411 liters (L)
1 quart (qt.)	= 0.946353 liter (L)
1 ounce (oz.)	= 29.573530 cubic centimeters (cm^3)
1 cubic centimeter (cm^3)	= 0.000264 gallon (gal.)
1 liter (L)	= 0.264172 gallon (gal.)
1 liter (L)	= 1.056688 quarts (qt.)
1 cubic centimeter (cm3)	= 0.033814 ounce (oz.)

Weight Measures

1 pound (lb.)	= 0.453592 kilogram (kg)
1 pound (lb.)	= 453.59237 grams (g)
1 ounce (oz.)	= 28.349523 grams (g)
1 ounce (oz.)	= 0.028350 kilogram (kg)
1 kilogram (kg)	= 2.204623 pounds (lbs.)
1 gram (g)	= 0.002205 pound (lb.)
1 kilogram (kg)	= 35.273962 ounces (oz.)
1 kilogram (kg)	= 1,000 grams (g)
1 gram (g)	= 0.035274 ounce (oz.)
1 gram (g)	= 1,000 milligrams (mg)
1 metric ton	= 1,000 kilograms (kg)

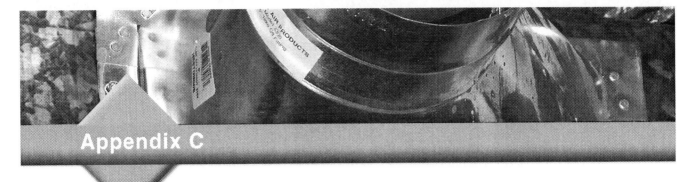

Area and Volume Formulas

Area Formulas

Right Triangle:	$a = \frac{1}{2}(bh)$	(one-half the product of the base times the height)
Rectangle:	$a = lw$	(length times width)
Square:	$a = s^2$	(side times side)
Parallelogram:	$a = bh$	(base times height)
Trapezoid:	$a = \frac{1}{2}h(b1+b^2)$	(one-half the height times the sum of the lengths of the two parallel sides)
Cube:	$s = l^2 \times 6$	(length times length times 6)
Circle:	$a = \pi r^2$	(pi times the square of the radius)
Arc:	$l = \dfrac{n}{360}(\pi d)$	(the product of the arc divided by 360 and the diameter times pi)
Ellipse:	$a = 0.7854ab$	(0.7854 times the narrowest diameter times the widest diameter)
Rectangular Solid:	$s = 2lw + 2lh + 2wh$	(two times length times width plus two times length times height plus two times width times height)

Cylinder

Lateral area:	$s = \pi dh$	(pi times diameter times height)
Surface area:	$s = \pi dh + 2 \times 0.7854d^2$	(pi times diameter times height plus two times .7854 times diameter squared)
Sphere (surface):	$s = \pi d^2$	(pi times diameter squared)
Semicircular-Sided Figure:	$a = \pi r^2 + dl$	(pi times the radius squared plus diameter times length)

Volume Formulas

Rectangular Solid:	$v = lwh$	(length times width times height)
Cube:	$v = s^3$	(side times side times side)
Right Triangular Solid:	$v = \frac{1}{2}(lwh)$	(one-half the product of length times width times height)
Cylinder:	$v = (\pi r^2)h$	(area of the circular base times the height)
Sphere:	$v = \pi d^2$	(pi times the square of the diameter)

Coefficient: A number or mathematical sign (plus or minus) that appears directly in front of a variable or grouping symbol.

Complement: In a right triangle, a non-right angle that, when added to the other non-right angle, equals 90 degrees.

Cosine: In a right triangle, the ratio between the side adjacent to an acute angle and the hypotenuse.

Fractional percentage: A percentage that is less than one percent.

Function: For an angle in a right triangle, the ratio of the sides and hypotenuse.

Hypotenuse: The side of a right-angled triangle that is opposite the right angle.

Lateral area: In a cylinder, the area of the surface that is perpendicular to the two circular sides.

Ratio: The relationship in quantity, amount, or size between two or more things.

Sine: In a right triangle, the ratio of the side opposite an acute angle and the hypotenuse.

Tangent: In a right triangle, the ratio of the side opposite an acute angle to the side adjacent to the angle.

Additional Resources

This module is intended to present thorough resources for task training. The following reference works are suggested for further study. These are optional materials for continued education rather than for task training.

Algebra and Trigonometry, Third Edition. 2008. Judith Beecher et al. Boston, MA: Addison-Wesley Longman, Inc.

Applied Construction Math: A Novel Approach. 2006. National Center for Construction Education and Research. Upper Saddle River, NJ: Prentice Hall.

Basic Math, Algebra, and Geometry with Applications. 2004. Cheryl Cleaves and Margie Hobbs. Upper Saddle River, NJ: Prentice Hall.

Math on Call: A Mathematics Handbook. 2004. Andrew Kaplan et al. Wilmington, MA: Great Source Education Group.

NCCER makes every effort to keep these textbooks up-to-date and free of technical errors. We appreciate your help in this process. If you have an idea for improving this textbook, or if you find an error, a typographical mistake, or an inaccuracy in NCCER's Contren® textbooks, please write us, using this form or a photocopy. Be sure to include the exact module number, page number, a detailed description, and the correction, if applicable. Your input will be brought to the attention of the Technical Review Committee. Thank you for your assistance.

Instructors – If you found that additional materials were necessary in order to teach this module effectively, please let us know so that we may include them in the Equipment/Materials list in the Annotated Instructor's Guide.

Write: Product Development and Revision
National Center for Construction Education and Research
3600 NW 43rd St, Bldg G, Gainesville, FL 32606

Fax: 352-334-0932

E-mail: curriculum@nccer.org

Craft _____   Module Name _____

Copyright Date _____   Module Number _____   Page Number(s) _____

Description _____

_____

_____

_____

(Optional) Correction _____

_____

_____

(Optional) Your Name and Address _____

_____

_____

04202-08

Plans and Specifications

04202-08
Plans and Specifications

Topics to be presented in this module include:

Overview

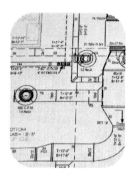

By the time that plans and specifications have made their way to the construction site, they have been subjected to considerable scrutiny. Their final form is a result of numerous negotiations between the building owner, architect, mechanical engineers, estimators, and other trades on site to arrive at a final system design to which everyone has agreed and signed-off. It is critical that you are able to interpret those plans and specifications accurately so you can create and install the system as described. Deviation from the plans or a misinterpretation could result in having to tear out and reinstall portions of a system which could seriously jeopardize the company's ability to turn a profit on the project. In this module, you will learn how to read and interpret building plans and specifications.

Objectives

When you have completed this module, you will be able to do the following:

1. Read selected plans to interpret project information.
2. Read and interpret section, elevation, and detail drawings.
3. Read specifications for information about selected parts of a construction job.

Trade Terms

As-built drawing	Member
Change order	Riser diagram
Elevation	Roof pitch
Footing	Schedule
Joist	Scupper
Joist hanger	Section
Legend	Submittal
Load capacity	

Required Trainee Materials

1. Pencil and paper
2. Calculator
3. Color highlighter
4. Set of plans provided with this module

Prerequisites

Before you begin this module, it is recommended that you successfully complete the following: *Core Curriculum; Sheet Metal Level One; Sheet Metal Level Two*, Module 04201-08.

This course map shows all of the modules in the second level of the Sheet Metal curriculum. The suggested training order begins at the bottom and proceeds up. Skill levels increase as you advance on the course map. The local Training Program Sponsor may adjust the training order.

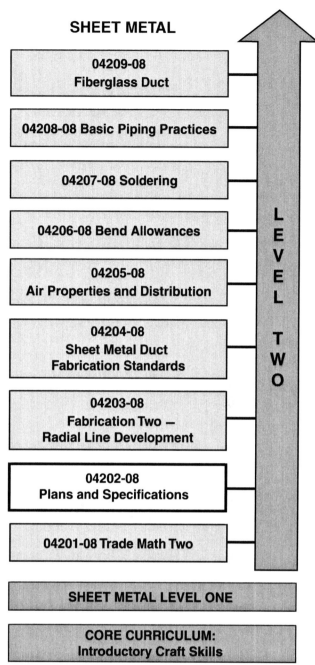

SHEET METAL

04209-08 Fiberglass Duct

04208-08 Basic Piping Practices

04207-08 Soldering

04206-08 Bend Allowances

04205-08 Air Properties and Distribution

04204-08 Sheet Metal Duct Fabrication Standards

04203-08 Fabrication Two — Radial Line Development

04202-08 Plans and Specifications

04201-08 Trade Math Two

SHEET METAL LEVEL ONE

CORE CURRICULUM: Introductory Craft Skills

LEVEL TWO

202CMAP.EPS

1.0.0 ◆ INTRODUCTION

Commercial construction is a complex process involving the work of many people, including architects, engineers, plumbers, carpenters, roofers, electricians, sheet metal workers, stonemasons, and landscapers. In addition, representatives of local government are involved in the construction process. This group includes people in the building department, who review the project documents and issue permits, record the ownership, and inspect the work to ensure its compliance with applicable codes. All of these people need to know what work is to be done, how it is to be done, and who is to do it. That information is recorded in the construction documents, which include contracts, specifications, and plans.

As a sheet metal worker, you may believe that you should focus on reading plans that pertain only to your trade. However, you will find that your work must be done in coordination with many other trades, so knowing how to read and understand plans used by other trades is essential for building a quality project on time and within budget.

At first glance, a complete set of commercial plans will seem confusing. With practice, however, you will soon learn to tell one type of plan from another, see how the different drawings in a set of plans relate to one another, and understand the lines and symbols in the drawings. This module gives you an overview of plans and specifications and includes exercises that will give you hands-on practice in working with these documents.

2.0.0 ◆ WORKING WITH PLANS

Plans for commercial projects include many drawings, **schedules**, notes, and **legends**. When working with commercial plans, keep the following points in mind.

- Read the title block (*Figure 1*). The title block contains information about the drawing such as the address of the building to be constructed, the scale, the drawing number, the date of the latest revision, and the name and address of the architect or engineer.
- Find the north arrow on the drawing (*Figure 2*). This arrow is included to show how the building is situated on the property. If you know where north is, you can accurately locate or describe the locations of walls and other parts of the building in relation to the north arrow.

CODE INFORMATION		REVISIONS
CODE USED: 1997 UBC		
OCCUPANCY GROUP: B		
TYPE OF CONSTRUCTION: TYPE V N		
BUILDING AREA:		**DRAWN BY** JLF/LSG
MAIN FLOOR: 5,890 SQ. FT. MECHANICAL ROOM: 324 SQ. FT.		**DATE** 5-30-00
TOTAL: 6,214 SQ. FT.		**PROJECT NO.** 9918

I HEREBY CERTIFY THAT THIS PLAN, SPECIFICATION, OR REPORT WAS PREPARED BY ME OR UNDER MY DIRECT SUPERVISION AND THAT I AM A DULY REGISTERED ARCHITECT UNDER THE LAWS OF THE STATE OF NORTH DAKOTA.

DATE: 5-30-2000   REG. NO. 578

DRAWING DESCRIPTION:

TITLE SHEET

A 0.1

PLOT SCALE: 1=1   FILE NAME: 9918_A_0-1.DWG

© 2000 Ritterbush – Ellig – Hulsing P.C. Architects – Planners

202F01.EPS

Figure 1 ◆ The title block.

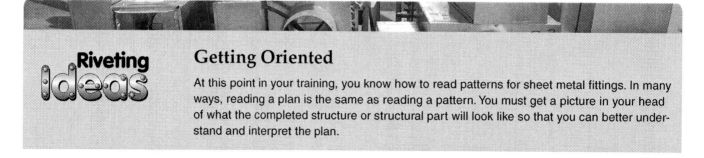

Riveting Ideas

Getting Oriented

At this point in your training, you know how to read patterns for sheet metal fittings. In many ways, reading a plan is the same as reading a pattern. You must get a picture in your head of what the completed structure or structural part will look like so that you can better understand and interpret the plan.

NOTE

Always verify the north direction on each drawing; it can change from drawing to drawing. The architect and/or engineer sometimes rotate the drawing layout to get more information on a sheet.

NORTH

202F02.EPS

Figure 2 ◆ North arrow.

- Drawings, schedules, legends, and notes in a set of plans are related to one another. It takes more than one type of drawing to show how a structure or part of a structure is to be built. Instructions and details that cannot be drawn are recorded in the schedules and notes. Because so much must be put into a drawing, symbols are used. These are defined in a legend. To fully understand what is to be built, you may have to look at more than one drawing and refer to notes, schedules, and legends to find the information you need.

- The architect may include addenda that clarify the project documents. These are usually written during the bidding process in response to questions from the contractor.

- If you do not recognize a symbol on a drawing, refer to the legend. You will find legends for many commonly used plan symbols in *Appendix A*.

- The thickness and type of lines used on plans may vary from one firm to another. However, in general, thicker lines show the overall outline of a structure, thinner lines indicate interior items (such as doors and windows), and dashed lines indicate items that are hidden from sight in the wall, floor, or ceiling (such as beams).

Specifications are contained in a large book organized so that each **section** applies to a specific trade or type of work. Plumbers, for example, will read the sections on plumbing but will probably not read the section that covers site work. It's a good idea, however, for every responsible person on the work site to review the sections of the specifications titled *Special Conditions* and *General Conditions*.

Sometimes a plan will contain what might be a mistake. You may, for example, add up the dimensions of the rooms inside a building and find that the total does not agree with the outside dimensions of the building, or a problem may crop up on the job site that requires an alteration to a drawing. The Request for Information (RFI) form has been developed to deal with these types of situations (*Figure 3*).

DID YOU KNOW?

Why Blueprints Were Called Blueprints

Blueprints, now an obsolete term, were plans that were hand-drawn on specially coated paper. They were then weighted with glass and set in a sunny spot. As a result of the development process, the coated paper turned blue, leaving the lines of the drawing white. Later, more advanced technologies enabled blueprints to be computer-generated on white paper with blue or black lines. You may be able to find blueprints in a museum or in the archives of some architectural firms.

Riveting Ideas

Clear Communication

Time is money, according to an old saying. That is certainly true in the construction business. All construction work must be scheduled so that it can proceed smoothly without delays. The heating, ventilating, and air conditioning (HVAC) workers, for example, cannot hang duct until the structural work is completed, and it's not desirable to have the plumbers running pipe at the same time that HVAC workers are installing duct. When a problem occurs that could delay the work of any trade, the general contractor wants to get it solved quickly. A clearly written RFI delivered in a timely fashion to all responsible people helps keep the job on schedule. Learn the process for handling RFIs at your work site so that you can do your part to keep the work on schedule.

DATE _____ RFI NO. _____

PROJECT NAME _____ PROJECT NO. _____

REQUEST: REF D.W.G.NO. _____ REV. _____ OTHER _____

_____
_____
_____
_____
_____
_____
_____
_____
_____
_____
_____
_____
_____
_____
_____

BY: _____ REPLY BY (DATE): _____

REPLY:

_____
_____
_____
_____
_____
_____
_____
_____
_____
_____
_____
_____
_____
_____
_____

DATE: _____

202F03.EPS

Figure 3 ◈ Request for Information.

Riveting Ideas

Contour Lines

A system of contour lines is used to describe the rise and fall of the ground at a site. Contour lines are spaced at regular intervals (usually one foot) which indicate the grade of the ground. Therefore, when you see contour lines that are close together, you know that the land grades steeply; the farther apart the lines, the more gradual the grade.

Most construction firms have their own forms and procedures for handling RFIs. As an apprentice, you probably will not be authorized to write an RFI and send it to the architect on your own. If you notice a problem or think that there may be a mistake on the plans, notify your supervisor. Your supervisor will complete the RFI and send it to the superintendent or project manager, who will review it and forward it to the general contractor. The general contractor will also review the RFI before sending it on to the architect or engineer. This system allows the supervisors and general contractor to be aware of all problems on the job site, to solve them, and to keep the work on schedule.

3.0.0 ◆ THE PLANS

For a commercial project, a complete set of plans will contain several types of drawings or plans. Each plan may, in turn, have its own set of additional drawings, notes, schedules, addenda, and legends. Plans may be organized into several broad categories:

- Civil
- Architectural
- Mechanical
- Structural
- Plumbing
- Electrical

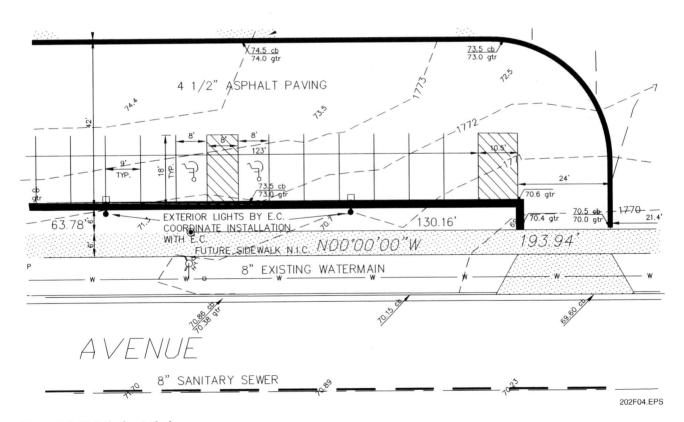

Figure 4 ◆ Detail of a civil plan.

3.1.0 Civil Plan

The civil plan is an aerial view that shows the location of the building or buildings to be built on the site (*Figure 4*). Civil plans are sometimes called site plans, survey plans, or plot plans. A civil plan may also show the contours of the ground surrounding the building site. Other information that may be shown on the civil plan includes the following:

- Landscape (showing existing trees and any that are to be removed)
- Water features (ponds, lakes, rivers, storm retention ponds)
- Sidewalks, driveways, curbs, and gutters
- Utilities
- Property dimensions
- Notes regarding grading
- A legal description of the property

All workers on the project must be able to accurately locate where their work is to be done on the building. For example, the project may call for connecting utilities to the southwest corner of a building or for locating a loading dock on the eastern side. To help workers orient themselves to the site, the civil plan includes an arrow that indicates which direction on the drawing is north.

3.2.0 Architectural Plan

The architectural plan includes several drawings that show how the various parts of the building should look from several views. These views include floor plans, **elevation** drawings, section drawings, and detail drawings. Also included in the architectural plan are schedules that describe the number, types, and sizes of such items as ceiling type and height, windows, and doors (*Figure 5*).

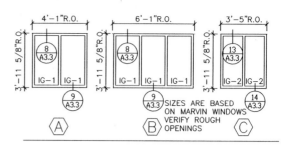

WINDOW TYPES

SCALE: 1/4" = 1'-0"

202F05.EPS

Figure 5 ◆ Detail of an architectural drawing showing window types.

3.2.1 Floor Plan

Like the site plan, a floor plan is an aerial view. Looking at a floor plan is similar to the view you would get if you could remove the roof of your building and look down from above (*Figure 6*). The floor plan contains a lot of detailed information that must be interpreted correctly by the different trades on a construction site. Included on a floor plan are the following items:

- Length and width of the building
- The outlines and dimensions (length and width) of rooms, hallways, elevator shafts, stairwells, doorways, doors, and windows
- The direction in which doors swing open, as well as the direction that any operable windows open
- Plumbing fixtures
- Inside utility connections
- Special features such as rolling shutters for front and rear entrances, built-in counters, or storage shelves

Scale

All drawings in a set of plans include a note showing the scale of the drawing. For example, the scale for a commercial drawing may read ⅛" = 1'0", which means that every ⅛-inch on the drawing is equal to 1 foot. Drawings are done to scale so that they will fit on paper of a manageable size. Drawings are drawn using architect's scales and engineer's scales. An architect's scale is usually triangular in shape, with graduations along its edges. An engineer's scale is three-sided and has several scales on it.

When working with scale, always use the drawing's dimensions. Don't use your own architect's scale or engineer's scale to measure the drawing itself because it may have been enlarged or reduced. A plan that is marked NTS (not to scale) means that the measurements give only relative positions and sizes. NTS measurements are not accurate enough for construction purposes.

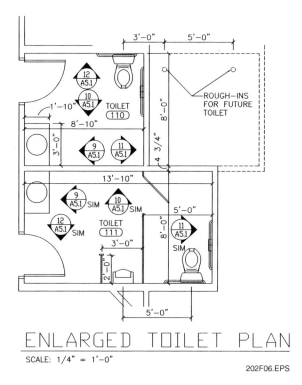

ENLARGED TOILET PLAN

SCALE: 1/4" = 1'-0"

202F06.EPS

Figure 6 ◆ Detail of a floor plan showing bathroom layout.

3.2.2 Elevation Drawings

The elevation drawing shows a face-on view of a building, room, or part of a building, such as a stairwell or elevator shaft. It is called an elevation because it shows height. A set of plans may include two types of elevation drawings: a presentation view and a working view. The presentation view gives a complete picture of how the proposed building will look (*Figure 7*). Working views are used by the construction team and contain construction details such as **roof pitch** and

the location of **footings**. Four exterior elevations are typically shown for one building. These views may be labeled north, south, east, and west or front, right side, left side, and rear. If two views are the same—for example, side views on a simple building—only three views are required. For irregularly shaped buildings, more than four views may be required. The following items are included on elevation views:

- Exterior style of the building, including steps, railings, and exterior lights
- Location of doors, windows, chimneys, covered entranceways, and decorative trim
- Exterior finish materials

An elevation view of the interior of a building shows the location of entranceways, doors, and windows and may also include other elements found inside a building such as stairways, ductwork, and plumbing fixtures.

3.2.3 Section Drawings

Sometimes the easiest way to understand how an object is put together is to take it apart. That is what, in effect, section drawings allow you to do (*Figure 8*). Section drawings show what an object or structure would look like if it were cut in half, so you can see how all the parts fit together. Most section views are vertical, but horizontal section views may be included. When the sectional view is of the long side of a building, it is called a longitudinal section. When the sectional view is of the short side of a building, it is called a transverse section. Section drawings are usually drawn to a larger scale than other drawings so details can easily be seen. In fact, section drawings often include detail drawings. Detail drawings are also

EAST ELEVATION

SCALE: 1/8" = 1'-0"

202F07.EPS

Figure 7 ◆ Detail of the presentation view.

Figure 8 ◈ Detail of a section drawing.

drawn to a larger scale. They focus on a detail or specific area of a building, piece of equipment, or an installation.

3.2.4 Schedules

Schedules are written descriptions that record the dimension, style, and materials used for such items as ceilings, windows, doors, hardware, mechanical equipment and controls, plumbing and electrical fixtures, and appliances (*Figure 9*). Each item on a schedule is identified by a number or letter and is referenced to the appropriate drawing in the set of plans.

3.3.0 Mechanical Plan

Engineers provide mechanical plans to support the architectural plans. Mechanical plans are drawn to show what is required to install a building's systems (*Figure 10*). These plans may include drawings of the following:

- Hot water heating or chilled water cooling distribution systems
- Sprinkler system
- Air distribution system
- Refrigeration system
- Control systems

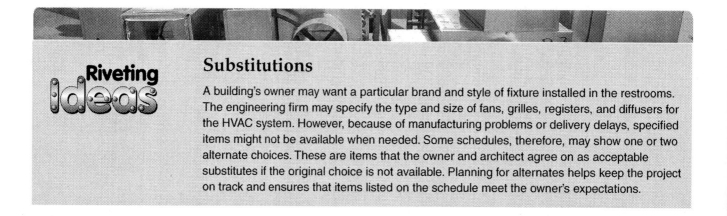

Riveting Ideas

Substitutions

A building's owner may want a particular brand and style of fixture installed in the restrooms. The engineering firm may specify the type and size of fans, grilles, registers, and diffusers for the HVAC system. However, because of manufacturing problems or delivery delays, specified items might not be available when needed. Some schedules, therefore, may show one or two alternate choices. These are items that the owner and architect agree on as acceptable substitutes if the original choice is not available. Planning for alternates helps keep the project on track and ensures that items listed on the schedule meet the owner's expectations.

SUPPLY AND EXHAUST FANS

MARK	TYPE	MFTR	MODEL NUMBER	SERVICE	CFM	S.P. IN.	WHEEL DIA. IN.	FAN RPM	FAN BHP	MOTOR HP	MAX SONES
EF-ED-1	EXH	GREENHECK		TRAUM1-851	1060						
EF-ED-2	EXH	GREENHECK		TRAUM1-850	1080						
EF-ED-3	EXH	GREENHECK		TRAUM1-846	1050						
EF-ED-4	EXH	GREENHECK		TRAUM1-844	1080						
EF-ED-5	EXH	GREENHECK		TRAUM1-843	1030						
EF-ED-6	EXH	GREENHECK		DECON1-833	270						
EF-ED-7	EXH	GREENHECK		ISO 1-863	420						
EF-ED-8	EXH	GREENHECK		ISO 1-868	260						
EF-ED-9	EXH	GREENHECK		ISO 1-873	260						
EF-ED-10	EXH	GREENHECK		GENERAL							
EF-ED-11	EXH	GREENHECK		GENERAL							
EF-ED-12	EXH	GREENHECK		TRIAGE/WAIT	2920						
EF-PARK-1	PROP VENT	GREENHECK		PARKING	9000						
EF-PARK-2	VENT	GREENHECK		PARKING	9000						

PARKING DECK VENTILATION FANS SHALL BE TWO SPEED

202F09.EPS

Figure 9 ◆ Schedule.

Mechanical plans also show all systems equipment such as pumps, heaters, compressors, boilers, fans, valves, and controls. General notes are also included in mechanical plans. These notes describe how the mechanical systems are to be installed and may include directions for the subcontractors. The notes may require, for example, that the HVAC subcontractor and the electrical subcontractor meet to coordinate the installation of the ductwork and wiring.

Before ordering the HVAC equipment, the HVAC subcontractor should confirm the equipment's electrical requirements. This information is included in the electrical drawings. If there is a difference, the HVAC subcontractor must write an RFI to clarify the correct electrical requirements.

3.4.0 Structural Plan

Included in the structural plan (*Figure 11*) are all of the drawings related to the framework of a structure, its foundation, and its roof. Most structural plans include section drawings, general notes, a foundation plan, and a roof-framing plan.

The section drawings are similar to architectural section drawings. However, in the structural plan, the section drawings focus on the structural framework. Details about the location of **joists** and **joist hangers**, connections between the framework **members**, and pitch (for roof plans) are included.

The general notes section includes applicable codes, a description of structural materials

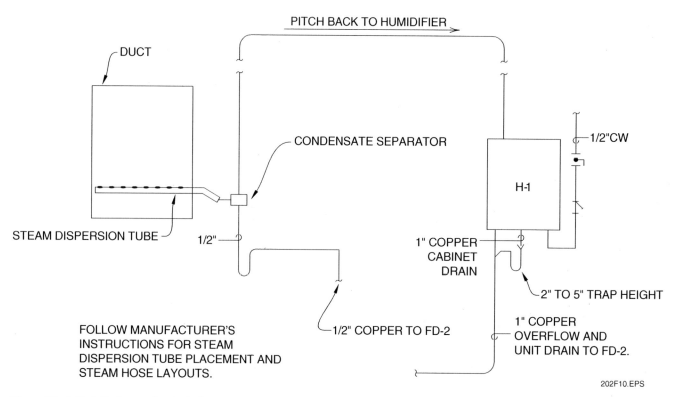

Figure 10 ◆ Detail of a mechanical plan.

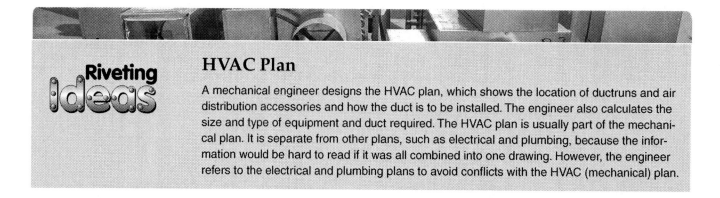

HVAC Plan

A mechanical engineer designs the HVAC plan, which shows the location of ductruns and air distribution accessories and how the duct is to be installed. The engineer also calculates the size and type of equipment and duct required. The HVAC plan is usually part of the mechanical plan. It is separate from other plans, such as electrical and plumbing, because the information would be hard to read if it was all combined into one drawing. However, the engineer refers to the electrical and plumbing plans to avoid conflicts with the HVAC (mechanical) plan.

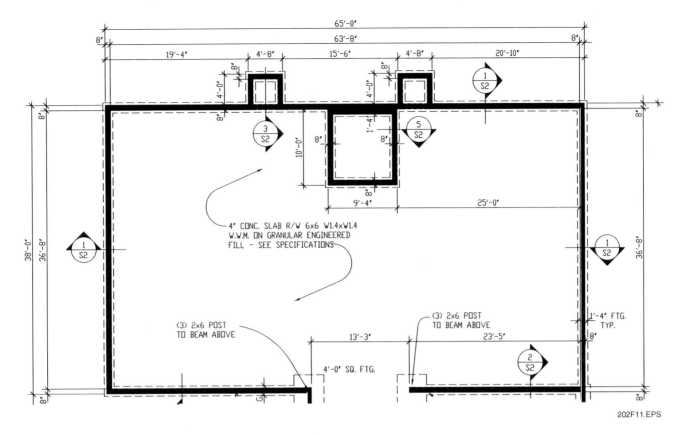

Figure 11 ◆ Detail of a structural plan.

required, and requirements for workers on the site. For example, the notes may require that all welders on the job have proof of having passed tests that verify they are qualified to use welding equipment.

The foundation plan is a drawing of the lowest level of a building. It includes the dimension and location of footings, steel girders, columns, beams, and other supporting structures. Related to the foundation plan is the structural floor plan, which includes the type and size of the framing and the underlying support for each floor in a building.

The roof plan includes a roof framing plan and a roof drainage plan. Mechanical equipment is often installed on the roofs of commercial buildings. Therefore, the framing must be designed to support not only the roof load but the concentrated loads of equipment installed there as well. Water from rain or melted snow can add to the load placed on the roof. Although the roofs on most commercial buildings appear to be flat, they include a pitch to allow water to drain. The roof drainage plan may include the location of drains, **scuppers**, downspouts, and/or gutters. These items may also be shown on the architectural drawings and details.

3.5.0 Plumbing Plan

The plumbing plan for a commercial building is similar to that of a residential building. It shows the layout of pipes and fittings that supply water to plumbing fixtures and carry away waste and gases (*Figure 12*). The plumbing plans for a manufacturing facility will also include plans for pipes that carry chemicals, gases, or hazardous waste products.

The plumbing plan looks similar to the floor plan. However, all details except for overall dimensions are removed so that the plumbing symbols can be easily read. Like the other plans in a set of plans, the plumbing plan also includes general notes, a schedule showing the type and sizes of plumbing fixtures, a legend, and detail drawings. On a multistory building, the plumbing plan will also include a **riser diagram** that shows the path of the piping from one floor to the next.

The plumbing plan may also show hot water heater vents, which are not always shown on the HVAC (mechanical) plan. The sheet metal contractor may be responsible for furnishing and installing these vents, as well as for installing combustion and relief ducts to the hot water heater.

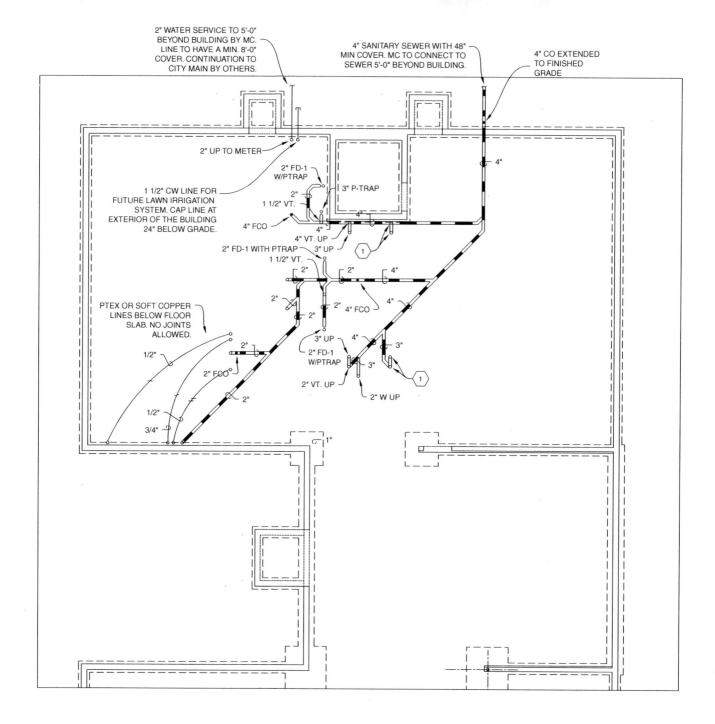

UNDERGROUND PIPING PLAN

1/8" = 1'-0"

2
M1

NORTH

202F12.EPS

Figure 12 ◆ Detail of a plumbing plan.

3.6.0 Electrical Plan

The electrical plan is similar to the plumbing plan. Instead of showing the path of pipes, however, it shows the path of the electrical wiring (*Figure 13*). In addition to details about **load capacities** and wire sizes, the plan includes the location for meters, distribution panels, fixtures, and switches. Like the plumbing plan, the electrical plan looks similar to the floor plan. All details except overall dimensions are removed so that the electrical symbols can be easily read. Notes and a legend describing the electrical symbols used on the plan are also included along with an electrical distribution riser diagram. The electrical plan will also include a mechanical equipment schedule.

4.0.0 ◆ SPECIFICATIONS

The specifications document spells out details such as:

- The type, color, and style of plumbing fixtures to be installed.
- The soundproofing standards the ductwork must meet.
- How walls, ceilings, and floors in mechanical areas are to be finished.

A sample specifications document is included in *Appendix B.*

Writing accurate and complete specifications is a serious responsibility and can be quite complicated. The specifications, combined with the plans and addenda, govern all of the decisions made during a construction project, and a single project can involve thousands of details.

Specifications are an important part of the construction contract. After the responsible parties agree to the contract, the general contractor provides copies of the specifications to all of the subcontractors on the job. Specifications include the following general information:

- Written descriptions of items that cannot be shown on the drawings
- Work standards, types of material to be used, and the responsibility of various parties to the contract
- Applicable building codes and zoning ordinances

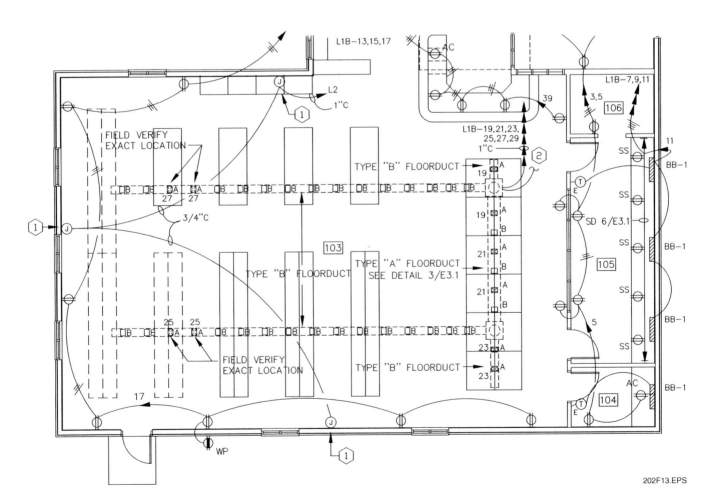

Figure 13 ◆ Detail of an electrical plan.

202F13.EPS

- Contract terms
- Permits and payment of fees
- Use and installation of utilities
- Supervision of construction

Even on a relatively small commercial job, a lot of information must be written down and organized so that all of the trades involved in a project can find the information that describes work for which they are responsible. The Construction Specifications Institute has developed a system to organize project information in a logical way. The system was updated in 2004 to accommodate more categories. Until the new CSI MasterFormat 2004 system works its way into the industry, you may encounter specifications following either of these two standards. Under the 2004 system, all construction work is divided into 49 divisions (*Table 1*), several of which are reserved for future use. These divisions, in turn, may have several sections and subsections (*Table 2*). Not all divisions will be included in every contract; however, most specifications include a section titled *Division 1—General Requirements*. In addition, many specifications will contain a section titled *Special Conditions*, which focuses on conditions specific to the job. Every responsible person involved with a construction project should review the general requirements and special conditions sections. Some jobs may still use the old 1995 specifications (*Table 3*).

5.0.0 ◆ SHOP DRAWINGS

Shop drawings show how a specific portion of the work is to be done, such as the fabrication and installation of ductruns. For large commercial jobs, a draftsman creates shop drawings based on a design drafted by an engineer. On smaller jobs, the draftsman may work from freehand sketches based on field measurements. Shop drawings, like

DID YOU KNOW?

CSI

The Construction Specifications Institute (CSI) is a membership organization located in Alexandria, Virginia. It works to meet the construction industry's need for a common system of organizing construction information. CSI's members include architects, engineers, builders, suppliers, building owners, and facilities managers. Founded in 1948, CSI has many chapters nationwide.

section and detail drawings, are drawn to a larger scale than the engineer's design.

An example of part of a shop drawing for ductwork installation is shown in *Figure 14*. Note that this shop drawing is not for the same building as shown throughout this module. A complete shop drawing for ductwork installation will include the following information:

- Overall dimensions
- Assembly instructions
- Installation instructions
- Exact layout of the ductwork
- Partitions, exterior walls, columns, beams, hanging ceilings, and any other potential obstructions to the ductwork
- Size and type of fittings
- Size and types of hangers and supports
- Notes that will assist in fabrication or installation

Depending on the contractor's preferred methods, either the draftsman or the shop foreman will create cut lists for each fitting. The cut list may be compiled from fabrication tickets created by sheet metal workers in the field or from looking at the construction drawings. The cut list may be a preprinted form generated by the draftsman (*Figure 15*), or it may be computer generated (*Figure 16*).

A number is assigned to the cut list, which matches a number on the shop drawing. Cut lists include the number and size of the fittings required and details about how the fittings are to be made.

The draftsman or sheet metal worker performs a number of tasks to ensure that the work will be done properly. For example, when drafting the shop drawings for ductwork, the draftsman does the following tasks:

- Checks the electrical, plumbing, and mechanical drawings to avoid conflicts during installation
- Consults available reference materials regarding approved HVAC equipment layout dimensions
- Checks the gauge specifications
- Notes where watertight duct construction is required, such as in rooms containing showers or dishwashing equipment
- Checks and verifies the proper location of dampers (including, but not limited to, fire, smoke, and volume dampers) and ensures that they are accessible
- Writes RFIs to the attention of the architect or designer regarding missing information or problems or to note where coordination of the work by other trades is critical

Table 1 Specifications Divisions, 2004 edition

MasterFormat™ 2004 Edition – Division Numbers & Titles

Division	Areas Covered	Division	Areas Covered

PROCUREMENT AND CONTRACTING REQUIREMENTS GROUP

Division 00 — Procurement and Contracting Requirements

SPECIFICATIONS GROUP

GENERAL REQUIREMENTS SUBGROUP

Division 01 — General Requirements

FACILITY CONSTRUCTION SUBGROUP

Division 02	Existing Conditions
Division 03	Concrete
Division 04	Masonry
Division 05	Metals
Division 06	Wood, Plastics, and Composites
Division 07	Thermal and Moisture Protection
Division 08	Openings
Division 09	Finishes
Division 10	Specialties
Division 11	Equipment
Division 12	Furnishings
Division 13	Special Construction
Division 14	Conveying Equipment
Division 15–19	Reserved

FACILITY SERVICES SUBGROUP

Division 20	Reserved
Division 21	Fire Suppression
Division 22	Plumbing
Division 23	Heating, Ventilating, and Air Conditioning
Division 24	Reserved
Division 25	Integrated Automation
Division 26	Electrical
Division 27	Communications
Division 28	Electronic Safety and Security
Division 29	Reserved

SITE AND INFRASTRUCTURE SUBGROUP

Division 30	Reserved
Division 31	Earthwork
Division 32	Exterior Improvements
Division 33	Utilities
Division 34	Transportation
Division 35	Waterway and Marine Construction
Division 36–39	Reserved

PROCESS EQUIPMENT SUBGROUP

Division 40	Process Integration
Division 41	Material Processing and Handling Equipment
Division 42	Process Heating, Cooling, and Drying Equipment
Division 43	Process Gas and Liquid Handling, Purification, and Storage Equipment
Division 44	Pollution Control Equipment
Division 45	Industry-Specific Manufacturing Equipment
Division 46	Reserved
Division 47	Reserved
Division 48	Electrical Power Generation
Division 49	Reserved

Table 2 Specifications Division 23 with Sample Subdivisions, 2004 edition

DIVISION 23 – HEATING, VENTILATING, AND AIR-CONDITIONING (HVAC)

Division	Areas Covered

23 00 00 HEATING, VENTILATING, AND AIR-CONDITIONING (HVAC)

23 01 00	**Operation and Maintenance of HVAC Systems**
23 01 10	Operation and Maintenance of Facility Fuel Systems
23 01 20	Operation and Maintenance of HVAC Piping and Pumps
23 01 30	Operation and Maintenance of HVAC Air Distribution
23 01 30.51	HVAC Air Duct Cleaning
23 01 50	Operation and Maintenance of Central Heating Equipment
23 01 60	Operation and Maintenance of Central Cooling Equipment
23 01 60.71	Refrigerant Recovery/Recycling
23 01 70	Operation and Maintenance of Central HVAC Equipment
23 01 80	Operation and Maintenance of Decentralized HVAC Equipment
23 01 90	Diagnostic Systems for HVAC
23 05 00	**Common Work Results for HVAC**
23 05 13	Common Motor Requirements for HVAC Equipment
23 05 16	Expansion Fittings and Loops for HVAC Piping
23 05 19	Meters and Gages for HVAC Piping
23 05 23	General-Duty Valves for HVAC Piping
23 05 29	Hangers and Supports for HVAC Piping and Equipment
23 05 33	Heat Tracing for HVAC Piping
23 05 48	Vibration and Seismic Controls for HVAC Piping and Equipment
23 05 53	Identification for HVAC Piping and Equipment
23 05 63	Anti-Microbial Coatings for HVAC Ducts and Equipment
23 05 66	Anti-Microbial Ultraviolet Emitters for HVAC Ducts and Equipment
23 05 93	Testing, Adjusting, and Balancing for HVAC
23 06 00	**Schedules for HVAC**
23 06 10	Schedules for Facility Fuel Service Systems
23 06 20	Schedules for HVAC Piping and Pumps
23 06 20.13	Hydronic Pump Schedule
23 06 30	Schedules for HVAC Air Distribution
23 06 30.13	HVAC Fan Schedule
23 06 30.16	Air Terminal Unit Schedule
23 06 30.19	Air Outlet and Inlet Schedule
23 06 30.23	HVAC Air Cleaning Device Schedule
23 06 50	Schedules for Central Heating Equipment
23 06 50.13	Heating Boiler Schedule
23 06 60	Schedules for Central Cooling Equipment
23 06 60.13	Refrigerant Condenser Schedule
23 06 60.16	Packaged Water Chiller Schedule
23 06 70	Schedules for Central HVAC Equipment
23 06 70.13	Indoor, Central-Station Air-Handling Unit Schedule
23 06 70.16	Packaged Outdoor HVAC Equipment Schedule
23 06 80	Schedules for Decentralized HVAC Equipment
23 06 80.13	Decentralized Unitary HVAC Equipment Schedule
23 06 80.16	Convection Heating and Cooling Unit Schedule
23 06 80.19	Radiant Heating Unit Schedule
23 07 00	**HVAC Insulation**
23 07 13	Duct Insulation

Table 3 Specifications Divisions with Sample Subdivisions, 1995 edition

Division	Areas Covered	Division	Areas Covered

Division 1—General Requirements
01100 Summary
01200 Price and Payment Procedures
01300 Administrative Requirements
01400 Quality Requirements
01500 Temporary Facilities and Controls

Division 2—Site Construction
02050 Basic Site Materials and Methods
02100 Site Remediation
02200 Site Preparation
02300 Earthwork
02400 Tunneling, Boring, and Jacking

Division 3—Concrete
03050 Basic Concrete Materials and Methods
03100 Concrete Forms and Accessories
03200 Concrete Reinforcement
03300 Cast-in-Place Concrete
03400 Precast Concrete

Division 4—Masonry
04050 Basic Masonry Materials and Methods
04200 Masonry Units
04400 Stone
04500 Refractories
04600 Corrosion-Resistant Masonry

Division 5—Metals
05050 Basic Metal Materials and Methods
05100 Structural Metal Framing
05200 Metal Joists
05300 Metal Deck
05400 Cold-Formed Metal Framing

Division 6—Wood and Plastics
06050 Basic Wood and Plastic Materials
and Methods
06100 Rough Carpentry
06200 Finish Carpentry
06400 Architectural Woodwork
06500 Structural Plastics

Division 7—Thermal and Moisture Protection
07050 Basic Thermal and Moisture Protection
Materials and Methods
07100 Dampproofing and Waterproofing
07200 Thermal Protection
07300 Shingles, Roof Tiles, and Roof Coverings
07400 Roofing and Siding Panels
07600 Architectural Sheet Metal

Division 8—Doors and Windows
08050 Basic Door and Window Materials and
Methods
08100 Metal Doors and Frames
08200 Wood and Plastic Doors
08300 Specialty Doors
08400 Entrances and Storefronts

Division 9—Finishes
09050 Basic Finish Materials and Methods
09100 Metal Support Assemblies
09200 Plaster and Gypsum Board
09300 Tile
09400 Terrazzo

Division 10—Specialties
10100 Visual Display Boards
10150 Compartments and Cubicles
10200 Louvers and Vents
10240 Grilles and Screens
10250 Service Walls

Division 11—Equipment
11010 Maintenance Equipment
11020 Security and Vault Equipment
11030 Teller and Service Equipment
11040 Ecclesiastical Equipment
11050 Library Equipment

Division 12—Furnishings
12050 Fabrics
12100 Art
12300 Manufactured Casework
12400 Furnishings and Accessories
12500 Furniture

Division 13—Special Construction
13010 Air-Supported Structures
13020 Building Modules
13030 Special Purpose Rooms
13080 Sound, Vibration, and Seismic Control
13090 Radiation Protection

Division 14—Conveying Systems
14100 Dumbwaiters
14200 Elevators
14300 Escalators and Moving Walks
14400 Lifts
14500 Material Handling

Division 15—Mechanical
15050 Basic Mechanical Materials and Methods
15100 Building Services Piping
15200 Process Piping
15300 Fire Protection Piping
15400 Plumbing Fixtures and Equipment
15500 Hydronic Piping
15750 Heat Transfer
15780 Packaged Air Conditioning Units
15850 Air Handling Equipment
15880 Air Distribution and Sheet Metal
15900 Temperature Control Systems
15990 Testing, Adjusting, and Balancing

Division 16—Electrical
16050 Basic Electrical Materials and Methods
16100 Wiring Methods
16200 Electrical Power
16300 Transmission and Distribution
16400 Low-Voltage Distribution

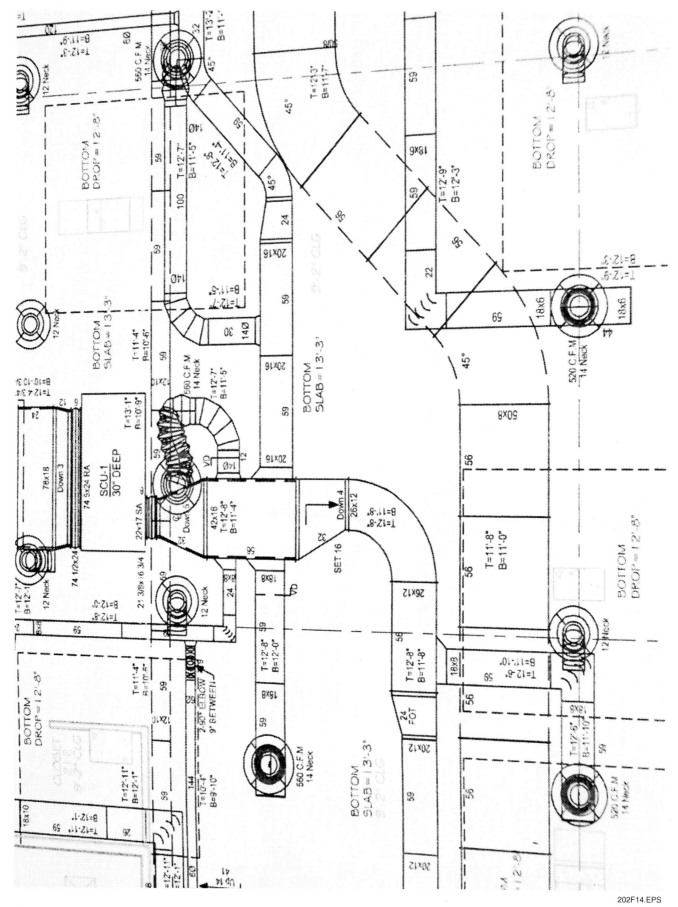

Figure 14 ◆ Section of a shop drawing showing duct installation.

202F14.EPS

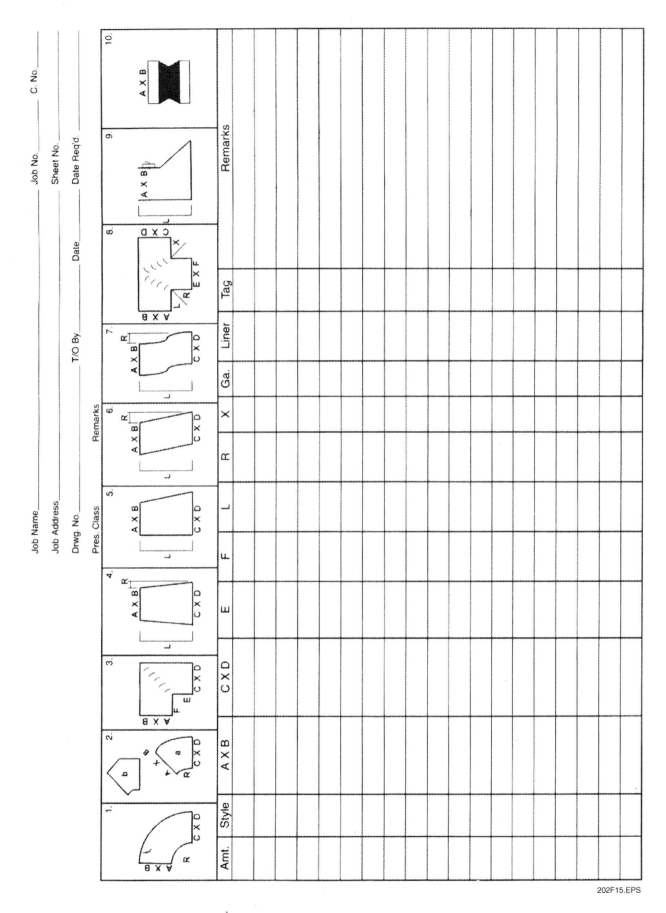

Figure 15 ◆ Draftsman-generated cut list.

202F15.EPS

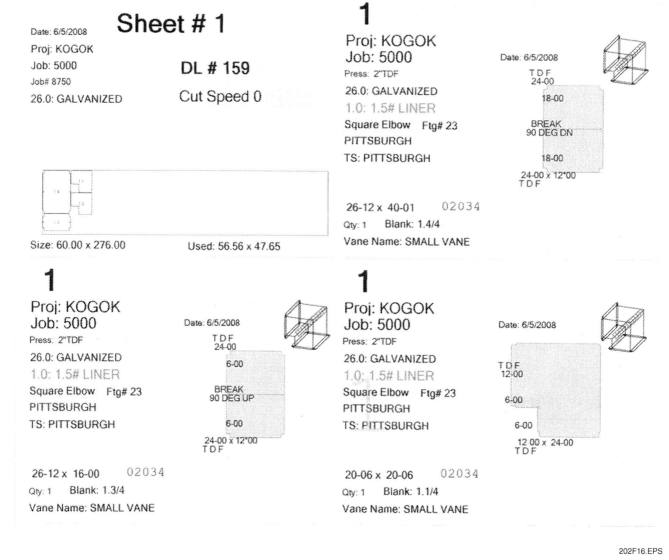

Figure 16 ◆ Computer-generated cut list.

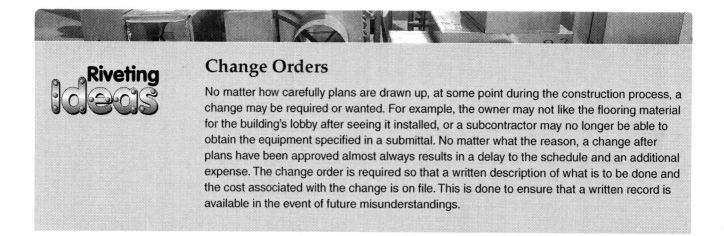

Change Orders

No matter how carefully plans are drawn up, at some point during the construction process, a change may be required or wanted. For example, the owner may not like the flooring material for the building's lobby after seeing it installed, or a subcontractor may no longer be able to obtain the equipment specified in a submittal. No matter what the reason, a change after plans have been approved almost always results in a delay to the schedule and an additional expense. The change order is required so that a written description of what is to be done and the cost associated with the change is on file. This is done to ensure that a written record is available in the event of future misunderstandings.

6.0.0 ◆ SUBMITTALS

Submittals are descriptions and drawings of special equipment or accessories that are to be supplied and installed by the subcontractor. These drawings and descriptions come from the equipment manufacturer. After successfully bidding on a job, a subcontractor gives copies of submittals to the general contractor for the project. The submittal should meet the standards noted in the specifications. *Figure 17* shows a typical submittal.

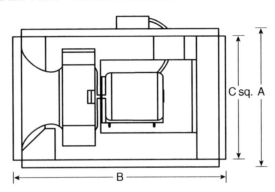

SUBMITTAL SHEET HV 4-8

IN-LINE CENTRIFUGAL DUCT FANS

MODEL VIDB – DIRECT DRIVE

UNIT SIZE	A	B	C	INLET OR OUTLET AREA	WHEEL DIA.	WT.
06	13⅞"	20"	12"	.979 SQ. FT.	10¾"	30
08	13⅞"	20"	12"	.979 SQ. FT.	10¾"	40
10	13⅞"	20"	12"	.979 SQ. FT.	10¾"	40
12	17⅞"	27⅜"	16"	1.750 SQ. FT.	11¹³⁄₁₆"	75
15	*21⅞"	31"	20"	2.740 SQ. FT.	14⅞"	90
18	*21⅞"	33⅜"	26"	4.650 SQ. FT.	17¹³⁄₁₆"	140

* "A"–1" Larger on access door sides

JOB NAME AND LOCATION	SUBMITTED BY

202F17A.EPS

Figure 17 ◆ A submittal for a duct fan (1 of 2).

SUBMITTAL SHEET HV 4-8

IN-LINE CENTRIFUGAL DUCT FANS

```
Centrifugal in-line duct fans shall be Carnes Company,
Model VIDB direct drive or Model VIBS belt drive as shown
on the plans and schedules. Fans shall be constructed of
heavy-gauge steel and electrocoated acrylic enamel finish
over phosphate primer. Wheels 12 inches in diameter and
larger shall have median foil blades to ensure quiet
operation. The motor-drive compartment shall be isolated
from the air stream and externally ventilated. Bearings
shall be lubricated and sealed for minimum maintenance
and designed for 200,000 hours of operation. Internal
parts shall be accessible for inspection or service
without disturbing inlet or outlet ductwork. Fans shall
be furnished with a mounted safety disconnect. Single-
phase motors shall have integral overload protection.
V-belt drives shall be adjustable. Horsepower and noise
levels shall not exceed the values shown. Oversized
motors are not acceptable. Performance ratings shall be
certified for air and sound.
```

202F17B.EPS

Figure 17 ◆ A submittal for a duct fan (2 of 2).

The submittal is reviewed by the general contractor, building's owner, architect, and any code enforcement authorities. Once all concerned parties accept the submittal, the subcontractor may install the equipment. An accepted submittal may not be changed without a written **change order** signed by all responsible parties.

7.0.0 ◆ AS-BUILT DRAWINGS

As-built drawings show completed mechanical installations. Sheet metal contractors prepare these drawings after the project is completed. As-built drawings can be used as a reference in the following situations:

- When alterations or additions are to be made to the building
- When modifications must be made to make room for an additional mechanical installation
- When a piece of mechanical equipment must be relocated

The draftsman or sheet metal worker uses separate symbols or notes to identify what equipment was removed and what equipment was relocated. These changes are recorded on the architectural plan in a different color so that it's clear where the changes differ from the original plan. The changes then become part of the permanent files for the building's mechanical systems.

8.0.0 ◆ EXERCISES

In this section, you will get hands-on practice in working with plans and specifications. There are five exercises in this section. To complete them, you will use the following drawings from the set of plans provided with this module:

- A 0.1 (title sheet)
- C-1, C-2 (site survey and site plan)
- A 2.1 (floor plan schedules and door details)
- A 3.1 (building elevations)
- A 3.2 (building sections)
- A 3.3 (wall sections, details)
- A 5.1 (interior elevations, casework, and details)
- A 6.1 (reflected ceiling plan)
- S1 and S2 (structural plans and details)
- M1 and M2 (mechanical plans)
- E 1.1, E 2.1, E 3.1, E 4.1 (electrical plans, details, and schedules)

You will also do an exercise using the specifications in *Appendix B*. This exercise is designed to give you practice in reading specifications and finding the required information.

Study the plans referenced for each exercise to answer the questions. You may have to refer to the plan symbols in *Appendix A* for some of the questions. On the work site, you will probably study plans and drawings as a team to ensure that all of the workers on your part of the project understand what tasks they are to do and how they are to do them.

The plans for these exercises are drawings for a light commercial construction building. The building is the new home for the Germans From Russia Heritage Society, located in Bismarck, North Dakota. The building will provide workrooms and offices for staff and visitors, as well as storage space for the society's records, historical artifacts, and library.

Drawing A 2.1 is the building's floor plan. This drawing includes two schedules: a room finish schedule and a door schedule. It also has elevation drawings of the different types of doors, frames, and windows used in the building. Subcontractors who want to bid on the job will use the information in the schedules to calculate the cost of materials, the number of workers that will be needed, and how long the job will take.

Each schedule is keyed to the floor plan. Compare the floor and finish schedule to the floor plan. All of the rooms on the floor plan are numbered. Notice that areas that are not normally considered rooms—such as the entry and vestibule—are also numbered. This numbering system makes it easy to locate exactly where a finish material should be installed. For example, the vault (112) will have vinyl composite tile with a 4" resilient base, concrete painted block walls, and a painted concrete ceiling. Information about the type and color of tile and baseboard will be found in the specifications.

Each item on the window schedule is also keyed to a corresponding number on the plan. This project requires only three types of windows, which are labeled A, B, and C on both the floor plan and the window type schedule.

The door schedule is organized like the window schedule. The door for each room and entrance on the plan is numbered and listed, by number, on the door schedule. You can see, for example, that the door (1081) into the workroom (108) is a wood door with the framing details shown in the detail drawing 1/A 2.1. Depending on the purpose of the plans, not all of the information may be filled in on the schedule. For example, the fire rating and remarks columns of the door schedule are blank.

Sometimes a set of plans will contain alternates. The owner of the building may want to see how changes to the building will affect the design and cost of the project. The alternate drawing will help the owner see how the changes will affect the design. If alternates are requested, the bid will include two costs, one for the entire project and one in which the alternate plans are included.

8.1.0 Exercise 1: Title Page

Study the title page in the package of plans included with this module to answer the following questions.

1. What is the name of the building to be built?

   _____

2. Where is the site located?

   _____

3. What is the project number?

   _____

4. How many sheets are included in the building packet?

   _____

5. What are the areas of construction addressed by this set of plans?

   _____

6. Does this set of plans conform to an accepted code? If so, which one?

_____

_____

8.2.0 Exercise 2: Site Survey and Site Plan

Study drawing C-1, Site Survey, and C-2, Site Plan. Then answer the following questions.

1. What is to be built on the site?

_____

2. Where can you find the address for this project? What is the address?

_____

_____

3. Who is responsible for obtaining necessary permits?

_____

4. What type of pipe and joints should be used for the sewer service?

_____

5. What is the name of the engineering firm that did the survey work?

_____

6. What is the highest existing grade of the property lot where this building will be built?

_____

8.3.0 Exercise 3: Architectural Plans

Study drawings A 2.1, Floor Plan Schedules and Door Details; A 3.1, Building Elevations; A 3.2, Building Sections; A 3.3, Wall Sections, Detail; A 5.1, Interior Elevations, Casework, and Details; and A 6.1, Reflected Ceiling Plan. Then answer the following questions.

1. To what scale is the floor plan drawn?

_____

2. Where is door 1012 located? What type of material is it made from?

_____

3. What type of flooring will be used in the multipurpose room? Where would you find this information?

_____

4. How will service personnel gain access to the mechanical room 201?

_____

5. What type of material will be used to face the interior walls of the toilets 110 and 111?

_____

6. What are the dimensions for the library room 103?

_____

7. What is the floor area, excluding walls, of the library room 103?

_____

8. What are the rough-in measurements for window type A?

_____

9. How is the mechanical room soundproofed?

_____

10. What is the roof pitch for the building?

_____

11. What is the distance from the top of the concrete floor to the top of the concrete footing?

_____

12. What type of insulation is used in the foundation walls?

_____

13. What type of framing will be found around the windows?

_____

8.4.0 Exercise 4: Structural Plans

Study drawings S-1, Foundation and Framing Plans and Notes, and S-2, Foundation & Framing Details. Then answer the following questions.

1. What are the dimensions of the mechanical space?

   _____

2. Using Truss C allows for the mechanical room in the attic. How much room will be available (height, width, and length) according to the plans?

   _____

3. How many penetrations through the mechanical room floor are there?

   _____

4. How thick are the wall footings?

   _____

8.5.0 Exercise 5: Mechanical Plans

Study drawings M-1, Piping Plan, and M-2, HVAC Plan. Then answer the following questions.

1. On plan M-1, detail drawing 8, what does NTS stand for?

   _____

2. What does 2" FD-1 WITH PTRAP stand for? Where can an example of its use be found on the M-1 plan?

   _____

3. How many cleanouts are shown in the underground piping plan? What size are they and where are they located?

   _____

4. How many hot water takeoffs come off the hot water line?

   _____

5. How many cold water takeoffs come off the cold water line after the meter location?

   _____

6. What size are the floor drains in the toilets?

   _____

7. The freeze-proof wall hydrants are to be installed 24" AFG. What does AFG stand for?

   _____

8. What is the largest return grille to be installed, and where is it located?

   _____

Riveting Ideas

Handling Plans

Folded plans are like road maps. Once you unfold one, it can seem impossible to refold it properly. Once you unfold several prints, you will have a lot of big sheets of paper to handle. In no time, you'll be surrounded by paper, and you'll find yourself spending more time sorting through the prints than actually reading them. To make it easier to handle plans, do the following:

- As you finish working with each plan, try to refold it along the fold lines and set it aside with the title block facing up so that you can easily see which plan it is.
- When working with two or more plans, unfold and stack them in order. Then turn them as you would the pages of a large book. This method ensures that the print you just finished looking at is on your left in case you need to refer to it again.
- Before looking for information on a plan, check to make sure that you have unfolded the right one.
- Take a moment to orient yourself to the plan. Look at the overall drawing before focusing in on the details.
- Think about where you might logically find something. For example, if you are looking for information related to mechanical equipment, you will most likely find it in the basement plan or in a room labeled Mech. If you are looking for the size and materials for a door, you will find this information on the door schedule.

9. How many exhaust fans will be installed? Where will they be located?

_____

10. How many louvers will be installed on the building's exterior?

_____

11. What is the largest duct in the building? What will it be used for?

_____

12. What separates the fresh air intake duct from the furnace return duct?

_____

13. What is the BtuH input for the furnace labeled F-1a? What is the total BtuH input for the two furnaces that are twinned (F-1a and F-1b)?

_____

14. What are the BtuH outputs of those furnaces?

_____

15. What is the efficiency rating of those furnaces?

_____

16. What is the output of the F-1a and b furnaces in cubic feet per minute (cfm) when in the heating mode?

_____

17. What are the dimensions of the ductwork after the 32" × 18" ductwork housing the H-1 (humidifier) distribution tube?

_____

18. What is the cfm of the diffusers located in the library ceiling?

_____

19. How are the dampers that control the fresh air intake/return air mix controlled?

_____

20. Please copy and complete the table below.

Furnace	Input BtuH	Output BtuH	Condensing Unit Tonnage	Liquid Line Dimensions	Suction Line
F-1a	120,000				
F-1b		112,000			
F-2			1.5 ton		
F-3				½ inch	
F-4					¾ inch

1. The _____ contain(s) the address of the building to be constructed, the scale, drawing number, date of the latest revision, and information about the architect or engineer.
   a. legend
   b. notes
   c. specifications
   d. title block

2. Mistakes found on a plan should be recorded on a _____.
   a. Request for Specification
   b. Request for Information
   c. change order
   d. cut list

3. Contour lines are found on the _____ plan.
   a. civil
   b. floor
   c. structural
   d. foundation

4. The _____ plan shows how the various parts of the building should look from several views.
   a. architectural
   b. electrical
   c. mechanical
   d. civil

5. A drawing of what an object or structure looks like from the inside is called a(n) _____ drawing.
   a. detail
   b. elevation
   c. section
   d. as-built

6. A mechanical plan includes _____ and _____ systems.
   a. plumbing; electrical
   b. air distribution; refrigeration
   c. roof-framing; irrigation
   d. piping; wiring

7. The path of water piping in a multistory building is shown in a _____.
   a. load capacity chart
   b. section drawing
   c. riser diagram
   d. drainage plan

8. Information about applicable building codes and zoning ordinances can be found in the _____.
   a. plans
   b. specifications
   c. schedules
   d. construction indexes

9. The _____ has developed a system to organize project information in a logical way.
   a. Construction Specifications Institute
   b. Building Officials and Contractors Association
   c. National Electrical Contractors Association
   d. International Code Council

10. Like section and detail drawings, _____ are drawn to a larger scale than the engineer's design.
    a. site plans
    b. elevation drawings
    c. shop drawings
    d. plot plans

11. A draftsman or sheet metal worker prepares _____ for each fitting to be made.
    a. transverse sections
    b. survey plans
    c. schedules
    d. cut lists

12. A draftsman will _____ to ensure that the work will be done properly.
    a. determine the work schedule
    b. check gauge specifications and verify the location of dampers
    c. perform an OSHA safety inspection
    d. modify a building code to fit the structural requirements

13. _____ are descriptions and drawings of equipment or accessories to be provided and installed by the subcontractor.
    a. Change orders
    b. As-built drawings
    c. Submittals
    d. Bids

14. The abbreviation AHU often stands for _____.
    a. air heating unit
    b. air handling unit
    c. adjusted heating unit
    d. air handler, underground

15. The abbreviation EX often stands for _____.
    a. line extension
    b. emergency exit
    c. exit light
    d. electric line

Summary

A construction project is a complex process involving many people who represent the various construction trades and local government. Without a clear set of plans and specifications, assigning and organizing the work efficiently would be impossible. Plans and specifications spell out every detail of the construction job from preparing the site to putting in the finishing touches. However, even the best plans can run into problems, so clear and open communication is also an important part of the construction process.

As a sheet metal apprentice, you must learn the basics of plan reading so that you can interpret the drawings that show how to accurately install an air distribution system and its accessories. However, you must also be able to read the prints used by other trades because you and your co-workers will have to coordinate your work with them.

In this module, you learned that reading plans is similar to reading patterns for sheet metal fittings. Your ability to see the finished object in your head—whether it's a fitting or a building—is an important skill in the construction industry. You also learned the importance of good communication. With so many details involved in a construction project, mistakes can and will happen, and this module explained how to clearly communicate problems to the people authorized to handle them.

Notes

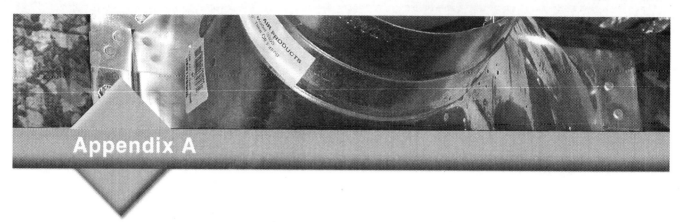

Plan Symbols and Abbreviations

The illustrations and table included in this appendix will familiarize you with commonly used plan abbreviations (*Table A-1*) and symbols (*Figures A-1* through *A-10*). You will find similar symbols on most plans. Note, however, that some architectural and engineering firms may use slightly different abbreviations and symbols. For example, see if you can find some differences between the abbreviations in *Table A-1* and the plans provided. Also, compare the plan lines in *Figure A-1* with other forms of plan lines in *Figure A-2*. As you can see, the abbreviations and symbols are similar, though not exactly alike. Generally, it is best to refer to the symbols and legends included in the full set of architectural plans with each project.

Table A-1 Standard Plan Abbreviations

ADD.	addition	MEZZ	mezzanine
AGGR	aggregate	MO	masonry opening
AHU	air heating unit	MECH	mechanical
L	angle	N	north
B	bathroom	NO.	number
BR	bedroom	OC	on center
BM	bench mark	OPP	opposite
BRKT	bracket	O.D.	outside diameter
CFM	cubic feet per minute	PNL	panel
CLK	caulk	PSI	pounds per square inch
CHFR	chamfer	PWR	power
CHWR	chilled water return	REINF	reinforce
CHWS	chilled water supply	RH	right-hand
CND	conduit	RFA	released for approval
CO	cleanout	RFC	released for construction
CU FT	cubic foot, feet	RFD	released for design
DIM.	dimension	RFI	released for information
DR	drain	SA	shock absorber
DWG	drawing	SHTHG	sheathing
ELEV	elevation	SQ	square
ESC	escutcheon	STR	structural
EX	exit	SYM	symbol
FAB	fabricate	THERMO	thermostat
FD	floor drain	TYP	typical
FLGE	flange	UNFIN	unfinished
FLR	floor	VEL	velocity
GR	grade	VTR	vent to roof
GYP	gypsum	WV	wall vent
HDW	hardware	WHSE	warehouse
HTR	heater	WH	weep hole
" or IN.	inch, inches	WDW	window
I.D.	inside diameter	WP	working pressure
LH	left-hand		

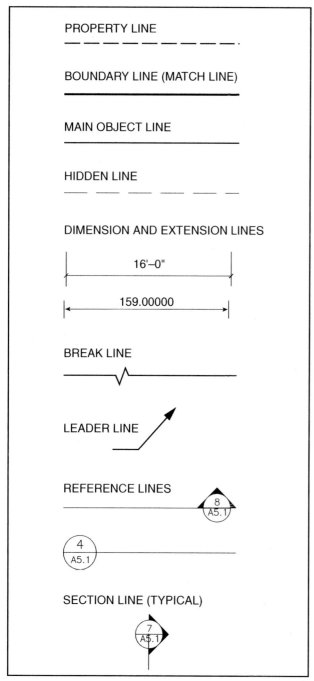

PROPERTY LINE

BOUNDARY LINE (MATCH LINE)

MAIN OBJECT LINE

HIDDEN LINE

DIMENSION AND EXTENSION LINES

16'–0"

159.00000

BREAK LINE

LEADER LINE

REFERENCE LINES

8
A5.1

4
A5.1

SECTION LINE (TYPICAL)

7
A5.1

202A01.EPS

Figure A-1 ◆ Lines used in plans.

LIGHT FULL LINE	
MEDIUM FULL LINE	
HEAVY FULL LINE	
EXTRA HEAVY FULL LINE	
CENTERLINE	
HIDDEN LINE	
DIMENSION LINE	4"
SHORT BREAK LINE	
LONG BREAK LINE	
MATCH LINE	
SECONDARY LINE	
PROPERTY LINE	

202A02.EPS

Figure A-2 ◆ Common plan lines.

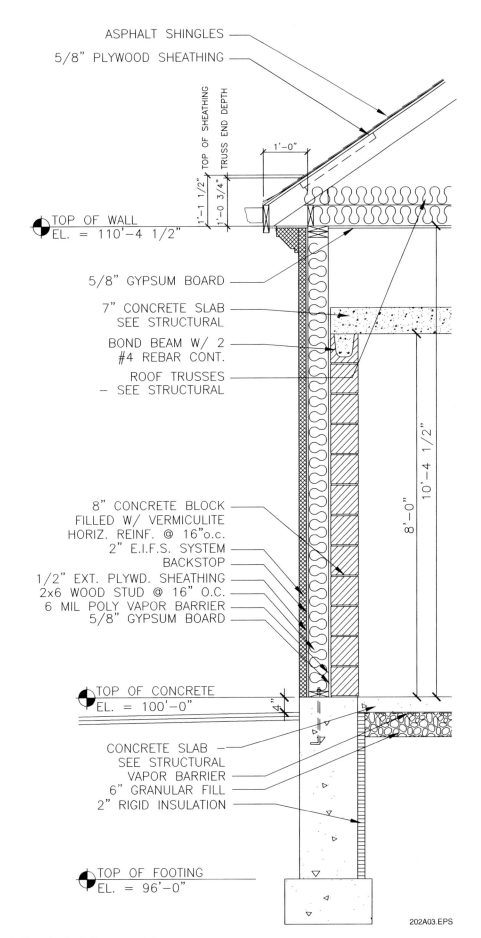

ASPHALT SHINGLES

5/8" PLYWOOD SHEATHING

TOP OF SHEATHING

TRUSS END DEPTH

1'-0"

1'-1 1/2"

1'-0 3/4"

TOP OF WALL
EL. = 110'-4 1/2"

5/8" GYPSUM BOARD

7" CONCRETE SLAB
SEE STRUCTURAL

BOND BEAM W/ 2
#4 REBAR CONT.

ROOF TRUSSES
— SEE STRUCTURAL

10'-4 1/2"

8'-0"

8" CONCRETE BLOCK
FILLED W/ VERMICULITE
HORIZ. REINF. @ 16"o.c.

2" E.I.F.S. SYSTEM
BACKSTOP

1/2" EXT. PLYWD. SHEATHING

2x6 WOOD STUD @ 16" O.C.

6 MIL POLY VAPOR BARRIER

5/8" GYPSUM BOARD

TOP OF CONCRETE
EL. = 100'-0"

4"

CONCRETE SLAB
SEE STRUCTURAL

VAPOR BARRIER

6" GRANULAR FILL

2" RIGID INSULATION

TOP OF FOOTING
EL. = 96'-0"

202A03.EPS

Figure A-3 ◆ Exterior building materials.

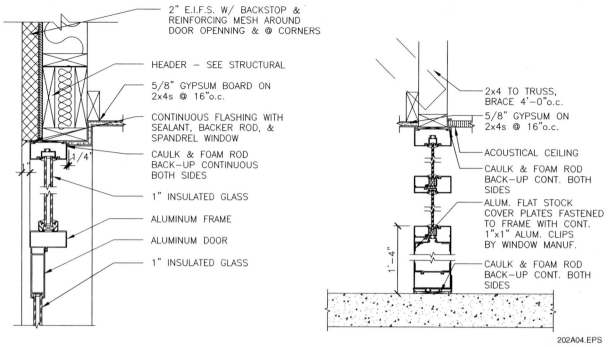

2" E.I.F.S. W/ BACKSTOP & REINFORCING MESH AROUND DOOR OPENNING & @ CORNERS

HEADER – SEE STRUCTURAL

5/8" GYPSUM BOARD ON 2x4s @ 16"o.c.

CONTINUOUS FLASHING WITH SEALANT, BACKER ROD, & SPANDREL WINDOW

CAULK & FOAM ROD BACK-UP CONTINUOUS BOTH SIDES

1" INSULATED GLASS

ALUMINUM FRAME

ALUMINUM DOOR

1" INSULATED GLASS

2x4 TO TRUSS, BRACE 4'-0"o.c.

5/8" GYPSUM ON 2x4s @ 16"o.c.

ACOUSTICAL CEILING

CAULK & FOAM ROD BACK-UP CONT. BOTH SIDES

ALUM. FLAT STOCK COVER PLATES FASTENED TO FRAME WITH CONT. 1"x1" ALUM. CLIPS BY WINDOW MANUF.

CAULK & FOAM ROD BACK-UP CONT. BOTH SIDES

202A04.EPS

Figure A-4 ◈ Interior building materials.

WINDOWS

DOUBLE–PANE   TRIPLE–PANE

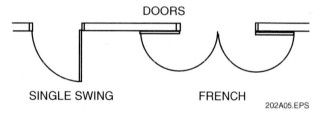

DOORS

SINGLE SWING   FRENCH   202A05.EPS

Figure A-5 ◈ Window and door symbols.

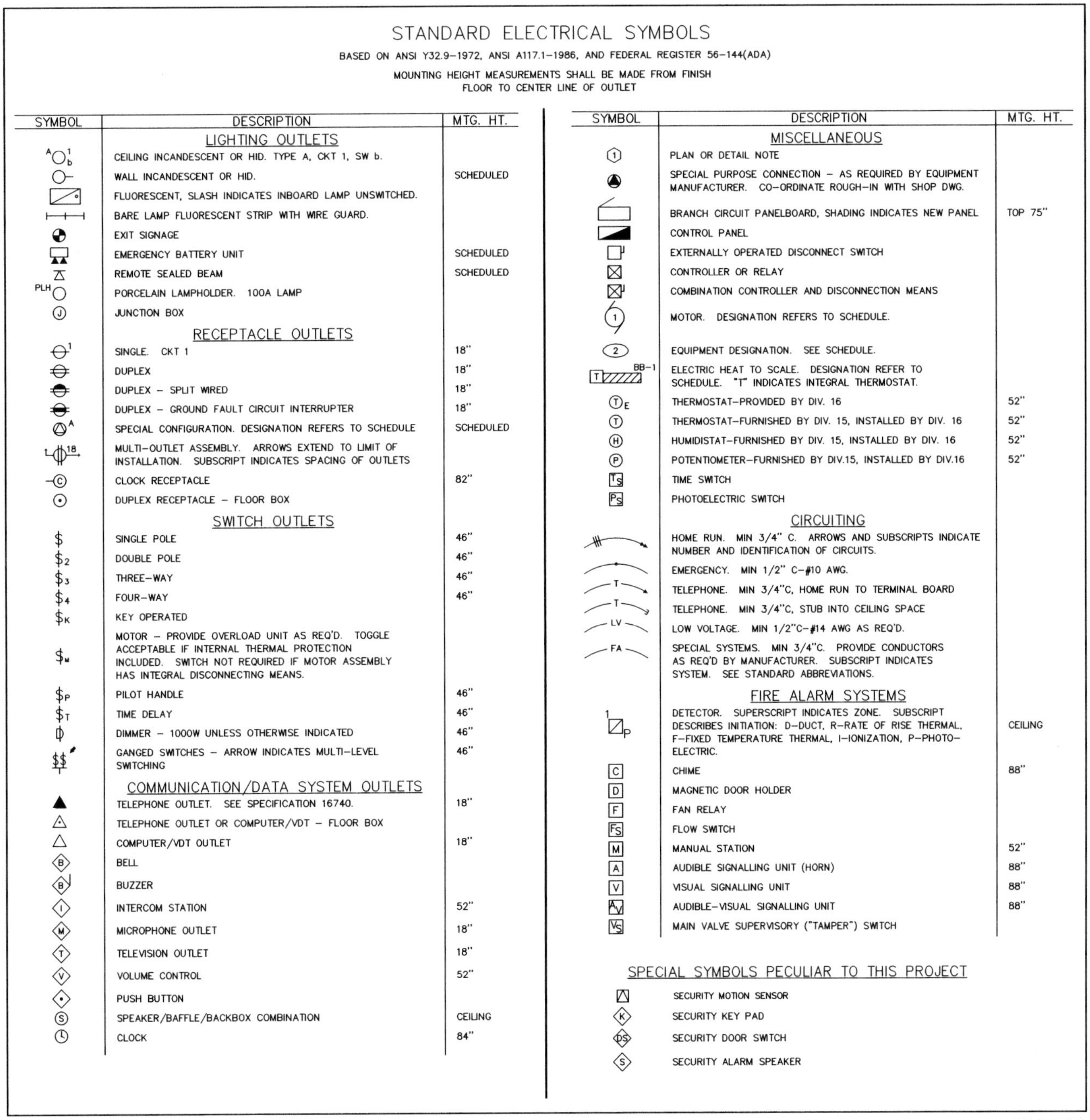

STANDARD ELECTRICAL SYMBOLS

BASED ON ANSI Y32.9–1972, ANSI A117.1–1986, AND FEDERAL REGISTER 56–144(ADA)

MOUNTING HEIGHT MEASUREMENTS SHALL BE MADE FROM FINISH
FLOOR TO CENTER LINE OF OUTLET

SYMBOL	DESCRIPTION	MTG. HT.
LIGHTING OUTLETS		
	CEILING INCANDESCENT OR HID. TYPE A, CKT 1, SW b.	
	WALL INCANDESCENT OR HID.	SCHEDULED
	FLUORESCENT, SLASH INDICATES INBOARD LAMP UNSWITCHED.	
	BARE LAMP FLUORESCENT STRIP WITH WIRE GUARD.	
	EXIT SIGNAGE	
	EMERGENCY BATTERY UNIT	SCHEDULED
	REMOTE SEALED BEAM	SCHEDULED
	PORCELAIN LAMPHOLDER. 100A LAMP	
	JUNCTION BOX	
RECEPTACLE OUTLETS		
	SINGLE. CKT 1	18"
	DUPLEX	18"
	DUPLEX – SPLIT WIRED	18"
	DUPLEX – GROUND FAULT CIRCUIT INTERRUPTER	18"
	SPECIAL CONFIGURATION. DESIGNATION REFERS TO SCHEDULE	SCHEDULED
	MULTI-OUTLET ASSEMBLY. ARROWS EXTEND TO LIMIT OF INSTALLATION. SUBSCRIPT INDICATES SPACING OF OUTLETS	
	CLOCK RECEPTACLE	82"
	DUPLEX RECEPTACLE – FLOOR BOX	
SWITCH OUTLETS		
	SINGLE POLE	46"
	DOUBLE POLE	46"
	THREE-WAY	46"
	FOUR-WAY	46"
	KEY OPERATED	
	MOTOR – PROVIDE OVERLOAD UNIT AS REQ'D. TOGGLE ACCEPTABLE IF INTERNAL THERMAL PROTECTION INCLUDED. SWITCH NOT REQUIRED IF MOTOR ASSEMBLY HAS INTEGRAL DISCONNECTING MEANS.	
	PILOT HANDLE	46"
	TIME DELAY	46"
	DIMMER – 1000W UNLESS OTHERWISE INDICATED	46"
	GANGED SWITCHES – ARROW INDICATES MULTI-LEVEL SWITCHING	46"
COMMUNICATION/DATA SYSTEM OUTLETS		
	TELEPHONE OUTLET. SEE SPECIFICATION 16740.	18"
	TELEPHONE OUTLET OR COMPUTER/VDT – FLOOR BOX	
	COMPUTER/VDT OUTLET	18"
	BELL	
	BUZZER	
	INTERCOM STATION	52"
	MICROPHONE OUTLET	18"
	TELEVISION OUTLET	18"
	VOLUME CONTROL	52"
	PUSH BUTTON	
	SPEAKER/BAFFLE/BACKBOX COMBINATION	CEILING
	CLOCK	84"

SYMBOL	DESCRIPTION	MTG. HT.
MISCELLANEOUS		
	PLAN OR DETAIL NOTE	
	SPECIAL PURPOSE CONNECTION – AS REQUIRED BY EQUIPMENT MANUFACTURER. CO-ORDINATE ROUGH-IN WITH SHOP DWG.	
	BRANCH CIRCUIT PANELBOARD, SHADING INDICATES NEW PANEL	TOP 75"
	CONTROL PANEL	
	EXTERNALLY OPERATED DISCONNECT SWITCH	
	CONTROLLER OR RELAY	
	COMBINATION CONTROLLER AND DISCONNECTION MEANS	
	MOTOR. DESIGNATION REFERS TO SCHEDULE.	
	EQUIPMENT DESIGNATION. SEE SCHEDULE.	
	ELECTRIC HEAT TO SCALE. DESIGNATION REFER TO SCHEDULE. "T" INDICATES INTEGRAL THERMOSTAT.	
	THERMOSTAT–PROVIDED BY DIV. 16	52"
	THERMOSTAT–FURNISHED BY DIV. 15, INSTALLED BY DIV. 16	52"
	HUMIDISTAT–FURNISHED BY DIV. 15, INSTALLED BY DIV. 16	52"
	POTENTIOMETER–FURNISHED BY DIV.15, INSTALLED BY DIV.16	52"
	TIME SWITCH	
	PHOTOELECTRIC SWITCH	
CIRCUITING		
	HOME RUN. MIN 3/4" C. ARROWS AND SUBSCRIPTS INDICATE NUMBER AND IDENTIFICATION OF CIRCUITS.	
	EMERGENCY. MIN 1/2" C–#10 AWG.	
	TELEPHONE. MIN 3/4"C, HOME RUN TO TERMINAL BOARD	
	TELEPHONE. MIN 3/4"C, STUB INTO CEILING SPACE	
	LOW VOLTAGE. MIN 1/2"C–#14 AWG AS REQ'D.	
	SPECIAL SYSTEMS. MIN 3/4"C. PROVIDE CONDUCTORS AS REQ'D BY MANUFACTURER. SUBSCRIPT INDICATES SYSTEM. SEE STANDARD ABBREVIATIONS.	
FIRE ALARM SYSTEMS		
	DETECTOR. SUPERSCRIPT INDICATES ZONE. SUBSCRIPT DESCRIBES INITIATION: D–DUCT, R–RATE OF RISE THERMAL, F–FIXED TEMPERATURE THERMAL, I–IONIZATION, P–PHOTO-ELECTRIC.	CEILING
	CHIME	88"
	MAGNETIC DOOR HOLDER	
	FAN RELAY	
	FLOW SWITCH	
	MANUAL STATION	52"
	AUDIBLE SIGNALLING UNIT (HORN)	88"
	VISUAL SIGNALLING UNIT	88"
	AUDIBLE–VISUAL SIGNALLING UNIT	88"
	MAIN VALVE SUPERVISORY ("TAMPER") SWITCH	

SPECIAL SYMBOLS PECULIAR TO THIS PROJECT

SYMBOL	DESCRIPTION
	SECURITY MOTION SENSOR
	SECURITY KEY PAD
	SECURITY DOOR SWITCH
	SECURITY ALARM SPEAKER

202A06.EPS

Figure A-6 ◈ Electrical symbols.

STANDARD ABBREVIATIONS

AC	ABOVE COUNTER (MIN 4" ABOVE BACKSPLASH)	MC	MECHANICAL CONTRACTOR
AFF	ABOVE FINISH FLOOR	MCA	MINIMUM CIRCUIT AMPACITY
AFG	ABOVE FINISH GRADE	MCB	MAIN CIRCUIT BREAKER
AHU	AIR HANDLING UNIT	MLO	MAIN LUG ONLY
ATC	AUTOMATIC TEMPERATURE CONTROL	NC	NURSES CALL
BOF	BOTTOM OF FIXTURE	NFDS	NON-FUSIBLE DISCONNECT SWITCH
CKT	CIRCUIT	NR	NOT REQUIRED
CP	CONTROL PANEL	P	PAGING/BACKGROUND MUSIC
CU	CONDENSING UNIT	PA	PUBLIC ADDRESS
CUH	CABINET UNIT HEATER	PRV	POWER ROOF VENTILATOR
DT	DUST TIGHT	RF	RELIEF FAN
EC	ELECTRICAL CONTRACTOR	RGSC	RIGID GALVANIZED STEEL CONDUIT
EF	EXHAUST FAN	RT	RAIN-TIGHT
EM	EMERGENCY	RTU	ROOFTOP UNIT
EP	EXPLOSION PROOF	S	SECURITY
EWC	ELECTRIC WATER COOLER	SD	SEE DETAIL
EWH	ELECTRIC WATER HEATER	SS	SURGE SUPPRESSION
F	FUSED	SW	SWITCH
FA	FIRE ALARM	TOP	TOP OF PANEL
FACP	FIRE ALARM CONTROL PANEL	T	TELEPHONE
FDS	FUSIBLE DISCONNECT SWITCH	TSSW	TWO SPEED SEPARATE WINDING
FLA	FULL LOAD AMPERES	TTB	TELEPHONE TERMINAL BOARD
FVNR	FULL VOLTAGE NON REVERSING	VT	VAPOR TIGHT
FVR	FULL VOLTAGE REVERSING	W	WALL MTG. (46" AFF FORWARD REACH)
IC	INTERCOM		(52" AFF SIDE REACH)
IG	ISOLATED GROUND	WP	WEATHERPROOF
IL	INTERLOCK	WT	WATER TIGHT
LV	LOW VOLTAGE	WM	WIREMOLD

SPECIAL ABBREVIATIONS PECULIAR TO THIS PROJECT

FARA	FIRE ALARM REMOTE ANNUNCIATOR
SSCP	SECURITY SYSTEM CONTROL PANEL
DR	DATA RACK

202A07.EPS

Figure A-7 ◈ Abbreviations used on electrical drawings in the plans.

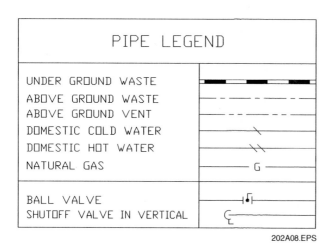

202A08.EPS

Figure A-8 ◈ Pipe legend.

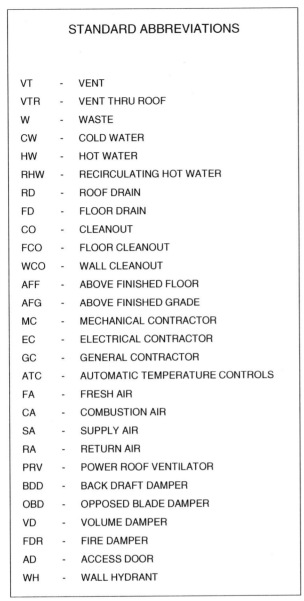

STANDARD ABBREVIATIONS

VT	-	VENT
VTR	-	VENT THRU ROOF
W	-	WASTE
CW	-	COLD WATER
HW	-	HOT WATER
RHW	-	RECIRCULATING HOT WATER
RD	-	ROOF DRAIN
FD	-	FLOOR DRAIN
CO	-	CLEANOUT
FCO	-	FLOOR CLEANOUT
WCO	-	WALL CLEANOUT
AFF	-	ABOVE FINISHED FLOOR
AFG	-	ABOVE FINISHED GRADE
MC	-	MECHANICAL CONTRACTOR
EC	-	ELECTRICAL CONTRACTOR
GC	-	GENERAL CONTRACTOR
ATC	-	AUTOMATIC TEMPERATURE CONTROLS
FA	-	FRESH AIR
CA	-	COMBUSTION AIR
SA	-	SUPPLY AIR
RA	-	RETURN AIR
PRV	-	POWER ROOF VENTILATOR
BDD	-	BACK DRAFT DAMPER
OBD	-	OPPOSED BLADE DAMPER
VD	-	VOLUME DAMPER
FDR	-	FIRE DAMPER
AD	-	ACCESS DOOR
WH	-	WALL HYDRANT

202A09.EPS

Figure A-9 ◈ Standard plumbing abbreviations.

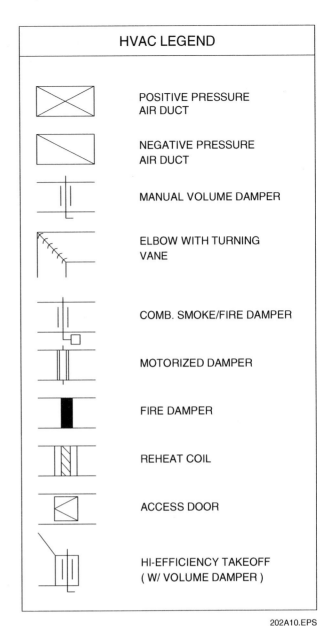

HVAC LEGEND

POSITIVE PRESSURE AIR DUCT

NEGATIVE PRESSURE AIR DUCT

MANUAL VOLUME DAMPER

ELBOW WITH TURNING VANE

COMB. SMOKE/FIRE DAMPER

MOTORIZED DAMPER

FIRE DAMPER

REHEAT COIL

ACCESS DOOR

HI-EFFICIENCY TAKEOFF (W/ VOLUME DAMPER)

202A10.EPS

Figure A-10 ◈ HVAC symbols.

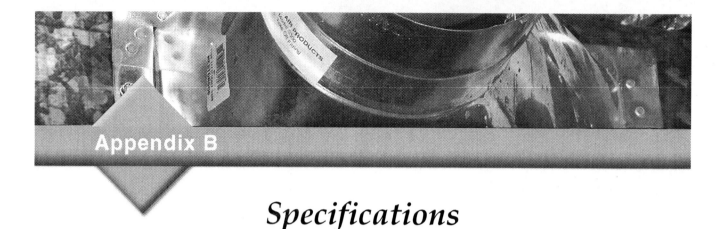

Specifications

The following is an excerpt from the booklet for the Germans From Russia Heritage Society project. It begins with a section titled *Special Conditions* and then goes on to include the construction divisions that are required for this project.

Special Conditions

Germans From Russia Heritage Society Office Building Project

Bismarck, North Dakota

- Where contradictions between plans and specifications exist, the plans shall take precedence.

- The heating and air-conditioning equipment noted in Section 15 is not to be a part of the general contract, but will be furnished and installed by the Germans From Russia Heritage Society. The Contractor shall be responsible for installing the support frames and platforms and shall work with the Germans From Russia Heritage Society equipment installer by providing crane service and helping set the rooftop units as noted on the drawings. The Germans From Russia Heritage Society's heating and air-conditioning installer will direct the setting of the rooftop units.

- **SAFETY**—All required exits must be kept usable throughout the construction period. General Contractor will provide lighted enclosed walkways through new construction areas as required by government authorities having jurisdiction.

- **COORDINATION OF WORK**—All work shall be coordinated with the proper Germans From Russia Heritage Society representative to minimize disruption to offices on the east wall of the building. These offices shall remain in operation until the offices are ready for occupancy.

- **TEMPORARY PARTITION**—After the ceiling and floor covering of the new addition have been built and before the existing office walls are removed, the Contractor shall install a temporary, full-height, braced 2" × 6" stud wall where shown on the plan. The Owner and the Contractor will coordinate the distance of the temporary wall from the existing wall.

- **TEMPORARY INTERIOR ENCLOSURES**—Areas shall be entirely enclosed with temporary partitions and separated from the remainder of the building. Partitions shall go from the floor to the ceiling or above, as necessary, to control dust and noise. The Germans From Russia Heritage Society representative or the Architect shall approve construction and installation of the temporary partitions.

- **SHOP DRAWINGS**—The following shop drawings and submittals are required. The Contractor shall submit five copies of each to the designated Germans From Russia Heritage Society representative.
  - Folding acoustical partitions
  - Exterior canopy system
  - Glass and glazing in the storefront
  - Electrical submittals
  - Plumbing submittals
  - Millwork, including doors
  - Hardware schedule
  - Ceiling tile sample
  - Overhead doors and dock equipment
  - Toilet partitions
  - Sprinkler drawings approved by Mutual Insurance Company

- **SALVAGED MATERIALS**—Salvaged materials from demolition of the area shall remain the property of the Owner. If the Owner declines ownership, these materials shall become the property of the Contractor, who shall be responsible for their disposal. These items include but are not limited to the following:

- Unit heaters
- Doors and hardware
- Electrical panels
- Switches, wiring, and light fixtures
- Water heaters
- Plumbing fixtures and trim

General Conditions

- **1-1 GENERAL**—The work to be done includes the furnishing of all labor, materials, and equipment necessary to complete the renovation of an existing building as shown and described in the drawings, specifications, and contract documents.

  - Before submitting a quote, the Contractor shall be satisfied as to the nature and location of the work; the materials, tools, equipment; other facilities required before and during the work; and general and local conditions that can affect the work. The Contractor accepts the work site as found, unless otherwise specifically stated in the quote for the work. The Owner shall make no compensation to the Contractor for any bid errors.

  - The Contractor's quote shall include the cost of all utility tie-ins, including tap fees for sewage, drainage, water supply, sprinkler system supply, natural gas, electricity, and so forth whether or not such work is on the Owner's property. Required performance bonds, permits, and deposits shall also be included in the quote. The Germans From Russia Heritage Society will not honor any extras for these items.

- **1-2 DEFINITIONS**—The words Germans From Russia Heritage Society shall mean the Germans From Russia Heritage Society Properties, Inc., Bismarck, North Dakota, acting through its duly authorized representatives.

  - The word Owner shall mean the Germans From Russia Heritage Society Properties, Inc., Bismarck, North Dakota.

  - The word Architect shall mean Sam Spearing, Architects and Engineers, who prepared contract documents for the project.

  - The word Contractor shall mean the person, persons, partnership, company, firm, or corporation entering into the contract for the performance of the work required, and the legal representative of said party or agent appointed to act for said party in the performance of the work.

  - The word Contract shall mean, collectively, all of the covenants, terms, and stipulations contained in the various portions of the contract, to wit: bid letters, bids, specifications, plans, and shop drawings as well as any addenda, letter of authorization or instruction, and any change orders that may be originated by the Germans From Russia Heritage Society.

- **1-3 CORRELATION AND INTENT OF DOCUMENTS**—The contract documents are complementary to the specifications, and what is called for by one shall be binding as if called for by all. It is the intent of the contractual documents to have a complete operating facility constructed with all services that are within the contract connected and in operating condition.

- **1-4 DRAWINGS AND SPECIFICATIONS**—This specification is for general use and may apply to more than one building system such as masonry buildings, pre-engineered metal buildings, and so forth. Check drawings for items that are included in the work. Drawings shall govern when variations from these specifications are found.

  - The Contractor shall promptly notify the Germans From Russia Heritage Society and the Architect regarding any apparent contradictions, ambiguities, errors, discrepancies, or omissions in the plans or specifications. No extras shall be allowed for any such items that the Contractor fails to report to the Germans From Russia Heritage Society and the Architect before the contract is awarded.

  - All drawings, prints, specifications, or other documents furnished by the Germans From Russia Heritage Society or prepared by the Contractor or subcontractors specifically for the work shall become the sole property of the Germans From Russia Heritage Society and shall be returned to the Germans From Russia Heritage Society at completion of the work.

  - When necessary, the Contractor shall furnish shop drawings in sufficient quantity, and after approval by the Germans From Russia Heritage Society these drawings shall be considered part of the contract documents. Checking and approval by the Germans From Russia Heritage Society shall not relieve the Contractor of any responsibility for errors, omissions, or discrepancies on such shop drawings.

- **1-5 INSURANCE**—The Contractor shall carry insurance as listed herein and furnish a certification of insurance before construction begins. The certification must state that the insurance will not be cancelled while the specified work is in progress without ten (10) days prior notice to the Germans From Russia Heritage Society.

Type and Limits of Insurance

Workmen's Compensation
   Statutory Amount

Contractor's Public Liability
Bodily Injury (including death)
   Each Person $250,000
   Each Accident $500,000

Property Damage
   Each Accident $100,000
   Aggregate $100,000

- The Contractor shall take out and maintain throughout the course of construction a builders' "all risks" insurance policy in the amount of the contract. This policy shall name the Owner and the Contractor as insured parties. Subcontractors' loss shall be adjusted with payment to the Owner and the Contractor. Until work is fully completed and accepted by the Owner, the Contractor shall promptly repair any damage to the work. The original of said policy shall be delivered to the Owner before work begins and returned to the Contractor after the work is finished for cancellation and refund of premium, if any.

- **1-6 BOND**—The Contractor shall furnish, where required by the Owner, an approved performance bond in an amount equal to one hundred percent (100%) of the contract price. The Contractor shall verify the amount with the Owner before bidding. The bond shall contain the following paragraph:

   And the said surety, for value received, hereby stipulates and agrees that no change, extension of time, alteration, or addition to the terms of the contract or to the work to be performed thereunder or the specifications accompanying the same shall in any way affect its obligations on this bond, and it does

hereby waive notice of any change, extension of time, alteration, or addition to the terms of the contract or to the specifications.

- **1-7 TAXES**—The Contractor shall pay all costs of Social Security payments, unemployment insurance, sales tax, and any other charges imposed by federal, state, and local governments.

- **1-8 PERMITS AND REGULATIONS**—The Contractor shall obtain all permits required for the work, including necessary temporary permits; give all notices; and pay all fees. All equipment, materials, and workmanship shall comply with the requirements of federal, state, and municipal codes and ordinances and underwriters' rules and specifications. Proposals shall be based on the plans and specifications with any exceptions required by codes and regulations being noted in writing by the Contractor so that an equitable adjustment can be made in the contract price. Otherwise, it shall be construed that the Contractor is willing to comply with such codes, ordinances, rules, and regulations without additional cost to the Owner. Any workmanship done or equipment and materials installed contrary to the above-noted regulations shall be removed and replaced at the Contractor's expense.

- **1-9 SUPERINTENDENCE AND EMPLOYEES**— The Contractor shall provide competent superintendence satisfactory to the Owner at all times. The superintendent shall not be changed without the Owner's consent, unless that person proves to be unsatisfactory to the Contractor and ceases to be employed by the Contractor. The Contractor shall be totally responsible for the employment, supervision, welfare, and compensation of employees and shall be responsible for any work performed by employees or subcontractors.

- **1-10 SUBCONTRACTS**—The Owner shall have the right to approve or disapprove the subcontractors to be used for the work.

- **1-11 SEPARATE CONTRACTS**—The Owner reserves the right to let other contracts in connection with the work. The Contractor shall give other contractors reasonable opportunity for the storage of their materials and the execution of their work. The Contractor shall properly connect and coordinate work with other contractors. When the work of any other contractor is involved in the execution and results of the Contractor's work, the Contractor shall inspect the work and report any discrepancy to the Owner.

- **1-12 DISCHARGE OF LIENS**—If at any time there is evidence that a claim chargeable to the Contractor may become a lien against the premises, the Owner may retain out of the contract price an amount sufficient to indemnify itself against any lien and against all costs and expenses (including attorney's fees) that the Owner may incur in connection with any claim or lien arising out of any action related thereto. If payment of the contract price has been made to the Contractor, the Contractor shall reimburse the Owner for all monies, costs, expenses, and disbursements (including attorney's fees) that the Owner may be compelled to pay to discharge any claim or lien against the premises. When requesting the final payment for completion of the contract, the Contractor shall submit to the Owner an executed and notarized Contractor's affidavit stating that all subcontractors and suppliers have been paid in full. Upon the Owner's request, the Contractor shall supply lien waivers from all subcontractors and material suppliers.

- **1-13 GUARANTEE**—The Contractor shall guarantee materials and workmanship against defects for a minimum period of one (1) year from the date of final acceptance. Upon notification from the Owner, the Contractor agrees to promptly repair or replace any defects and all resulting damage to the satisfaction of the Owner and at no cost to the Owner.

- **1-14 PAYMENTS TO THE CONTRACTOR**—At thirty (30) day intervals, the Contractor may submit to the Owner requests for payment for the work completed to date, less ten percent (10%) of the amount due, less the amount of payments previously made.

- **1-15 CREDITS AND EXTRAS**—When additions to or deletions from the work covered in the contract are required by the Owner, a change order shall be executed by the Owner and the Contractor including the costs of the changes involved.

- **1-16 INSPECTIONS**—The Germans From Russia Heritage Society and its representatives shall, at all times, have access to the project, and the Contractor shall give the Germans From Russia Heritage Society sufficient advance notice of when the work will be ready for inspection at the following stages of construction:
  - Inspection of soil condition after excavation for footings
  - Inspection of steel reinforcing before placement of concrete
  - Inspection after structural steel is in place
  - Inspection of electrical and plumbing rough-in before floor slab is poured
  - Inspection of roofing application as roofing contractor begins work
  - Inspection of electrical and plumbing rough-in before walls and ceiling are installed
  - Final electrical, mechanical, and plumbing inspections
  - Inspection of driveway base material after it has been set up prior to paving
  - Final inspection of complete job

- **1-17 CLIMATIC CONDITIONS**—When so directed by the Germans From Russia Heritage Society, the Contractor shall suspend work that may be subject to damage by climatic conditions.

- **1-18 TEMPORARY SERVICES**—The Contractor shall pay for all fuel, electrical current, and water required for construction purposes. The Contractor shall also provide temporary heat and temporary toilet, if required, as directed by the Owner.

- **1-19 SUBSTITUTIONS**—The Contractor shall base the bid on the cost of the materials and products specified. If the Contractor desires to substitute any equal material of another brand or manufacturer, said substitution shall be requested in writing at least five (5) days prior to the bid due date. Requests from subcontractors will not be considered. Samples or technical data on the proposed product or material shall accompany these requests.

- **1-20 MEASUREMENTS, LINES, AND GRADES**—The Contractor shall be responsible for the accuracy of all lines and grades of the work. The Contractor shall do all fieldwork necessary to lay out and maintain the work. No extra charge or compensation will be allowed due to differences between actual dimensions and the measurements indicated on the drawings; any difference found shall be submitted to the Owner for consideration before proceeding with the work. The Owner will provide a survey of the property including reference points, property corners, and benchmarks.

- **1-21 OCCUPANCY PRIOR TO FINAL ACCEPTANCE**—The Owner reserves the right to take possession and use any completed or partially completed portion of the project, providing it does not interfere with the Contractor's work. Such possession or use of the project shall not be considered as final acceptance of the project or any portion thereof.

- **1-22 SAFETY**—The Contractor shall provide and continuously maintain adequate safeguards, such as railings, temporary walks, lights, and so forth, to prevent accidents, injuries, or damage to any person or property. The Contractor shall adhere to the requirements of the federal Occupational Safety and Health Act as it relates to the work covered in the contract.

- **1-23 USE OF PREMISES**—The Contractor shall occupy, use, and permit others to use the premises only for the purpose of completing the work to be performed under the contract with the Owner. Storage and other uses required by the Contractor shall be in areas designated by the Owner.

- **1-24 USE OF ADJOINING PREMISES**—The Contractor shall confine operations to the area contained within the property lines as shown on the plot plan. The Contractor may use public streets and alleys as permitted by the jurisdictional authorities. No equipment, forms, materials, scaffold, or persons shall encroach or trespass on any adjoining property, unless the Contractor obtains prior written consent of the landowner.

- **1-25 PROTECTION OF ADJACENT WORK AND PROPERTY**—The Contractor shall protect all adjacent work and property such as structures and plantings from all damage resulting from the operations. Should it be necessary to remove or trim an existing planting, the Contractor shall secure all permits and approvals and pay any and all costs associated with such work. The Contractor shall check all projects, offsets, footings, and so forth and determine that there are no encroachments on the adjoining property. Where encroachments occur as a result of the work performed under these plans and specifications, the Contractor shall remove such encroachments and pay associated costs with no expense to the Owner.

- **1-26 CLEANING UP**—The Contractor shall keep the project clean at all times. No accumulation of waste material or rubbish shall be permitted. At the completion of the work, the Contractor shall remove all rubbish, tools, and surplus materials and shall leave the work area broom clean and ready for use, unless otherwise specified.

- **1-27 CONSTRUCTION-RELATED DOCUMENTS REQUIRED BY THE GERMANS FROM RUSSIA HERITAGE SOCIETY**—Prior to the payment by the Germans From Russia Heritage Society of the final contract, or the first month's rent for facilities built by developers other than the Germans From Russia Heritage Society, the following documents shall have been placed on record with the Germans From Russia Heritage Society:

  - A letter from a registered consulting engineering firm attesting to the adequacy of the asphalt paving design
  - An executed copy of the roofing guarantee
  - A completed maintenance data sheet (forms supplied by the Germans From Russia Heritage Society)
  - Copies of manufacturers' and shippers' bills of lading for structural steel columns, beams, bar joists, and steel decking
  - Copies of soil density tests
  - Copies of concrete cylinder tests
  - Copies of asphalt pavement tests (base thickness and topping density)
  - The final letter of acceptance of the fire protection sprinkler system from Mutual Insurance Company
  - Copies of manufacturers' warranties on all heating and air-conditioning equipment

Division 15 Heating, Ventilating, and Air Conditioning

SCOPE OF WORK:
- This section, together with applicable paragraphs of Section 15010, essentially outlines the work to be accomplished by the Mechanical Contractor for this project. This section and each of the sections to which it applies is subject to the requirements of The Instructions to Bidders, General Conditions, Special Conditions, and Division 1 of these complete Specifications.
- This Contractor shall furnish all labor, equipment, and materials necessary for a complete ventilation system as shown on the Drawings and herein specified.
- Provide Shop Drawings on material and equipment in this Section as per Paragraph 15010-09.

SHEET METAL DUCTS (RECTANGULAR LOW PRESSURE):
- All low pressure rectangular ductwork, except as noted on the Drawings or in the Specifications, shall be of galvanized sheet steel. Duct construction, duct joints, duct seams, locks, etc., shall be in accordance with the *SMACNA HVAC Duct Construction Standards*, latest edition.
- Ducts shall conform to dimensions on the Drawings unless locations of structural members prohibit. In case of a change in dimension, cross-sectional areas shall be maintained. All

ducts shall be straight and smooth on the inside with finished joints. No inside standing seams will be permitted in ductwork. All rectangular ductwork shall be cross-broken in sheet width 18" and larger. No single thickness partitions shall be allowed between ducts.

Changes in size throughout shall be of perfect rectangular cross sections with a pitch of approximately 1 to 4. Abrupt changes or offsets will not be accepted. The Architect/ Engineer reserves the right to make minor changes in run of certain ducts without extra cost to the Owner if necessary to avoid unforeseen structural or other interferences.

- All low pressure ductwork shall be constructed to a 2" pressure class unless noted otherwise. The Engineer reserves the right to order any open seams or joints made tight by caulking or taping. No open joints at corners or elsewhere will be allowed. All vertical ducts or risers shall be self-supporting. Where space permits, elbows shall have a centerline radius equal to 1½ times their width. Shorter radius and square throat elbows shall be used where required to fit restricted areas only and where required by the duct connection details on the Drawings. All short radius and square throat elbows shall be provided with turning vanes.

- All ducts shall be substantially supported with hangers to the ceiling or otherwise, depending upon local conditions, placing supports not over eight feet apart along the length of the duct. All hangers shall be made of one inch wide galvanized iron strips of 16-gauge, riveted or bolted to sides of duct. The upper ends of hangers shall be securely fastened to ceiling or masonry by means of expansion shields or concrete inserts.

- Ductwork shall be fabricated in accordance with the following schedule:

– Up to 12" width	26-gauge
– To 30" width	24-gauge
– To 60" width	22-gauge
– To 90" width	20-gauge
– 90" width and up	18-gauge

- All right elbows shall be provided with proper size Tuttle and Bailey "Ductruns" or equivalent. Ducts and plenums over 60" shall be reinforced with 1½" × 1½" × ⅛".

- On all ductwork with perimeter greater than 120", the "Ductmate" connection system shall be used to join ducts. Where this system is used, the duct seams and connections do not have to have duct sealer applied. Ductmate shall be fabricated from 20 gauge galvanized steel and be joined with No. 440 tape between joints.

- Ductwork shall be as high as possible in all rooms to maintain proper headroom. Where two or more ducts cross each other, they must be arranged in such a manner as to maintain the greatest possible clearance underneath. The Ventilating Contractor shall not cover any electrical outlets or junction boxes. This Contractor shall consult with the Plumbing, Heating, Temperature Control and Electrical Contractors to avoid interference with piping runs, etc.

- All ductwork, where internal insulation is to be applied, shall be increased in size so the net internal dimension is that indicated on the Drawings.

- Watertight drip pans shall be provided below all power vent, relief vent, gravity vent, fresh air and exhaust openings through roof, either built into the ductwork or, if no duct is installed, independently suspended below opening. All fresh air, exhaust air, and relief air louver plenums shall have watertight seams on bottom portion of plenums. All fresh air ductwork and other ductwork, as noted on the Plans, shall be caulked watertight.

FLEXIBLE CONNECTIONS:

- Flexible connections shall be provided on inlet and outlet of all fans and air handling units. Connections shall be a 20-oz. woven glass fabric double coated with neoprene similar to "Vent Glass." Fabric shall be waterproof and fire retardant. Ductmate PRO Flex or Duro Dyne equivalent will be acceptable.

- Use 1" × 1" × ⅛" angles to clamp the duct fabric to the ductwork, fan section, and fan discharge openings and to the fans or air handlers. Use ⁵⁄₁₆" stove bolts or self-tapping screws on approximately 6" centers. Use #14 gauge 1" wide bands to bolt fabrics to round openings. These connections shall not be less than 6" wide and shall be made with slack in the fabric.

- Flexible connections may be eliminated from the inlet and outlet of fan housings that have the entire fan and motor assembly isolated for vibration.

DAMPERS:

- Balancing Dampers: Install volume dampers where shown on the Drawings. Dampers shall be fitted with adjustable quadrants or regulators. Quadrants shall be plainly marked to indicate position of damper. All dampers will be tight fitting and reinforced to prevent vibration. Quadrants shall be Ventlok No. 555 or 635, ⅜" square rod quadrants. Use No. 605 or 607 end bearing as required. Use No. 637, 638, or 639 elevated regulators for ducts with external insulation.

- Back-draft Dampers: Furnish and install aluminum, counter-balanced backdraft dampers in ductwork to all power roof ventilators and elsewhere as indicated on the Drawings. Dampers shall be multi-bladed with gasketed edges of felt or polyurethane foam for noiseless operation. Dampers shall be Arrow Type 500-ACB or Bent Products Model 3100 with blade-to-blade linkage and adjustable counterweight or equivalent by Cesco, Penn, Nailor Industries, Air Balance, Ruskin, or Safe Air.
- Automatic Dampers: These dampers, except as identified below, are to be furnished under the Automatic Temperature Control Paragraph of this Division. Dampers so furnished, but which are to be field-mounted, shall be carefully installed by this Contractor.

ACCESS DOORS:
- Furnish and install access doors where shown on the Drawings. In addition, whether shown on the Plans or not, install an access door to allow inspection of each fire damper or smoke damper not accessible by removal of a register or diffuser. Doors shall be complete with hinges and cam locks and shall open outward. Access doors for fire dampers shall have a label with letters not less than ½" in height reading "Fire Damper."
- Unless noted to be of different size on the Drawings, each access door shall be 12" × 12", Air Balance, Inc. Model FSA100 with 1" thick insulation. Equivalent units by Censco, Kees, Air Balance, Nailor Industries, Vent Products, Safe Air, or Phillips are acceptable.
- Ductmate "sandwich" doors in 12" × 8" size will be acceptable.
- Access doors at humidifiers shall be Air Balance Model FSA1000-G-1, or equivalent, hinged with glass see-thru porthole, 3" min. dia.

DUCT SEALER:
- All exhaust air, supply air, mixed air and fresh air ductwork shall have joints sealed with mastic.
- Acceptable sealers include Durodyne S-2, Permatite (Class II), 3M Duct Sealer 800, or 3M Duct Sealer 900.

CONCRETE BASES:
- Bases under mechanical equipment shall be furnished and installed by the General Contractor, but this Contractor shall be responsible for coordinating with the General Contractor as to size and location.

INTERNAL DUCT LINER AND INSULATION:
- The following ducts shall be lined with fiberglass duct liner.

  - All return ductwork for furnace system F-1a & F-1b and F-4, ½" thick.
  - All return ductwork for furnace systems F-4, ½" thick.
  - All air transfer ducts, ½" thick.
- Duct liner shall be equivalent to Certain-teed Ultra-lite duct liner. ½" thick liner shall be 2.0 lb. density and 1" thick shall be 1.5 lb. density. Coating on air side shall meet flame spread and smoke requirements of *NFPA 90A*.
- All ducts requiring insulation shall be lined by carefully adhering the liner in a continuous piece to clean, flat metal sheets with a quick tacking, fire retardant adhesive. Coated side of duct liner shall face the air stream. All exposed edges shall be coated with adhesive. Ducts over 24" width or breadth shall have liner additionally secured with spot welded pins and washers, one pin and washer to each 2 square feet of duct liner.
- Where duct liner insulation is installed in ducts, the ducts shall be increased in size to accommodate the insulation such that the net duct dimension is as shown on the Drawings.
- Duct linings shall meet *UL #181 Erosion Test* and the linings, vapor barriers, and adhesive shall meet the flame spread rating of not more than 25 and smoke developed rating of not more than 50.
- Insulation shall be as manufactured by Certain-teed, Owens Corning, Knauf, or Schuller International.

AIR EXTRACTORS:
- Furnish and install air extractors where shown in branch duct takeoffs and at all registers except last register in run which shall be supplied through "Ductruns." Extractors shall be Tuttle and Bailey "Vectrol" VLR, Nailor Industries, Krueger EX8/EX-88 or equivalent. Air extractors shall have external adjustment capacity.

HIGH EFFICIENCY TAKEOFFS:
- All round takeoffs, where connections are made to flexible duct, shall be made with high efficiency rectangular openings with 45-degree pitch on body, rectangle to round transition with built-in volume damper as manufactured by Sheet Metal Connectors Inc.
- Equivalent fittings by other manufacturers will be accepted.

AIR FILTER GAUGES:
- Gauges shall be Dwyer Magnehelic™ Gauges, Catalog No. 2001-AF, with range of 0 to 0.1 inches of water. Each gauge shall be installed with necessary stops and tubing.

- Install gauges as follows: Across F-1a, F-1b, F-2, F-3 and F-4 filters.
- Equivalent gauges by Orange Research are acceptable.

AIR DUCT THERMOMETERS:

- Air duct thermometers shall be Trerice, Catalog No. B83204, bi-metal, dial thermometers with 3-inch stainless steel case, 4-inch stem length, No. 065-0015, 3" O. flange for duct mounting. Thermometer range –40°F to 160°F.
- Install thermometers as follows: on discharge ducts from F-1a, F-1b, F-2, F-3 and F-4.
- Equivalent thermometers by Tel-Tru, Weiss, Weksler, or U.S. Gauge are acceptable.

FLEXIBLE HOSE:

- Insulated flexile duct used in low pressure heating/cooling applications shall be Thermaflex, Type M-KE, UL 181, Class 1, air duct constructed of polymeric liner bonded to corrosion-resistant steel wire helix, fiberglass insulating blanket and outer vapor barrier of fiberglass reinforced metalized film laminate. Wiremold Type WK, ATCO UPC 90, Flexmaster III, Hart & Cooley NT-25 VM, or Certainteed equivalent are acceptable.
- Flexible hose shall be sealed to the spin-in take-off and neck of the diffuser or air device with duct tape and secured with a Thermaflex "Duct Strap" or radiator type hose clamp.

LOUVERS:

- Air intake louvers shall be Airolite Type K638 stationary extruded aluminum louvers. Exhaust louvers shall be Type K609. Louver shall be 4" thick, fabricated of 12-gauge aluminum with custom Kyne finish, 70% fluoropolymer coatings, resistance to ultraviolet deterioration and chemical attack. Color selection by Architect. Louver shall be furnished with ½" square mesh, 16 gauge aluminum bird screen on louver interior.
- Furnish extruded aluminum sill for each louver.
- Louvers by Cesco, Arrow, American Warming and Ventilating, Louvers and Dampers, Inc., Industrial Louver, Dowco, Penn, Ruskin, and Air Control Products will be acceptable.

REFRIGERATION PIPING:

- This Contractor shall furnish and install all refrigeration piping and accessories as shown on the Drawings. This work shall be accomplished by trained refrigeration mechanics whose normal activity is installation and service of refrigeration systems.
- Insulation of refrigerant piping shall be as specified in Section 15180.

- Refrigerant piping shall be ACR Type "L" hard drawn, hydrated, copper tubing with forged copper fittings joined with Silfos solder.
- Furnish and install Sporlan thermostatic expansion valve of proper type, Sporlan "Catch-All" filter/dryer and Sporlan "See-All" combination moisture and liquid indicator on each liquid line. Furnish and install a Sporlan or Alco hot gas bypass valve and bypass piping where shown on the Drawings.
- All tubing must be clean and dry. If there is any evidence of dust, moisture, or corrosion, the tubing must be cleaned out by drawing a swab soaked in methyl alcohol or Virginia Number 10 degreasing solvent (a product of the Virginia Chemical Company) through the tubing as many times as necessary to thoroughly clean the tubing. To eliminate the formation of copper oxide on the inside of the tubing, all air must be flushed from the inside of the refrigerant lines before soldering and a slow stream of dry nitrogen must be passing through the tube during the soldering process.
- After the refrigeration piping has been installed and a leak test has been completed, a vacuum of 500 microns shall be pulled on the system by use of a vacuum pump built especially for that purpose for a minimum of six hours and held without leak. After the system has been charged with refrigerant and the proper amount of oil, it shall be placed in operation. After 24 hours operation, the refrigerant charge and oil level shall be rechecked for proper amounts.
- The Drawings show a piping arrangement based on the specified condensing unit and cooling coil. If some other arrangement of piping is necessary to accommodate an approved equal condensing unit the Mechanical Contractor elects to install, he shall install the required piping and any necessary accessories, such as solenoid valves or expansion valves, at no additional cost to the Owner.
- Workmen performing work on refrigeration systems shall be certified to do so. Certification shall involve a third party test focused on the environmental science surrounding the issue of refrigerants and their effect on the environment, sensible operating and maintenance practices and a clear understanding of the requirements for leak prevention, recovery of refrigerants, and safe disposal of refrigerants. Only certified technicians will be allowed to install the refrigeration piping systems and charge them with refrigerant or to reclaim refrigerant from existing systems. Certification of technicians shall

meet all requirements as set forth by Section 608 of the Clean Air Act Amendments. Agency providing testing and certification is The Refrigeration Service Engineering Society (RSES) or National Association of Plumbing Heating and Cooling Contractors (NAPHCC). Contractor shall utilize refrigerant reclaim unit and approved containers for used refrigerant when removing the refrigerant charge from new or used equipment. Flashing refrigerant gas to atmosphere is strictly prohibited.

THROWAWAY FILTERS:

- Throwaway filters not furnished with equipment shall be nominal 1" thick, pleated media, equivalent to American Air Filter 30/30. In addition to the initial installation, furnish three (3) sets.
- Filters shall feature woven media blended from synthetic and cotton fibers to provide a nominal 25% efficiency based on ASHRAE 52-76. Cam Farr 30/30, Airguard DPE 40, or Continental CPLT 32 are acceptable. Where indicated on the Drawings, filters shall be installed in a field fabricated frame with slide channels to hold the filters in place. Provide a hinged access door in the ductwork for filter removal.

REGISTERS, GRILLES AND DIFFUSERS:

- Registers, grilles, and diffusers, as scheduled on the Drawings, are those of the Price Manufacturer. Equivalent units by Titus, Nailor Industries, Krueger, J & J, Reliable Metal, Anemostat, Hart and Cooley, or ACP will be acceptable.

CONDENSING UNITS:

- Condensing units scheduled on the Drawings shall be complete with service valves, service ports, low pressure switch, internal pressure relief valve, and compressor overload protection, prewired control panel, operating charge of HCFC-22, and crankcase heater.
- Options shall include time delay, low ambient lockout set at 50°F, control transformer, if required, liquid line filter/dryer, and 5-year warranty on the compressor.
- Unit casing shall be louvered steel with baked enamel finish and have condenser fan discharge grille and condenser coil hail guards.
- Units scheduled on the Drawings are Trane. Equivalent units by Carrier, Bryant, or Rheem are acceptable.

FURNACES:

- Furnish and install condensing, sealed combustion, natural gas-fired furnace as scheduled on the Drawings.

- Burner controls shall include hot surface ignition and vent proving switch.
- Unit enclosure shall have baked enamel finish, access doors for service, and throwaway filters. Furnish three spare sets for each furnace.
- Fans will be forward curved, direct-drive, dynamically and statically balanced. The multispeed motor shall be factory lubricated and have internal overload protection.
- Each unit shall be furnished with direct expansion cooling coil matched to condensing unit, condensate drain pan, coil cabinet, and concentric vent/intake termination kit.
- Units shall come complete with controls including 24-volt transformer, indoor fan relay, line voltage terminal block, low-voltage terminal strip, and auxiliary board for interface with commercial thermostat provided and installed by ATC. Provide twinning kit to furnaces F-1a and F-1b.
- Units scheduled on the Drawings are Trane. Equivalent units by Carrier, Bryant, or Rheem are acceptable if they can operate in tandem with one thermostat and are equal in performance.

HI-EFFICIENCY FURNACE VENT/INTAKE PIPING:

- Vent piping to be Hart & Cooley "Ultravent" or a material as approved by the furnace manufacturer. Intake piping to be Schedule 40 PVC with DWV fittings joined with solvent cement. Vent and intake piping shall be sized as per manufacturer's recommendations. Terminate vent and intake with manufacturer's vent kit in accordance with Manufacturer's instruction.

SMALL CABINET FANS:

- Small cabinet fans shall have galvanized steel housing lined with ½" insulation. Fan wheels shall be direct drive with motor and fan mounted with resilient isolators. Type CSP fans shall be inline with inlet and outlet collars. Fan wheels shall be centrifugal and fans shall feature backdraft dampers.
- Motor disconnect shall be internal of the cord and plug type.
- Fans scheduled on the Drawings are Greenheck. Equivalent fans by Pace, Acme, Jenn-Fan, Penn, or Carnes will be acceptable.

HUMIDIFIER H-1:

- Furnish and install a self-contained microprocessor, electronically controlled wall-mounted steam humidifier having a capacity as scheduled on Drawings. Furnish humidifier with stainless steel duct distributor pipe of length as scheduled on Drawings. Furnish and install rubber steam hose of adequate length for connection or duct distributor pipe.

- Humidifier shall have locking backed enamel 14 gauge aluminum cabinet with all adjusting controls hidden from general tampering. Cabinet shall house the cleanable stainless steel steam generator cylinder. Electrode wires shall be fastened with a nut to prevent loosening of the connection. Friction connections are not acceptable. Humidifier shall be capable of operating on water in the range of 60 to 1500 micromhos conductivity.
- Humidifier shall discharge steam at 212°F and atmospheric pressure. Duct distributor pipes shall be usable in the vertical or horizontal position.
- Steam shall be generated in a stainless steel cleanable evaporation tank. The evaporation tank will be easily removable from the unit. Electronic level sensing assembly, heating elements and manual reset high temperature safety cut-out switch will be secured to the top cover of the evaporation tank. For servicing, the cover will be easily removable from the evaporation tank by means of spring latches. After removal of the top cover, the evaporation tank will become an easily cleanable empty vessel.
- All electrical wiring will be detachable between the top cover and the electrical cabinet by quick connectors.
- The evaporation tank shall have a safety overflow connection and the drain port, which will be located on the side wall of the evaporation tank. This will minimize the risk of blockage caused by sediment buildup in the bottom of the tank. The overflow and drain port will be detached for servicing by means of a single quick disconnect assembly.
- The supply water to the unit shall be controlled by a solenoid valve and the drain shall be operated by a motorized ball valve which will permit solid mineral particles to pass through the drain without being obstructed.
- The control circuit shall be 24 VAC. The unit shall be managed by a microprocessor controlled P.C. board that shall have the following features:
  - Water level display LEDs.
  - Adjustable drain cycles every 4, 8, and 24 hours or total drain cycles override.
  - Push button for "Start-up" commissioning cycle and "Reset" override button for normal operation.
  - After 1,000 hours of operation (humidifier "ON") the front panel of the humidifier will display "CHECK" light to indicate service required, but the unit shall continue to operate manually.
  - After 72 hours of no demand, the humidifier will go to "End of Season" mode draining the unit.
  - The humidifier will accept ON/OFF or modulating signals from a Building Automation System or a remote controller.
- The control modulating signal shall be 0–10 VDC or 2–10 VDC or 4–20 mA to modulate 0–100% of the capacity. Modulation will be done by silent SSR's using zero voltage crossing. The SSR's will be backed up by an electro-mechanical contactor.
- The front panel display on the humidifier shall include indicator lights to annunciate "POWER," "FILL," and "STEAM" humidity demand, "DRAIN" cycle, and "CHECK" system warning. It shall also include a manual three position rocker switch for "Automatic Operation," "Unit Off," and "Manual Drain."
- The steam distribution will be done by "Multi-steam" system distribution designed for short absorption distances and low air duct temperature.
- The unit shall have CSA-NTRL certification.
- The unit shall be supplied with the appropriate controls and safety ON/OFF duct high-limit humidistat controls for MODULATING operation.
- Appropriate inspection of the installation and start-up will be done by the manufacturer's agent (72 hours advance notice).
- Humidifiers to be furnished complete with distribution hose, distribution pipes, proportional plus integral humidistat, flow switch, and duct mounted high limit switch.
- Humidifier models scheduled on Drawings are by Neptronic.

HUMIDIFIER H-2:
- Provide and install Aprilaire model 550 bypass type humidifier with 0.5 gph capacity, 24c-60Hz-0.7AMPs, ½" I.D. plastic hose drain, 6-inch round bypass opening, 24v Humidistat, Saddle valve, and Transformer.
- Unit dimensions are 13" wide × 12¾" high × 9" deep.

SYSTEM BALANCING COORDINATION:
- The Sheet Metal Contractor shall make any changes required for correct balance as recommended by the Balancing Contractor at no additional cost to the Owner. Such changes may encompass, but are not necessarily restricted to, pulleys, belts, duct-work dampers, or the addition of dampers and access doors. The Sheet Metal Contractor shall provide labor, equipment, and cost of performing necessary corrections.

- All ductwork and coils shall be cleaned and left free of loose insulation and construction debris.
- The Sheet Metal Contractor shall put all heating, ventilating, and air conditioning systems into full operation and shall continue the operation of same during each working day of testing and balancing. Correct operation of equipment and system components and cleanliness of ductwork shall be the responsibility of the Sheet Metal Contractor.
- Filters shall be cleaned or replaced just prior to final balancing.
- All fans shall be initially started, lubricated, and balanced to eliminate noise and vibration.

As-built drawing: A drawing of a completed installation that is used as a reference to make a drawing showing actual changes to the project.

Change order: A written order signed by the architect and the owner and given to the contractor after the execution of the contract that authorizes a change in the work.

Elevation: A drawing that shows the vertical elements of a building (exterior or interior) as a projection on a vertical plane. Elevation may also be the height of a point on land.

Footing: The concrete base on which a building's foundation walls are built.

Joist: One of a series of parallel beams of timber, reinforced concrete, or steel used to support floor and ceiling loads.

Joist hanger: A metal hanger or strap used to attach a joist to a beam.

Legend: A list that defines the pictorial symbols on a construction drawing.

Load capacity: The amount of load that a structural member, support, pipe, or an electric circuit can safely withstand without failing.

Member: In structural engineering, any part of a structure that is complete in itself.

Riser diagram: A diagram that shows the vertical path of plumbing, electrical wiring, or duct from one story of a building to another.

Roof pitch: The pitch of a roof generally expressed as the angle of pitch in degrees or as a ratio of the vertical rise to the horizontal run.

Schedule: A detailed list of components to be installed in a building such as doors, windows, flooring, and wall treatments.

Scupper: An opening in a wall or parapet (the part of a wall that extends above the roof) that allows water to drain from a roof.

Section: A drawing that shows how a building or part of a building would look if it were cut open to show the interior.

Submittal: A detailed drawing and description of an accessory or piece of equipment to be installed in a building.

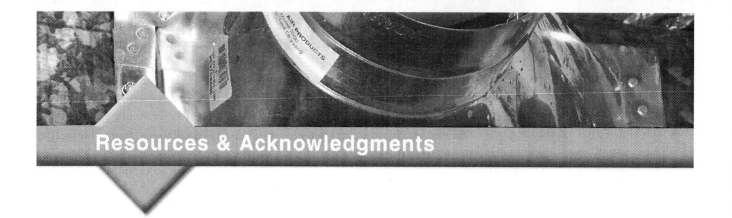

Additional Resources

This module is intended to present thorough resources for task training. The following reference work is suggested for further study. This is optional material for continued education rather than for task training.

Blueprint Reading for Construction, Second Edition. 2004. James A.S. Fatzinger. Upper Saddle River, NJ: Prentice Hall.

References

The Construction Specifications Institute, 99 Canal Center Plaza, Suite 300, Alexandria VA 22314

Acknowledgments

NCCER would like to thank the following companies for permitting use of the plans and specifications of the Germans From Russia Heritage Society:

Architects: Ritterbush-Ellig-Hulsing, P.C., Bismarck, ND

Mechanical & Electrical Engineers: Prairie Engineering, P.C., Bismarck, ND

Structural Engineers: Traeholt Associates, P.C., Bismarck, ND

Civil Engineers: Swenson, Hagen & Co., P.C., Bismarck, ND

Figure Credits

Kogok Corporation, 202F14, 202F15, 202F16
Carnes Company, 202F17A, 202F17B

NCCER makes every effort to keep these textbooks up-to-date and free of technical errors. We appreciate your help in this process. If you have an idea for improving this textbook, or if you find an error, a typographical mistake, or an inaccuracy in NCCER's Contren® textbooks, please write us, using this form or a photocopy. Be sure to include the exact module number, page number, a detailed description, and the correction, if applicable. Your input will be brought to the attention of the Technical Review Committee. Thank you for your assistance.

Instructors – If you found that additional materials were necessary in order to teach this module effectively, please let us know so that we may include them in the Equipment/Materials list in the Annotated Instructor's Guide.

Write: Product Development and Revision
National Center for Construction Education and Research
3600 NW 43rd St, Bldg G, Gainesville, FL 32606

Fax: 352-334-0932

E-mail: curriculum@nccer.org

Craft _____ Module Name _____

Copyright Date _____ Module Number _____ Page Number(s) _____

Description _____
_____
_____
_____
_____

(Optional) Correction _____
_____
_____

(Optional) Your Name and Address _____
_____
_____
_____

04203-08

Fabrication Two – Radial Line Development

04203-08
Fabrication Two –
Radial Line Development

Topics to be presented in this module include:

Overview

Radial line development is a method frequently used to lay out fittings such as cones and other tapered shapes. These fittings are described using lines that originate at their base and converge at a single common point, or apex. The base can be a circle, square, or rectangle. In this module, you will learn about and practice laying out tapered fittings using radial line development techniques.

Objectives

When you have completed this module, you will be able to do the following:

1. Describe the principles of radial line development used to determine layouts for sheet metal fittings.
2. Use the principles of radial line development for the layout of selected sheet metal fittings.
3. Layout and fabricate selected sheet metal fittings.

Trade Terms

Apex	Radial
Chord	Splitter cone
Gore	Stays
Pitch	Taper

Required Trainee Materials

1. Appropriate personal protective equipment
2. Pencil and paper

Prerequisites

Before you begin this module, it is recommended that you successfully complete the following: *Core Curriculum; Sheet Metal Level One; Sheet Metal Level Two*, Modules 04201-08 and 04202-08.

This course map shows all of the modules in the second level of the Sheet Metal curriculum. The suggested training order begins at the bottom and proceeds up. Skill levels increase as you advance on the course map. The local Training Program Sponsor may adjust the training order.

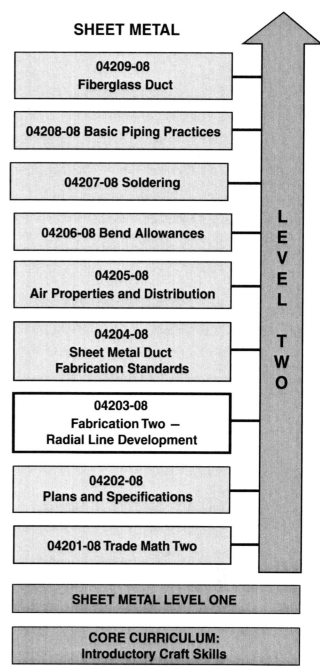

SHEET METAL

04209-08 Fiberglass Duct

04208-08 Basic Piping Practices

04207-08 Soldering

04206-08 Bend Allowances

04205-08 Air Properties and Distribution

04204-08 Sheet Metal Duct Fabrication Standards

04203-08 Fabrication Two — Radial Line Development

04202-08 Plans and Specifications

04201-08 Trade Math Two

SHEET METAL LEVEL ONE

CORE CURRICULUM: Introductory Craft Skills

LEVEL TWO

203CMAP.EPS

1.0.0 ◆ INTRODUCTION

When you create a pattern using **radial** line development, you will use many of the same procedures you learned in *Fabrication One – Parallel Line Development*. In that module, you used parallel line development to create patterns for parallel forms. In this module, you will learn how to use radial line development to make patterns for **tapered** fittings. These tapered fittings may have a circle for their base, such as a cone, or they may be shapes with a base that can be drawn within a circle, such as squares, hexagons, and octagons. What these tapered fittings have in common is that lines drawn from their corners end in a common center, called an **apex**, over the center of the base.

2.0.0 ◆ RADIAL LINE DEVELOPMENT PRINCIPLES

The radial line development method is best suited to fabricating fittings in which the lines drawn from the corner of the fitting end in an apex over the base of the fitting. You will use this method most often to make patterns for rounded, tapered fittings such as cones. You can also use it for other tapered shapes. Imagine unrolling a cone on a flat plane. You would see that the apex of the cone remains fixed at one point and that the cone unrolls in a circular path around the apex (*Figure 1*). Would the same thing happen with a pyramid shape? Yes, it would (*Figure 2*). If you unfold the pyramid, the apex remains fixed and the sides of the pyramid unfold in a circle around the apex. Thus, you can use the radial line method to develop a pattern for a pyramid with sides that are equally tapered.

In the radial line development method, you will extend the side lines of a tapered form until those lines intersect. The point at which those lines meet is the apex. Because you are working with tapered forms, those lines will extend from the base form on an angle or slope. Note that radial line development works best as the slope of the taper increases. If the slope of the taper is too shallow, you will have to increase the length of your extension lines, which puts the apex too far away to be practical.

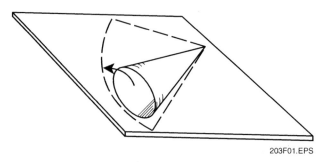

Figure 1 ◆ Unrolling a cone.

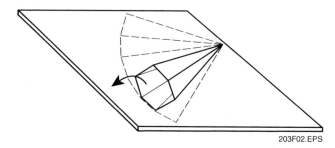

Figure 2 ◆ Unfolding a pyramid.

To understand why this is so, try this experiment. Draw a square. Next to it draw another square, but make the sides of the square slant inward very slightly. Now draw a third square, but make the sides slant inward at a greater slope. Extend the side lines of each shape upward until the lines intersect. As you will see, the extension lines on the first square will never meet. The extension lines on your second square will meet, but at a much higher point than those on the third square.

Before you can develop a pattern for a radial form, you must know two things about the fitting: the true lengths of the sides and the circumference or the perimeter of the base. In the following exercise, you will develop a pattern using the radial line development method (*Figure 3*). To complete this exercise, follow these steps:

Step 1 Draw the elevation view, showing the height of the apex.

Step 2 Draw the plan view. Use this view to determine the length of the stretchout.

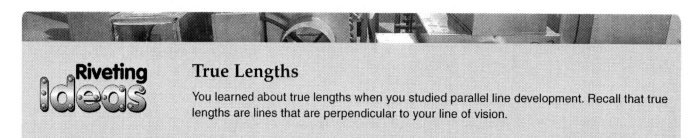

Riveting Ideas

True Lengths

You learned about true lengths when you studied parallel line development. Recall that true lengths are lines that are perpendicular to your line of vision.

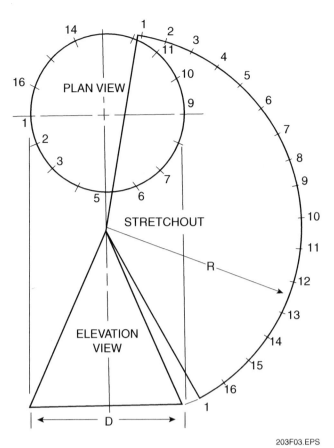

Figure 3 ◆ The radial line development method.

Step 3 Draw the stretchout arc with the radius equal to the true length of the edge of the fitting. Draw the stretchout arc so that it is long enough to contain each space in the plan view and in the same order.

Step 4 Get the true lengths of the sides from the elevation view, and place them on the stretchout to complete the pattern development.

A number of tapered shapes and fittings can be laid out using radial line development. These shapes and fittings include round cones, square or rectangular pyramids, funnels, round tapers, and roof jacks (*Figure 4*).

3.0.0 ◆ SAFETY

When you work on the tasks in this module, you will use tools that, if used incorrectly, can seriously injure you or your co-workers. You have already learned many safe work habits in other modules, but because safety is so important, you should take a moment to review these general safety rules before you do any of the tasks in this module:

- Always wear appropriate personal protective equipment.
- Be sure to read and follow safety instructions on sheet metal shop machines and welding equipment.
- Stay alert. You must tell your instructor if you are taking any medication that could make you drowsy.

> **WARNING!**
> In addition to safety gear that you must wear in the sheet metal shop, you must also pay attention to your clothing and accessories. Pieces of fabric, your hair, and even jewelry can become caught in machinery. This can damage the machine or cause you to be seriously injured. For example, if your hair becomes caught in the moving parts of a machine, you will be pulled into those moving parts with it. Don't wear loose-fitting clothing, ties, or scarves. Don't wear any jewelry, including watches. Securely tie back long hair away from your face.

- Always check welding equipment to ensure that hoses and connections are undamaged and in proper working order.
- If you are not sure about how to do a procedure, how to check equipment, or how to use any of the tools or equipment, you must ask your instructor for help.
- Keep your work area clean. Put tools away when you are finished with them. Keep cords free from tangles and out from underfoot.

ROUND
CONE

SQUARE AND
RECTANGULAR
PYRAMID

FUNNEL

ROUND TAPER
WITH END ON
SLANT

ROOF
JACK

203F04.EPS

Figure 4 ◆ Examples of shapes and fittings made using radial line development.

Riveting Ideas

Muscle Memory

As you use the tools of your trade, your muscles, especially those in your hands and arms, will develop a kind of memory. You'll know when this happens. One day you'll find that using your tools properly and safely is almost as natural as breathing. Take the time now to learn the right way to use your fabrication tools, because you are actually training your muscles to help you become a skilled professional.

4.0.0 ◆ RADIAL LINE DEVELOPMENT TASKS

There are twelve tasks in this module. The figures that accompany each task are not drawn to scale. The lines in the figures are identified with numbers and letters. Some of these numbers and letters are marked with a prime symbol ('). This mark helps you to identify which numbers and letters are being referred to in the figures.

The instructions given are for the radial line development method. Some of the fittings in this module can be developed using parallel line development or triangulation, and your instructor may have you practice these techniques as well.

Reading through the steps in each task can be confusing at first. As you work your way through each task, you'll get the hands-on experience that is the best teacher in fabrication work.

4.1.0 Layout Tools

At this stage in your apprenticeship training, you should be familiar enough with the tools of your trade to select the right tools for each task. For example, you know that to draw patterns on metal you use dividers. As you study each task, refer to the following list of layout tools and fabrication equipment and determine which you should use. If you are not sure which tools to use, ask your instructor.

- Scratch awls
- Pencils and felt-tipped markers
- Straightedge
- Flat steel square and combination square
- Prick punch and center punch
- Dividers
- Trammel points
- Marking gauge
- French curves
- Circumference rule
- ¼-inch tape measure

4.2.0 Fabrication Equipment

Your instructor will give you specifications for the gauge of sheet metal and details about any fabrication materials needed to complete the task.

- Hand seamer
- C-clamps and parallel clamps
- Sheet metal hammer and mallet
- Bench plate and stakes
- Sheet metal snips
- Chisel
- Rivet set and hand groover
- Drill bits and drill
- Files
- Hacksaw
- Bar folder
- Sheet metal brake
- Sheet metal shear
- Beading, crimping, turning, slip-roll forming, and burring machines
- Ring and circle shears
- Screwdrivers, wrenches, pliers
- Electric spot welder
- Gas welding equipment
- Arc welder
- Whitney hand (lever) punch

5.0.0 ◆ TASK 1: RECTANGULAR WEATHER CAP

In this task, you will lay out and fabricate a rectangular weather cap (*Figure 5*). The weather cap protects the square or rectangular vents in commercial buildings. Small caps are made in one piece; larger ones are often made in four pieces with standing seams on the corners. To lay out and fabricate this fitting, follow these steps:

Step 1 Draw the elevation and plan views.

Step 2 Draw half the elevation or side view.

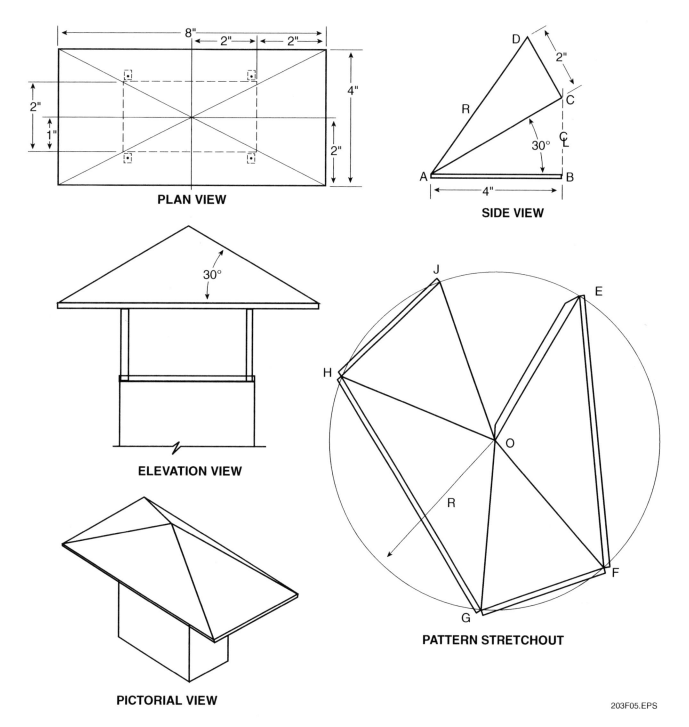

PLAN VIEW

SIDE VIEW

ELEVATION VIEW

PICTORIAL VIEW

PATTERN STRETCHOUT

203F05.EPS

Figure 5 ◆ Rectangular weather cap.

a. Draw the baseline AB and square a line upward from point B to establish the centerline of the drawing.

b. From point A, draw a line at 30 degrees or as established by your instructor. Establish point C.

c. From point C, draw a line perpendicular to line AC. Measure outward on this line from point C a distance equal to one-half the width of the weather cap (2 inches). Establish point D. The developed radius line (AD) is the true-length corner line and therefore is the radius of the pattern development circle.

Step 3 Draw the pattern stretchout.

a. Use point O as the center and distance AD as the radius, and scribe the circle for the pattern stretchout.

Identifying Lines

When you develop a pattern, you should number the lines on the side view and on the pattern. Doing this reduces the chance for error. It also saves time because you won't have to count or recount lines.

b. Draw a straight line from point O to the circumference of the circle, and establish point E.

c. Hold a rule so that **chords** equal to twice the length of AB can be formed in the circle to establish point F. Chords are lines drawn from one part of a circle's circumference to another, without passing through the center.

d. Using point F as the center and a distance equal to the width of the weather cap as the radius, establish point G on the circumference.

e. Repeat steps c and d to establish points H and J on the circumference of the circle.

f. Draw straight lines between the established points on the circumference to locate the bend lines and the cutout section.

g. Add allowances for the seams or joints, and locate the rivet holes for the **stays**. (Note: Some shops prefer that the stays be located on the corners; others prefer that they be located in the center of each side.) Stays are usually made from 1" × ⅛" band iron. The edge of the rectangular weather cap is usually bent down vertically about ½ inch, with a ¼-inch hem.

Step 4 Mark the pattern for cutting and bending. Cut and notch the pattern. Form the fitting and spot weld or rivet the seam.

Step 5 Clean up your work area.

6.0.0 ◆ TASK 2: SYMMETRICAL TAPERED DUCT

In this task, you will lay out and fabricate a symmetrical tapered duct (*Figure 6*). The symmetrical tapered duct is also called a round equal-taper joint. This is the most frequently used type of taper fitting. To lay out and fabricate this fitting, follow these steps:

Step 1 Draw the plan view and the elevation view.

Step 2 Draw the half profile view. (Note: Only half of a front profile is necessary to develop the full taper pattern.)

a. Draw a baseline CC' equal to one-half the diameter of the large end of the fitting. Square a line upward from point C' an undetermined distance to establish the centerline of the half profile.

b. From point C', measure up the distance equal to the height of the fitting, and establish point B'.

c. Square a line upward from point B' a distance equal to one-half of the diameter of the small end of the fitting, and establish point B.

d. Draw a straight line from point C through point B until it intersects with the centerline, and establish point A.

Step 3 Draw the pattern stretchout.

a. With distance AC as the radius and point A as the center, scribe an arc to establish the circumference of the

Pictorial Views

Each of the tasks in this module includes a pictorial view. The pictorial view is a simple line drawing that shows what the completed fitting looks like. Compare the pictorial view with the elevation and plan views and the pattern drawings. Practice visualizing how the pattern pieces come together into the finished fitting. The ability to mentally see the relationship between the pattern and the finished fitting is an important skill to develop.

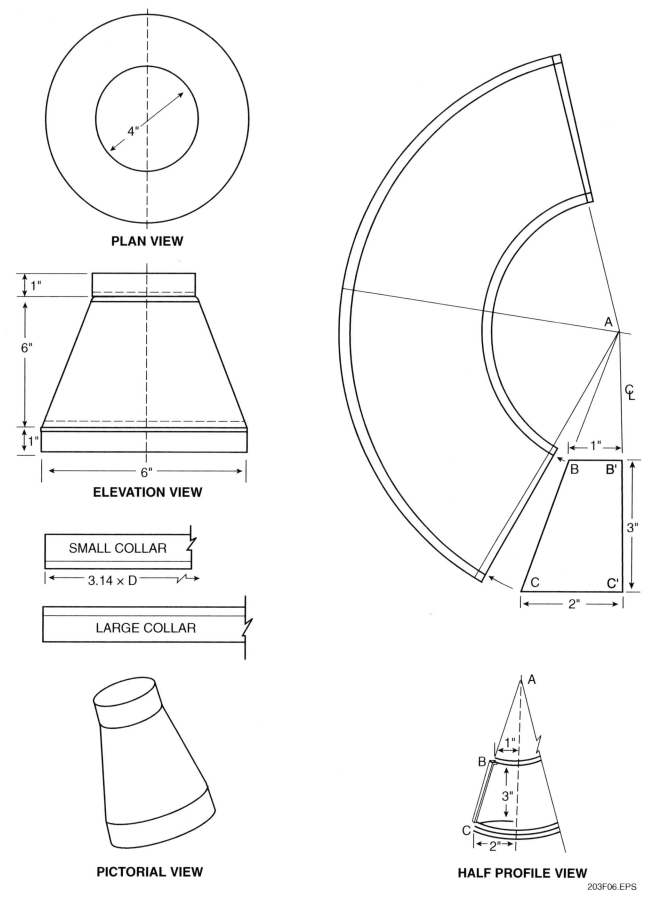

PLAN VIEW

ELEVATION VIEW

SMALL COLLAR

3.14 × D

LARGE COLLAR

PICTORIAL VIEW

HALF PROFILE VIEW

203F06.EPS

Figure 6 ◆ Symmetrical tapered duct.

large diameter of the fitting. Remember that circumference equals π × diameter, or 3.14 × 4.

b. With distance AB as the radius and A as the center, scribe an arc to establish the circumference for the small diameter of the fitting.

c. Draw straight lines from point C on the stretchout to point A. This will automatically establish the circumference of the small diameter. (Note: The half front profile can be included within arc CD without any wasted space.)

Step 4 Lay out the collars with the length of the stretchouts equal to the circumference of the large and small diameters of the arc. You can determine the width from the working drawing.

Step 5 Make allowances for seams and edges on the patterns.

Step 6 Cut out the patterns and form the shapes.

Step 7 Rivet or spot weld the seams and edges according to your instructor's directions.

Step 8 Clean up your work area.

7.0.0 ◆ TASK 3: ROOF PITCH STACK FLANGE

In this task, you will lay out and fabricate a stack flange for a roof **pitch** (*Figure 7*). This fitting may also be called a round taper with a mitered base or a roof jack on a pitch. Most fittings of this type are drawn according to pitch. Pitch describes the amount of rise in relation to a horizontal run. It may be measured in degrees, feet, or inches. For example, a pitch of ⁷⁄₁₂ inches is 7 inches of rise in 12 inches of horizontal measurement. To lay out and fabricate this fitting, follow these steps:

Step 1 Draw the elevation or side view.

a. Draw a vertical line to establish the centerline of the fitting.

b. At any distance on the centerline, establish point 16. From point 16 measure up 5 inches and establish point 0.

c. Draw a horizontal line through point 0 extending to the right and left of the centerline.

d. Draw a 30-degree line through point 16 or the pitch angle as directed by your instructor.

e. With point 0 as the center and one-half the diameter of the small end of the fitting as the radius, draw a half circle below the diameter line of the small end of the fitting.

f. Divide the half circle into six equal parts and mark the seven points of intersection 1, 2, 3, 4, 5, 6, and 7. Connect these points by vertical lines to line 1-7.

g. Draw lines at 30 degrees to the centerline through points 1 and 7 to intersect at the apex.

h. Draw lines from the apex through the intersection points of the perpendicular lines with small diameter line 1-7 to intersect the pitch line drawn through point 16, and establish points 13, 14, 15, 16, 17, 18, and 19.

i. From points 14, 15, 16, 17, and 18 on the pitch line, draw horizontal lines to the left to intersect the line from the apex to point G, and establish points F, E, D, C, B, and A.

j. Add 1 inch below point 13 to establish point 20. Add 1 inch above point 19 to establish point 21. (The size of the flange is 1 inch.)

Step 2 Draw the half plan view.

a. Draw vertical lines downward from points 20 and 21 on the pitch line to establish the length of the roof flange pattern, and locate points H and J.

b. Draw a horizontal line between points H and J. Measure down a distance equal to one-half the width of the roof flange, and draw line IK.

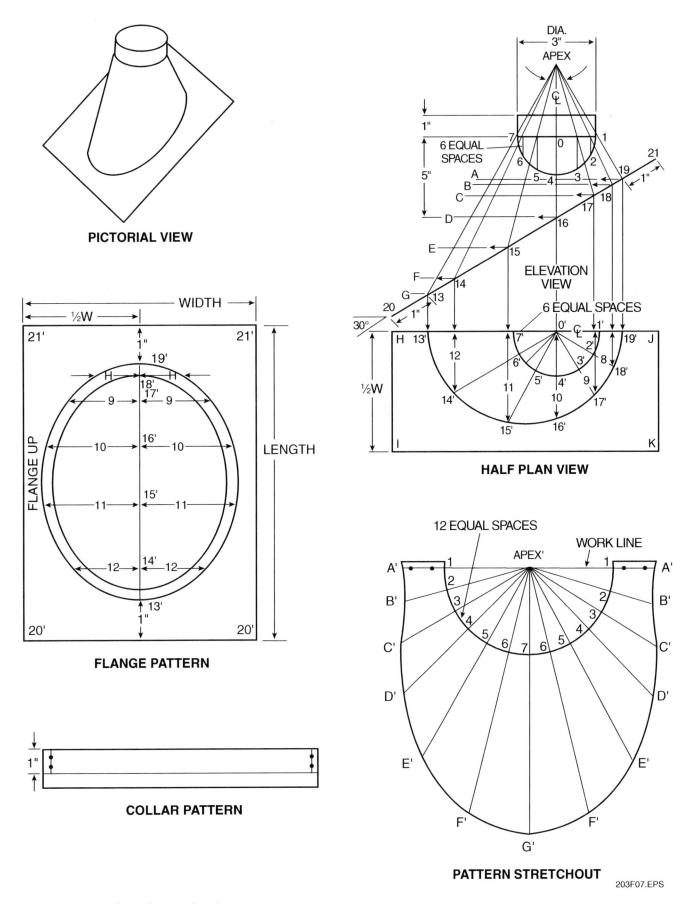

PICTORIAL VIEW

FLANGE PATTERN

COLLAR PATTERN

HALF PLAN VIEW

ELEVATION VIEW

PATTERN STRETCHOUT

203F07.EPS

Figure 7 ◈ Stack flange for a roof pitch.

c. Identify the intersection point of the vertical centerline and line HJ as point 0'. With point 0' as the center and one-half the diameter of the small end of the fitting as the radius, draw a half circle.

d. Divide the half circle into six equal parts and locate the seven points of intersection 1', 2', 3', 4', 5', 6', and 7'. Extend lines through these points from point 0'.

e. Project point 14 from the elevation downward to intersect line 0'-14'. Mark this line as distance 12. Follow the same procedure for points 15, 17, 18, and 19. Mark each distance on the lines as shown (11, 9, 8).

f. To find distance 0'-16', set your dividers to distance 16D in the elevation view; with 0' as the center, establish point 16' and mark this line as 10.

g. Draw the developed curve outline by connecting points 13', 14', 15', 16', 17', 18', and 19'.

Step 3   Draw the pattern stretchout.

a. Draw a horizontal work line and, at any point near the center of the work line, locate a center (apex') from which to develop radii.

b. With your dividers set to the distance between the apex and point 1 on the elevation view, use apex' on the pattern stretchout as the center, and locate point 1 on the work line. With this distance as the radius and the apex' as the center, scribe an arc in a circular manner equal to the circumference of the small end of the roof stack.

c. Divide the arc into 12 equal spaces and locate points 1, 2, 3, 4, 5, and 6 on each side of the arc centerline (point 7). Extend lines from the apex' through each of the points just established.

d. Working from the elevation view, transfer distance apex-A to apex'-A' on the stretchout pattern. Follow the same procedure for each of the other corresponding distances (apex-B to apex'-B', and so on).

e. Draw the stretchout pattern outline through the developed points.

Step 4   Lay out the flange pattern.

a. Draw a rectangle equal to the dimensions shown in the working drawing.

b. Find the center of the rectangle (bisect the centerline), and draw a centerline through the shape.

c. Measure up 1 inch from the bottom horizontal line, and establish point 13'.

d. Transfer lengths 13-14, 14-15, 15-16, 16-17, 17-18, and 18-19 from the pitch line of the elevation view above point 13'.

e. Use the half plan view as your reference, and transfer distance 12 on each side of the centerline at point 14'; distance 11 on each side of the centerline at point 15'; distance 10 on each side at point 16'; distance 9 on each side at point 17'; and distance 8 on each side at point 18'.

f. Draw the pattern outline through the developed points.

Step 5   Lay out the collar pattern.

Step 6   Recheck the dimensions and general pattern layout and mark the pattern for fabrication.

Step 7   Cut the sheet metal, and form the seams and flange. Make allowances for the seams and edges on the pattern. Form the pattern, set and lock the seams and spot weld, solder, or rivet the flange as directed by your instructor.

Step 8   Clean up your work area.

8.0.0 ◆ TASK 4: EXHAUST WEATHER CAP

In this task, you will lay out and fabricate an exhaust weather cap (*Figure 8*). When a cap is to be placed over an exhaust vent or discharge cap, the inside **splitter cone** should be equal to the diameter of the stack to properly deflect air. Band-iron brackets may be needed to support the cap. The number of brackets required depends on the size and the gauge of the metal. Generally, three brackets will be enough to support a medium-sized cap. To lay out and fabricate this fitting, follow these steps:

Step 1   Draw the elevation view. The diameter of the cap should be twice the diameter of the stack. The distance from the top of the stack to the bottom of the cap line should be about one-half of the stack diameter. The pitch of the cap is usually 30 degrees. The diameter of the inside splitter cone should be equal to the diameter of the stack. The tip of the splitter cone should begin at the base of the hood line.

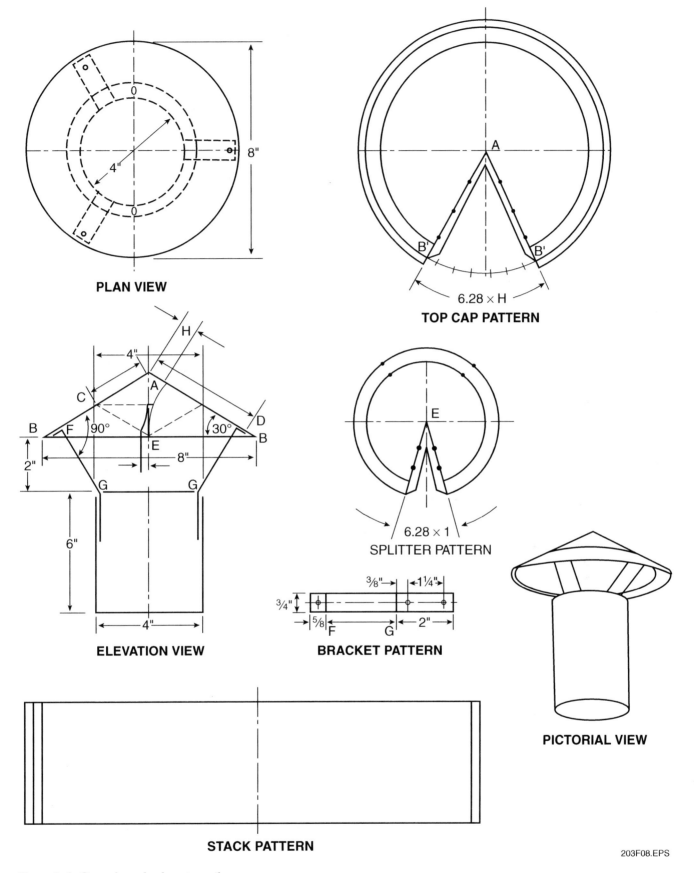

PLAN VIEW

TOP CAP PATTERN

$6.28 \times H$

ELEVATION VIEW

SPLITTER PATTERN

6.28×1

BRACKET PATTERN

PICTORIAL VIEW

STACK PATTERN

203F08.EPS

Figure 8 ◈ Cone-shaped exhaust weather cap.

Step 2 Draw the plan view.

Step 3 Lay out the top cap pattern.

    a. Draw a vertical line, intersect it at the center with a horizontal line, and establish point A as the center.

    b. With distance AB as the radius and point A as the center, scribe an arc to establish the circumference circle for the top cap.

    c. Draw a straight line to intersect the outer circumference line at point B.

    d. Set your dividers at a radius equal to distance H on the elevation view. Multiply this distance by 6.28 (2π) to find the cutout for a full pattern. Measure this distance along the circumference, and establish point B'. Mark the cutout line and make a seam allowance.

    e. With distance AD from the elevation view as the radius and point A as the center, scribe an arc to establish the location for the brackets to be fastened. Divide this line into three, four, or six parts, and mark the positions for the holes to be drilled.

    f. With distance AC as the radius and point A as the center, scribe another arc to establish the line for riveting the splitter cone. Divide this line, and mark it in a similar manner.

Step 4 Lay out the inside splitter cone.

    a. Use distance CE as the radius and point E as the center, and scribe an arc to establish the circumference of the splitter cone circle.

    b. Scribe a second arc ¼-inch outward from the circumference circle to allow for riveting laps.

    c. Set your dividers at radius C to centerline AE and scribe an arc across line BFE to establish distance I. Multiply this distance by 6.28 to establish the cutout for the splitter cone pattern.

    d. Mark the riveting line for drilling.

Step 5 Lay out the bracket pattern. This distance FG is equal to FG on the elevation view.

Step 6 Lay out the stack pattern to equal the circumference of the stack (3.14 × 4) for the length of the stretchout. The width of the stretchout should equal the height of the stack.

Step 7 Make seam allowances on each of the patterns that require them. Mark the brackets for bending.

Step 8 Check the dimensions, cut out and notch the pattern, and form the patterns. Assemble the parts and rivet, spot weld, or solder into place according to your instructor's directions.

Step 9 Clean up your work area.

9.0.0 ◆ TASK 5: ROOF PEAK GRAVITY VENTILATOR

In this task, you will lay out and fabricate a roof peak gravity ventilator (*Figure 9*). Today, most roof gravity ventilators are mass-produced, but you may still be required to lay out and fabricate ventilators for custom installations. To lay out and fabricate this fitting, follow these steps:

Step 1 Draw the elevation and plan views.

Step 2 Lay out the stretchout for one-half of the taper stack section.

    a. Draw a vertical centerline and label the very bottom of line A.

    b. After point A is established, measure up the height of the stack (ridge to top = 5 inches), and establish point 1'.

    c. Draw a horizontal line from point 1' to the right of the centerline a distance of 2 inches (the radius of the top of the stack), and establish point 4.

    d. With your dividers set at radius 1'-4 and with 1' as the center, scribe a quarter circle. Divide the circle into four equal parts and identify the spaces as 1, 2, 3, and 4.

    e. From points 2 and 3 on the quarter circle, draw vertical lines upward to intersect line 1'-4.

    f. Extend a 30-degree line downward and to the right from point 4 to represent the slant of the stack. From point A, draw a 30-degree line downward and to the right to intersect the line just drawn from point 4, and establish point DD'. This represents the pitch of the stack pattern. Extend the 30-degree line D4 upward to intersect the centerline at point X.

    g. Draw straight lines downward from point X through the points established in Step 2e, and establish points B and C on the pitch line AD. From points A, B, and C, draw horizontal lines to the right from the pitch line and establish points A', B', C', and D' on line 4D.

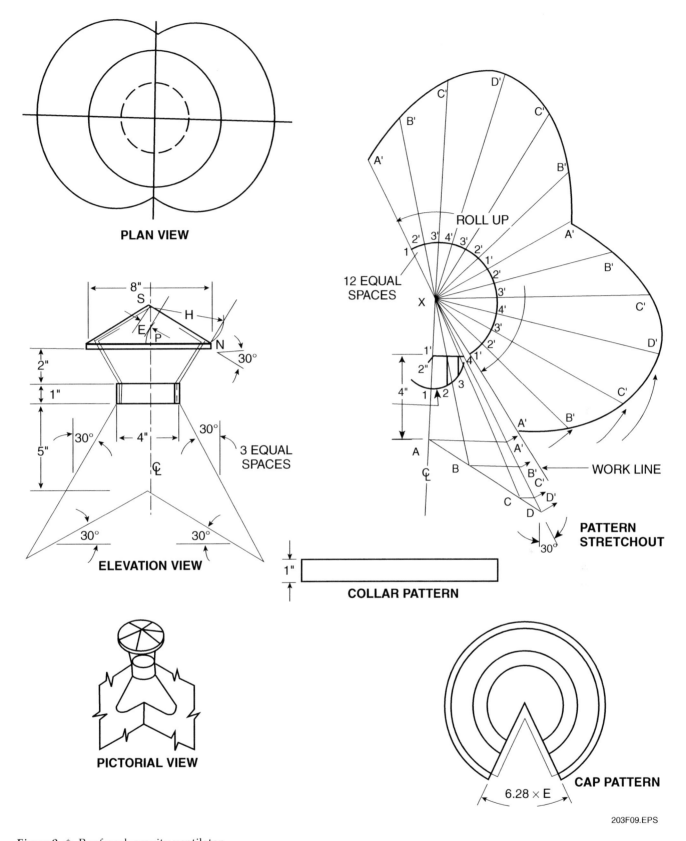

PLAN VIEW

ELEVATION VIEW

PICTORIAL VIEW

COLLAR PATTERN

ROLL UP

12 EQUAL SPACES

WORK LINE

PATTERN STRETCHOUT

CAP PATTERN

$6.28 \times E$

3 EQUAL SPACES

203F09.EPS

Figure 9 ◈ Roof peak gravity ventilator.

h. With distance X4 as the radius and point X as the center, scribe a semicircular arc around point X. Draw a work line downward and to the right of line XD. Where this line intersects the arc just drawn, establish work line point 1'. Calculate the circumference of the small end of the stack pattern (3.14 × 4), and measure this product around the semicircular arc. Establish the end of this distance as point 1' on the work line. Divide arc 1'-1' into 12 equal parts. Label the intersections of these parts with the X4 arc as points 1', 2', 3', and 4'. Extend straight lines through each of these points from point X to an undetermined length.

i. With point X as the center and distance XA' as the radius, scribe an arc across line 1'-1' to establish points A'. With point X as the center and distance XB' as the radius, scribe an arc across lines X2' to establish points B'. Follow the same procedure for points XC' and XD'.

j. When all 12 of the points have been established on the outer perimeter of the stretchout, draw the pattern outline through the developed points.

Step 3 Make a working drawing for the cap.

a. On the elevation view, find points P, N, S, and H.

b. With distance NP as the radius and point N as the center, scribe an arc across the 30-degree line of the top of the cap, and establish distance E.

Step 4 Lay out the cap pattern.

a. With distance H as the radius and any point on the layout sheet as the center, scribe an arc to establish the diameter of the cap.

b. Find the segment for the cutout (6.28 × distance E), and establish points U and V on the cap pattern.

c. Scribe an arc to locate the rivet or bolt holes for the brackets.

Step 5 Recheck the dimensions of the layouts, and measure out the seam and edge allowances.

Step 6 Lay out the top collar pattern.

Step 7 Cut and notch the pattern.

Step 8 Form the pattern and edges, and set and lock the seam. Attach the brackets.

Step 9 Clean up your work area.

10.0.0 ◆ TASK 6: PIPE INTERSECTING A TAPER

In this task, you will lay out and fabricate a pipe (round duct) intersecting a taper at a given angle (*Figure 10*). This fitting may also be called a round tee intersecting a taper at a 45-degree angle. This fitting is used to connect two pipes of different diameters whose axes lie at angles to each other. When making this type of connection, you must make the fitting so that it has minimum resistance to airflow. To lay out and fabricate this fitting, follow these steps:

Step 1 Draw the working view.

a. Draw a horizontal centerline and at the left end of the line, establish point Q.

b. Measure to the right from point Q a distance equal to the height of the taper (7 inches), and establish point O.

c. Draw lines perpendicular to and extending through points Q and O on the horizontal centerline. Measure 3 inches above and below point Q, and establish points S and T. Measure 1½ inches above and below point O, and establish points R and U.

d. Draw straight lines between T and U and between R and S.

e. Measure 2¼ inches to the right from point Q to establish point 10 (the point of intersection for the 45-degree centerline).

f. From the point of intersection of the 45-degree line and line TU, measure out (on the 45-degree line) 2 inches, and establish point W.

g. Draw a line perpendicular to the 45-degree centerline at point W and extending to the sides of point W. With your dividers set at the radius equal to one-half the diameter of the round tee (1¼ inches) and with point W as the center, scribe a half circle around point W, and establish points Y and Z. Divide the half circle into six equal spaces, and from the points draw lines parallel to the 45-degree centerline. Label these lines B, C, D, E, F, and G.

h. Draw a line perpendicular to line YZ downward to intersect line TU. Draw line H as shown. Set the point of your dividers where H intersects the horizontal line of the taper. Set the distance at ½ H as the radius, and scribe

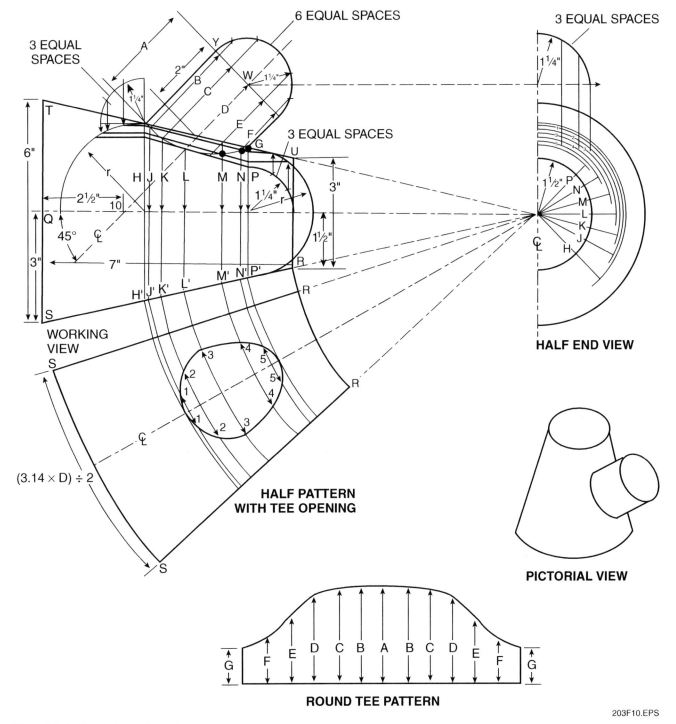

Figure 10 ◆ Pipe intersecting a taper.

an arc to the left of the intersection of H with line TU.

i. With the intersection of lines YH and TU as the center and 1¼ inches as the radius, scribe a quarter arc to the left. Divide this 90-degree arc into three equal spaces. From these spaces, draw vertical lines to intersect the arc previously drawn. Draw horizontal lines from the intersecting points to intersect with line H.

j. With point P as the center and distance PG as the radius, scribe a long arc.

k. With point P as the center and 1¼ inches as the radius, scribe a 90-degree arc and divide it into three equal spaces. From each point, draw a vertical line upward to intersect the arc PG.

l. From the intersection points on the arc, draw horizontal lines to the left to intersect line P. Draw straight lines to connect the points on line H and line P.

Where the 45-degree lines B, C, D, E, and F intersect with the straight lines from line H to line P, mark the intersection line of the round tee and the taper.

m. Draw vertical lines downward from the intersection points just established to intersect line RS, and establish points H', J', K', L', M', N', and P'.

Step 2 Draw the half end view. The half end view is needed to establish the opening in the taper. You can fabricate the taper and the round tee and then place the tee on the taper and mark the cutout.

a. Extend lines RS and TU from the working view to the right to intersect with the horizontal centerline at point X.

b. Project a horizontal line to the right from point W on the working view to intersect with the vertical centerline of the half end view, and establish point W'.

c. With W' as the center and 1¼ inches as the radius, draw a quarter circle, and divide it into three equal parts. Draw vertical lines downward from the points just established.

d. With point X as the center and distance QT and distance OU as radii, draw the half circles to represent the large end and the small end of the taper.

e. With point X as the center and distances H, J, K, L, M, N, and P as radii, scribe arcs within the half pattern as indicated. Mark the intersection points of the lines and the arcs as 1, 2, 3, 4, and 5.

Step 3 Lay out the round tee pattern.

a. Calculate the circumference of the tee section (3.14 × 2), and draw a horizontal line equal to that distance. Divide the line into 12 equal parts.

b. Use the working view and transfer lengths A, B, C, D, E, F, and G as perpendicular lines upward from the horizontal line just drawn.

c. Draw the pattern outline through the developed points.

Step 4 Lay out the half pattern for the taper and the opening for the intersecting tee.

a. With point X as the center and XS as the radius, scribe an arc to establish the length of the taper pattern SS. Calculate the circumference of the large end (3.14 × 6), and establish one-half this distance on line SS for a half pattern. (Note: The half pattern is used to conserve space.)

b. From the bottom point S on the half pattern, draw a straight line to point X.

c. With X as the center and distance XR as the radius, scribe an arc to establish the small end of the taper (half pattern). Draw straight lines between points S and R on the half pattern.

d. Bisect distance SS, and draw a straight line to point X to establish the centerline of the half pattern stretchout.

e. With X as the center and distances XH, XJ, XK, XL, XM, XN, and XP as radii, scribe arcs from the working view across the centerline of the half pattern stretchout.

f. From the half end view, transfer the arc length from the vertical centerline to each of the points 1, 2, 3, 4, and 5 to each side of the centerline on the half pattern taper stretchout to establish the outline for the tee opening.

Step 5 Make seam allowances, and recheck the measurements for the taper stretchout and the intersecting tee.

Step 6 Cut and form the patterns.

Step 7 Bend and set the seams and connect the tee to the taper.

Step 8 Clean up your work area.

11.0.0 ◆ TASK 7: TAPERED OFFSET DUCT

In this task, you will lay out and fabricate a tapered offset duct (*Figure 11*). Collars are usually attached to the openings of a round offset taper joint. The collar should be wide enough for a secure connection with the connecting lengths of pipe. Soldering or welding the collars on metal taper joints is considered to be a good practice even though the specifications may not call for these procedures. The seams on the body of the taper section should be riveted or soldered to prevent buckling when the collars are attached. To lay out and fabricate this fitting, follow these steps:

Step 1 Draw the elevation view.

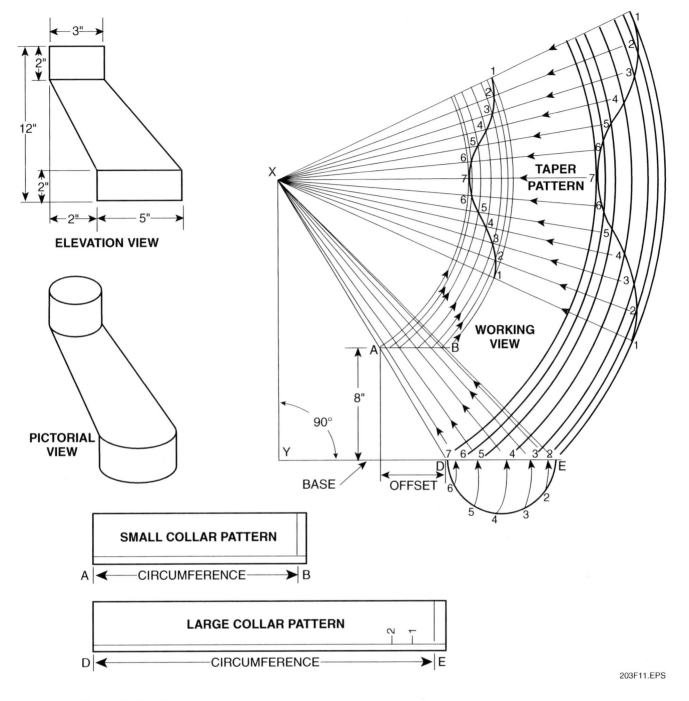

Figure 11 ◈ Tapered offset duct.

203F11.EPS

Step 2 Draw the working view.

  a. Draw a baseline of undetermined length.
  b. Establish the height of the pattern.
  c. Establish the diameter of the small end of the fitting (line AB).
  d. Establish the distance for the offset (line CD).
  e. Establish the diameter of the large end of the fitting (line DE).

  f. Extend lines AD and BE upward to establish point X.
  g. Draw a vertical line downward from point X to intersect the baseline and establish point Y.
  h. Bisect line DE, and with the bisect point as the center and one-half the diameter of the large end of the fitting as the radius. Draw a half circle below line DE. Divide the half circle into six equal parts.

i. With point Y as the center and point 6 on the half circle as the radius, scribe an arc from point 6 to intersect line DE. Scribe an arc from each of the other points on the half circle in the same manner.

j. Draw straight lines from each of the numbered points on line DE to point X.

Step 3   Lay out the collar patterns.

a. Lay out the collar patterns as shown. The lengths of the collars are equal to the circumference of the openings for the large and small ends of the fitting.

b. Divide one-fourth of the large collar pattern into equal spaces to use as a reference in the layout of the taper pattern stretchout.

Step 4   Lay out the taper stretchout pattern.

a. With point X as the center and each point of intersection on line AB and line DE as the radius, scribe arcs of undetermined length upward.

b. Determine point 1 on the perimeter of the outer arc at any convenient location above line DE.

c. With your dividers set at a distance equal to one of the equal spaces on the large collar pattern, use point 1 as the center, and scribe an arc on arc line 2 to establish point 2 on the taper pattern. With point 2 as the center, scribe an arc across arc line 3 to establish point 3. Follow the same procedure to establish each numbered point on the taper pattern stretchout.

d. Draw the pattern outline through the developed points.

e. Draw straight lines from each of the established points on the pattern outline to point X. The intersection of these lines with the arcs drawn from line AB automatically establishes the corresponding points on the pattern stretchout for the small end of the taper.

f. Draw the pattern outline through these developed points.

Step 5   Make seam allowances on each of the developed patterns. Recheck the measurements.

Step 6   Cut and form the patterns.

Step 7   Set and lock the seams. Crimp the small end of the fitting, if required.

Step 8   Clean up your work area.

12.0.0 ◆ TASK 8: TWO-WAY Y-BRANCH

In this task, you will lay out and fabricate a two-way Y-branch (*Figure 12*). Y-branches are used to complete the piping for blowpipe and exhaust systems. The design engineer may specify welded or soldered seams and connections to ensure efficient operation of the system and to eliminate the possibility of noxious gases seeping back into the conditioned space. An exhaust duct system transports air and gases, such as fumes from an air handler or heating system. A blowpipe system transports a product by means of a high-velocity air stream. These products could include grain, tobacco, or sawdust. To lay out and fabricate this fitting, follow these steps:

Step 1   Construct the working drawing.

Step 2   Draw the half front view.

a. Draw the horizontal baseline equal to the diameter of the large end of the fitting. Establish points 1 and 7.

b. Bisect line 1-7, and establish point 0. Draw a vertical line upward from point 0 equal to the height of the taper part of the fitting (8 inches). Draw a line parallel to line 1-7 and intersecting the vertical centerline at point X. Measure to the right from point X one-half the distance between the small ends of the Y-branch (2 inches), and establish point A. Measure to the right from point A, a distance equal to the small end of the Y-branch (3 inches), and establish point G.

c. Draw straight lines upward through points 7A and points 1G to intersect at point Y. Draw a vertical line downward from point Y to intersect the line extended through points 1-7 to establish point Y'.

d. With point 0 as the center and 3 inches as the radius, draw a half circle below line 1-7. Divide the half circle into six equal parts. Mark points 2, 3, 4, 5, and 6 on the half circle arc.

e. From point Y', draw lines to points 5 and 6 to intersect the vertical centerline at points 11 and 12 when they are extended to points 5 and 6 on the half circle.

f. With point Y' as the center and distance Y'2 as the radius, scribe the arc to establish point 2' on horizontal line 1-7. Draw a straight line from point 2' to Y. Establish point F where the straight line intersects line AG.

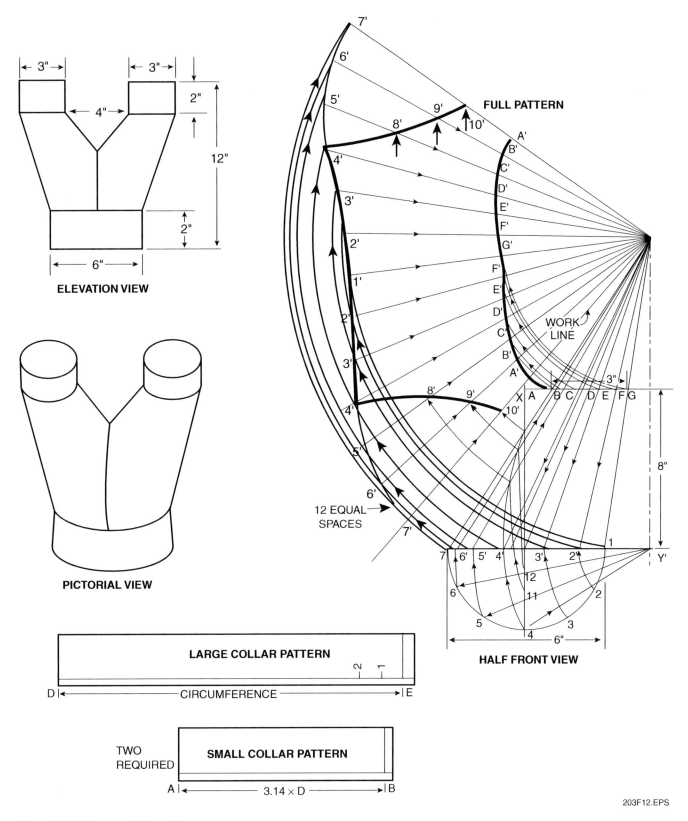

ELEVATION VIEW

PICTORIAL VIEW

FULL PATTERN

WORK LINE

12 EQUAL SPACES

HALF FRONT VIEW

LARGE COLLAR PATTERN

CIRCUMFERENCE

TWO REQUIRED

SMALL COLLAR PATTERN

$3.14 \times D$

203F12.EPS

Figure 12 ◈ Two-way Y-branch.

g. Follow the same procedure to establish points 3', 4', 5', and 6' and, consequently, points E, D, C, and B on line AG.

h. With Y' as the center and distance Y'11 as the radius, scribe an arc upward to intersect line 1-7. Draw a vertical line upward from this intersection point to intersect line Y5', and establish point 8. Follow the same procedure to establish point 9 on line Y'6. Establish point 10 where the vertical centerline intersects line A7.

Step 3 Lay out the full pattern for one leg.

a. Draw the work line downward from point Y at any convenient angle.

b. Use point Y as the center and radii Y1', Y2', Y3', Y4', Y5', Y6', and Y7' to scribe long arcs to the upper left of the work line.

c. Establish point 7' where the work line intersects arc Y7. With point 7' as the center and distance 1-2 on the half circle as the radius, intersect arc line 6' to establish point 6'. Follow the same procedure to locate points 5', 4', 3', 2', and 1'.

d. With point Y as the center and distance Y10 as the radius, scribe an arc to intersect lines Y7', and establish point 10'.

e. With point Y as the center and distance Y9 as the radius, scribe arcs to establish points 9' on lines Y6'.

f. With point Y as the center and distance Y8 as the radius, scribe arcs to establish points 8' on line Y5'. Locate points 4' on the pattern stretchout in a similar manner.

g. With point Y as the center and distance YA as the radius, scribe an arc to intersect lines Y7' to establish point A'.

h. With point Y as the center and distance YB as the radius, scribe an arc to intersect line Y6' to establish points B'. Follow a similar procedure to establish points C', D', E', F', and G'.

i. Draw the pattern outline through the developed points.

Step 4 Lay out the collars with the length of the stretchouts equal to the circumference of the large and small diameters of the arc. The width can be determined from the working drawing.

Step 5 Make allowances for seams and edges on the patterns.

Step 6 Recheck the measurements.

Step 7 Cut out the patterns and form the shapes.

Step 8 Rivet or spot weld the seams as directed by your instructor.

Step 9 Clean up your work area.

13.0.0 ◆ TASK 9: OFF-CENTER TAPERED DUCT

In this task, you will lay out and fabricate an off-center tapered duct (*Figure 13*). To lay out and fabricate this fitting, follow these steps:

Step 1 Draw the plan view to illustrate the amount of offset for the small end of the tapered fitting.

Step 2 Draw the front or elevation view.

a. Draw vertical lines downward from the plan view to locate the size of the small and large ends of the fitting and the amount of offset of the opening.

b. Establish horizontal line CD equal to the diameter of the small end of the fitting and line 1-7 equal to the diameter of the large end of the fitting.

c. Bisect line 1-7, and draw the half circle 1 to 7. Divide the half circle into equal parts as shown.

d. Draw a straight line from point 1 to C and another one from point 7 to D. Continue these two lines until they intersect at point X.

e. Square a line upward from line 1-7 to intersect point X; establish point B at the base of this line on line 1-7.

f. Use point B as the center, and scribe arcs from points 1, 2, 3, 4, 5, and 6 to intersect line 1-7. Draw straight lines from each of these intersecting points with line 1-7 to terminate at point X.

Step 3 Develop the taper pattern stretchout.

a. Use point X as the center to scribe an arc from each intersecting point on line 1-7 to establish the approximate area for the taper pattern.

b. At any convenient distance, draw a work line from point X toward the arc lines to establish point 1'.

c. Set your dividers to span any one of the equal spaces on the half circle, and with point 1' as the center, scribe an arc toward large arc 2' to establish point 2' on the arc.

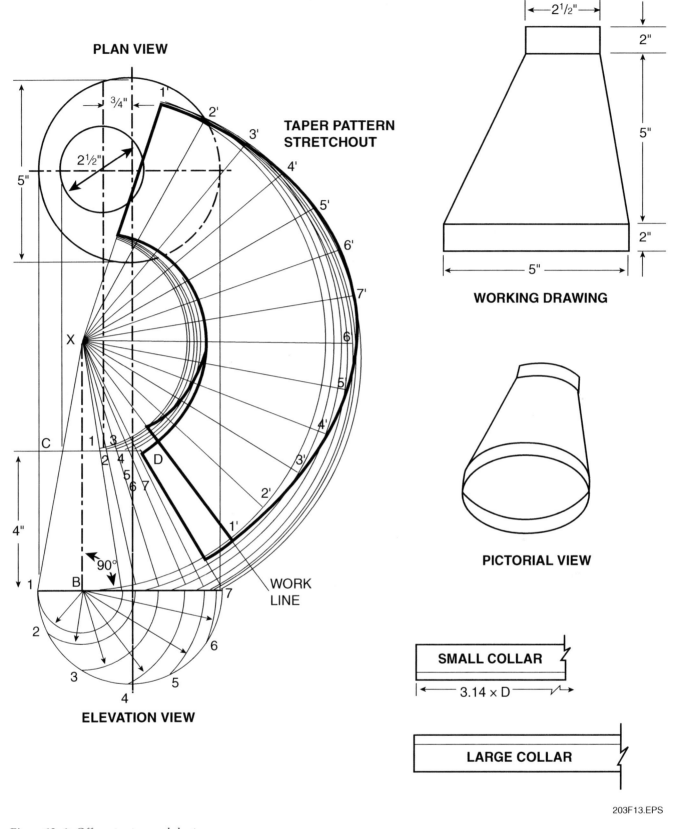

PLAN VIEW

3/4"

2 1/2"

5"

TAPER PATTERN STRETCHOUT

X

C

D

4"

90°

WORK LINE

ELEVATION VIEW

2 1/2"

2"

5"

2"

5"

WORKING DRAWING

PICTORIAL VIEW

SMALL COLLAR

3.14 × D

LARGE COLLAR

203F13.EPS

Figure 13 ◈ Off-center tapered duct.

d. Continue swinging your dividers from one arc space to the next until the full circumference of the taper pattern stretchout is set out as 1' to 7' and 7' to 1' as shown.

e. Draw the large end pattern outline through the established points.

f. Use point X as the center to scribe arcs upward from each intersecting point on line CD (points 1, 2, 3, 4, 5, 6, and 7) to establish the general outline for the small end of the taper.

g. Mark each of the points of intersection with their respective lines as established from the large end layout.

h. Draw the small end pattern outline through the established points.

Step 4 Lay out the collars with the length of the stretchouts equal to the circumference of the large and small diameters of the arc. You can determine the width from the working drawing.

Step 5 Make allowances for seams and edges on the patterns.

Step 6 Cut out the patterns and form the shapes.

Step 7 Rivet or spot weld the seams and edges according to your instructor's directions.

Step 8 Clean up your work area.

14.0.0 ◆ TASK 10: SQUARE-TO-SQUARE TAPERED DUCT

In this task, you will lay out and fabricate a square-to-square tapered duct (*Figure 14*). This is a trunk line fitting designed to carry air to a branch of the ductrun. This type of fitting is usually made in two pieces because it can be cut from the sheet metal economically. However, your instructor may have you make this fitting in one piece for practice. To lay out and fabricate this fitting, follow these steps:

Step 1 Draw the plan view.

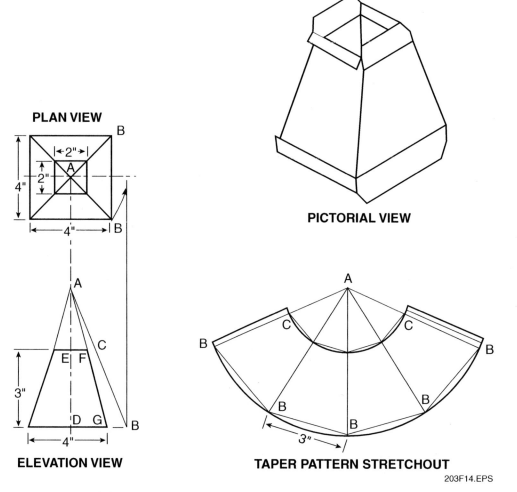

PLAN VIEW

PICTORIAL VIEW

ELEVATION VIEW

TAPER PATTERN STRETCHOUT

203F14.EPS

Figure 14 ◆ Square-to-square tapered duct.

Step 2 Draw the elevation view.

    a. Draw a horizontal baseline, and locate the diameter of the large end of the taper.

    b. Bisect the baseline, and establish point D.

    c. Square a line upward from point D to establish the centerline.

    d. Measure upward on the centerline, and locate the height of the fitting.

    e. Draw the top line at the given height, and set one-half of the diameter of the small end of the taper on each side of the centerline to establish points C and E.

    f. Draw a straight line upward from point G through point F, and establish point A on the centerline.

    g. Transfer the distance AB (slant line) from the plan view to the baseline point D to B to establish the true length of the corner on the taper pattern stretchout.

Step 3 Lay out the taper pattern stretchout.

    a. With point A (at any convenient location for the stretchout) as the center and distance AB as the radius, scribe a large arc to establish the outer circumference of the stretchout.

    b. Adjust your dividers to any side, equal to the diameter of the large end of the taper from the plan view, and scribe an arc across the outside arc on the taper pattern. Set this distance four times on arc B to draw the sides of the pattern.

    c. Draw straight lines connecting each point B on the pattern. Draw straight lines from each point B to point A to establish the sides of the pattern.

    d. With point A as the center and distance AC as the radius, scribe the arc at point C to establish the circumference of the small end of the taper.

    e. Draw straight lines connecting the intersecting points of arc C and the lines from A to B.

    f. Make allowances on the pattern for the S- and drive-clips.

Step 4 Recheck the dimensions and make seam allowances.

Step 5 Cut, notch, and form the pattern. Set and lock the seam. Complete the bends for the S- and drive-clips.

Step 6 Clean up your work area.

15.0.0 ◆ TASK 11: SHOE TEE INTERSECTING A TAPER ON CENTER

In this task, you will lay out and fabricate a shoe (or boot) tee intersecting a taper on center (*Figure 15*). To lay out and fabricate this fitting, follow these steps:

Step 1 Draw the elevation or front view.

Step 2 Construct the working drawing.

    a. Using the elevation view as a reference, draw a horizontal centerline.

    b. Establish point E' at the left of the horizontal centerline.

    c. From point E' on the centerline, measure a distance of 7 inches to the right, and establish point A'.

    d. From point A', measure a distance of 2⅝ inches to the left, and establish point O'.

    e. Draw lines perpendicular to the centerline through points E' and A'.

    f. Measure one-half the diameter of the large end of the fitting, and mark this distance above and below point E', establishing points E and Y'. Follow the same procedure for establishing points A and Z' above and below point A'.

    g. Draw straight lines through points E and A and through points Y' and Z' to intersect the centerline, establishing the apex.

    h. Square line R upward from point O' a distance equal to the height of the shoe portion of the tee from the center of the fitting (5 inches). Establish point O.

    i. Draw a horizontal line through point O, and establish points 1 and 7 (O7 and O1 will be equal to one-half the diameter of the tee, or 1 inch).

    j. Draw a 45-degree line downward from point 1 to intersect line AE, and establish point T and line U. Square a line downward from point 7 to intersect line AE, and establish point J and line Q.

    k. With point O as the center, draw a half circle above line 1-7. Divide the half circle into seven equal spaces, and number the spaces 1 through 7.

l. Draw vertical lines downward from points 2, 3, and 4 on the half circle to intersect line 1-7. From these intersection points with line 1-7, draw 45-degree lines downward and to the left an undetermined distance to establish lines V, W, and X. Draw vertical lines downward through the taper pattern from points 5 and 6 on the half circle to establish lines Y and N.

m. Square lines downward from points T and J where the shoe intersects the taper. Using the points at which these lines cross the centerline as circle centers, and with your dividers set at 1 inch or the radius of the shoe opening, swing quarter circle arcs. Divide each arc into four equal spaces. Label the points on the large end of the taper as E, F, G, and H and those on the small end of the taper as A, B, C, and D. With the same center, swing an arc from T on the left and from J on the right down to where they cross the centerline. Square lines upward from the 1-inch divided quarter circle to intersect the larger arc. Label these points. Square lines to the left and right to where they intersect line T and the centerline and line J and the centerline. Connect and draw lines to establish points as follows: from BF to cross N and V; from CG to cross Y and W; and from DH to cross R and X. Draw lines and curves through these established points to outline the miter and show how it sits on the taper.

n. Draw line L'P' at 90 degrees to line 1T (line L'P' can be positioned at any point on line 1T). Measure each distance from line 1-7 to points 4, 3, and 2 on the half circle and transfer this distance from line L'P' to the corresponding 45-degree lines and establish points 1', 2', 3', and 4'. Draw the developed curved line 1'-4' through these developed points.

o. Label the 45-degree lines from line L'P' to line 1-7 as U', V', W', and X'.

p. From the points where the shoe tee intersects with the taper pattern, draw vertical lines downward until they intersect with line Y'Z' of the taper pattern. Label the points as (from right to left) J, K, L, M, N, P, S, and T.

Step 3 Lay out the taper pattern stretchout.

a. Draw a vertical line of undetermined length, and establish the apex.

b. With the apex as the center and distance apex-Y' as the radius, draw an extended arc to include the circumference of the large end of the taper. Locate this distance, and label the arc line as E' and Y'.

c. With the apex as the center and distance apex-Z' as the radius, scribe an arc to establish the circumference of the end of the taper. Draw straight lines from E' and Y' to intersect at the apex, thereby establishing points A' and Z'.

d. Bisect arc E'Y', and establish the centerline. Mark the intersection of arc A'Z' and the centerline as point A'. Mark the intersection of arc E'Y' as point E'.

e. From the 3-inch arc on the working drawing, transfer arc chords F, G, and H to both sides of the centerline on arc E'Y'. Label the points F', G', and H'.

f. From the 1½-inch radius arc on the working drawing, transfer arc chords B, C, and D on each side of the centerline on arc A'Z'. Label the points B', C', and D'. Draw straight lines connecting points B' and F', points C' and G', and points D' and H'.

g. With the apex as the center and distance apex-J from the working view as the radius, scribe an arc between the edges of the taper pattern stretchout. Follow the same procedure for the apex of each arc to points K, L, M, N, P, S, and T. Mark the points of intersection with the previously determined corresponding chord lines.

h. Draw the cutout outline through the developed points.

Step 4 Lay out the shoe tee pattern.

a. Draw a horizontal line at any convenient location. Calculate the length of the line by multiplying the radius of the round portion of the tee by 3.14 (3.14 × 1). Establish points X and X on the horizontal line. Divide the line into six equal spaces. Square lines downward from each identified space division.

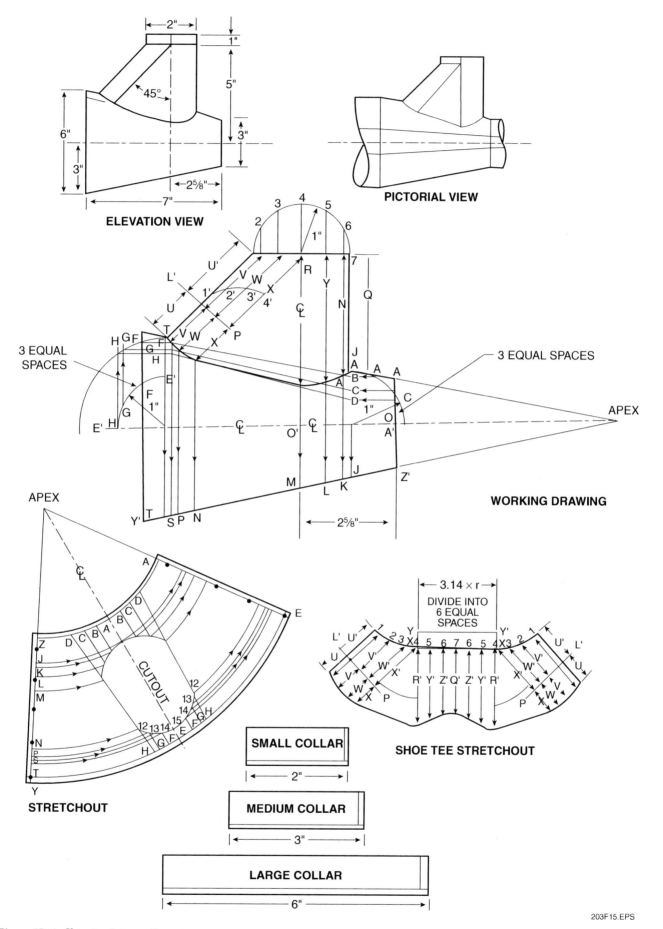

ELEVATION VIEW

PICTORIAL VIEW

WORKING DRAWING

3 EQUAL SPACES

3 EQUAL SPACES

APEX

APEX

STRETCHOUT

SHOE TEE STRETCHOUT

3.14 × r

DIVIDE INTO 6 EQUAL SPACES

SMALL COLLAR

2"

MEDIUM COLLAR

3"

LARGE COLLAR

6"

203F15.EPS

Figure 15 ◆ Shoe tee intersecting a taper on center.

b. Draw 45-degree lines downward and outward from both points X on each end of the horizontal line.

c. Transfer length X' from the working drawing to the 45-degree line on each side of the pattern. Square lines outward from this intersection with the 45-degree lines and establish lines L'P' on each side of the stretchout.

d. From the ellipse line on the working drawing, transfer chords 4'-3', 3'-2', and 2'-1' along line L'P' outward from the 45-degree lines on each side of the pattern. At each point of intersection with line L'P', draw lines perpendicular to and extending along both sides of lines L'P'.

e. Transfer the vertical lines Q, N, Y, and R from the working drawing to the pattern stretchout.

f. Working from both lines L'P', on each side of the pattern transfer lengths U', V', W', and X' to the upper side of the lines as indicated. Transfer lengths U, V, W, and X to the bottom of the lines.

g. Draw the pattern outline through the developed points.

h. Measure up from line XX the height of the tee collar, and draw lines to include the proper area within the rectangle.

Step 5 Lay out the half collar pattern. Draw a rectangle equal to the included area as developed in Step 4h. Make seam allowances.

Step 6 Lay out the collars for the small end and the large end of the taper.

Step 7 Recheck all dimensions. Make seam allowances on the patterns.

Step 8 Cut, notch, and form the patterns.

Step 9 Fabricate the fitting, set and lock the seams, and then rivet or solder them as directed by your instructor.

Step 10 Clean up your work area.

16.0.0 ◆ TASK 12: 90-DEGREE TAPERED ELBOW

In this task, you will lay out and fabricate a 90-degree tapered elbow (*Figure 16*). You can rivet, groove, or spot weld elbow seams. Many sheet metal workers prefer to use a grooved seam because it can be fabricated quickly and the grooving method produces a tighter seam. General guidelines for the layout procedure for an elbow fitting are as follows:

- The seams should run along the centerline of the elbow or on the heel or throat as the pattern is laid out.
- The seams should be staggered. The seam of the first **gore** should be on the side opposite the seam of the second gore, and so on.
- The gores should be run through the elbow rolls in the same position as they lie in the pattern.

To lay out and fabricate this fitting, follow these steps:

Step 1 Draw the elevation view.

Step 2 Construct the working drawing.

a. Calculate the length of the centerline 90-degree arc in the elevation view by multiplying the 6-inch radius by the constant 1.57 (½π), which equals 9.42 inches or approximately 9⁷⁄₁₆ inches. Construct this centerline in the working drawing. Establish point 1, measure upward from point 1, and establish point 9 (9⁷⁄₁₆ inches up from point 1).

b. Draw lines perpendicular to and extending on both sides of the centerline at points 1 and 9 equal to one-half the diameter of the large end of the fitting and one-half the diameter of the small end of the fitting, respectively. Establish points A and points K. Draw straight lines upward from points K through points A to intersect at point X.

c. Measure 5 inches (large end diameter) to the left and right of point 1, and establish points P and S. Square lines upward from points P and S.

d. Divide centerline 1-9 into eight equal spaces, and label the points of intersection with line 1-9 as 2, 3, 4, 5, 6, 7, and 8.

e. Draw a straight line from point S through point 2 to intersect the vertical line from point P, and establish point T.

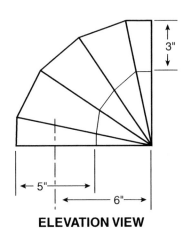

ELEVATION VIEW

3.14 × 5" DIVIDE INTO 8 EQUAL SPACES

PATTERN STRETCHOUT

1.57 × 6" DIVIDE INTO 8 EQUAL SPACES

WORKING DRAWING

203F16.EPS

Figure 16 ◈ 90-degree tapered elbow.

f. Draw a straight line from point T through point 4 to the right, and establish point U. Draw a straight line from point U through point 6 to the left, and establish point V.

g. Draw a straight line from point V through point 8 to the right, and establish point W.

h. Label the points of intersection of these lines with line AK on the right side of the working drawing as points B, E, F, and J. Label the corresponding points of intersection on the left side with line AK as points C, D, G, and H.

i. With point 1 as the center and 2½ inches (the radius of the large end) as the radius, draw a half circle below line KK. Divide the half circle into four equal spaces. Label the points L, M, and N.

j. Draw vertical lines upward from each of these points to intersect line KK and establish points 10 and 11.

k. Draw straight lines downward from point X to points 10 and 11.

l. Draw lines to the left from the intersecting points of lines X10 and X11 and the developed elbow section lines BC, DE, FG, and HJ, parallel to line KK until they intersect line AK on the left of the working drawing.

Step 3 Lay out the gore sections.

a. With X as the center and distance XK as the radius, draw a long arc to the left and upward from point K.

b. Calculate the circumference of the large end, and establish this length on the long arc (3.14 × 5 inches).

c. Divide the circumference into eight equal spaces. Label the spaces as L, M, N, and K (use each label twice).

d. Draw straight lines from each of these established points to point X.

e. With X as the center and distances XJ, X1, X2, X3, and XH as radii as developed on line AK (left side of the working drawing), scribe arcs to the left to intersect lines XK, XL, XM, and XN on the pattern stretchout.

f. With point X as the center and the other developed points on line AK as radii, scribe arcs to the left to intersect the corresponding lines.

g. Draw the pattern outline through the developed points. (Note: If the gore sections are to be welded, continue on to the next phase. If the sections are to be connected by seams, make the proper allowances by extending the pattern the proper amount for each gore.)

Step 4 Recheck the measurements. Cut the sections on the developed lines. Roll each gore section to the required diameter.

Step 5 Fasten the sections to form the tapered elbow, seams, and edges as instructed.

Step 6 Clean up your work area.

1. The _____ is a common center over the center of the base of a tapered figure.
   a. parallel form
   b. radial form
   c. vertex
   d. apex

2. If the pitch of the taper is too shallow, you will have to _____ the length of your extension lines.
   a. narrow
   b. increase
   c. widen
   d. decrease

3. Before you can develop a pattern for a radial form, you must determine the _____ and the _____.
   a. true lengths of the sides; circumference or perimeter of the base
   b. true lengths of the sides; area of the base
   c. circumference or perimeter of the base; location of the apex
   d. circumference or perimeter of the base; slope of the taper

4. Use the plan view to determine _____.
   a. the height of the apex
   b. the true length of the sides
   c. the length of the stretchout
   d. the area of the base

5. When developing a pattern, label the lines on the side view and on the pattern using _____.
   a. numbers
   b. letters
   c. lengths
   d. angles

6. When drawing half the elevation or side view of a rectangular weather cap, establish the centerline of the drawing by _____.
   a. drawing the baseline and squaring a line upward from one endpoint
   b. drawing a line at 30 degrees from one point
   c. determining the radius of the pattern development circle
   d. measuring outward a distance equal to one-half the width of the weather cap

7. When laying out a symmetrical tapered duct, the baseline CC' should be equal to _____ of the large end of the fitting.
   a. one-half the diameter
   b. the diameter
   c. twice the radius
   d. one-half the circumference

8. Lay out the collars of a symmetrical tapered duct with the length of the stretchouts equal to the _____ of the large and small diameters of the arc.
   a. radius
   b. area
   c. width
   d. circumference

9. In an exhaust weather cap layout, the distance from the top of the stack to the bottom of the cap line should be approximately _____.
   a. twice the stack diameter
   b. the same as the length of the stack
   c. one-half the stack diameter
   d. one-half the length of the stack

10. A pipe intersecting a taper is used to _____.
    a. connect to a round offset taper joint
    b. deflect air from an exhaust vent
    c. connect two pipes of different diameters
    d. complete the piping in a blowpipe system

11. When laying out a pipe intersecting a taper, the half end view is used to establish the opening in the _____.
    a. taper
    b. round tee
    c. intersecting tee
    d. intersecting elbow

12. _____ systems use a high-velocity air stream to transport products.
    a. Blowpipe
    b. Exhaust
    c. Loop perimeter
    d. Extended plenum

13. To lay out the full pattern for one leg in a two-way Y-branch, begin by _____.
    a. drawing the horizontal baseline
    b. establishing a point where the work line intersects the arc
    c. drawing the work line downward at an angle
    d. scribing an arc upward to intersect the work line

14. Square-to-square tapered ducts are usually fashioned from _____ separate pieces of metal.
    a. two
    b. three
    c. four
    d. five

15. To lay out the taper pattern stretchout for a square-to-square tapered duct, set your dividers to any side equal to the _____ from the plan view.
    a. diameter of the small end of the taper
    b. circumference of the small end of the taper
    c. diameter of the large end of the taper
    d. circumference of the large end of the taper

Summary

In this module, you built on the skills you learned in *Fabrication One – Parallel Line Development* to lay out and fabricate fittings using the radial line development method. This method is best suited to fabricating fittings in which the lines drawn from the corner of the fitting end in a common center, or apex, over the base of the fitting. Tapered ducts, weather caps, Y-branches, and elbows are all examples of the type of cone-shaped fittings developed using the radial line development method.

At this point in your training, you are seeing many of the principles you learned in earlier modules start to come together into objects that you can make and that have a specific purpose in the construction industry. Taking the time now to hone your skills and focus on pride of craftsmanship will pay off as you move up the apprenticeship ladder.

Notes

Trade Terms
Introduced in This Module

Apex: In a tapered fitting, the point where lines drawn from the corners cross each other. The apex is located over the center of the base.

Chord: A line drawn that goes from one part of a circle's circumference to another without passing through the circle's center.

Gore: A tapering wedge-shaped unit on a sheet metal fitting.

Pitch: A slope (usually of a roof) expressed as a ratio of vertical rise to horizontal run.

Radial: An arc defined by the sweep of the radius of a circle.

Splitter cone: The inverted cone in an exhaust weather cap that prevents gases from accumulating in the cone's peak.

Stays: The metal bands that attach the weather cap to the curb. Stays are usually located in the corners, but sometimes in the center, of the weather cap.

Taper: To narrow from the base to the apex.

Additional Resources

This module is intended to present thorough resources for task training. The following reference works are suggested for further study. These are optional materials for continued education rather than for task training.

Today's 40 Most Frequently-Used Fittings, 1996. Richard S. Budzik. Chicago, IL: Practical Publications.

Ultimate Sheet Metal Fabrication, 1999. Timothy Remus. Stillwater, MN: Wolfgang Productions.

NCCER makes every effort to keep these textbooks up-to-date and free of technical errors. We appreciate your help in this process. If you have an idea for improving this textbook, or if you find an error, a typographical mistake, or an inaccuracy in NCCER's Contren® textbooks, please write us, using this form or a photocopy. Be sure to include the exact module number, page number, a detailed description, and the correction, if applicable. Your input will be brought to the attention of the Technical Review Committee. Thank you for your assistance.

Instructors – If you found that additional materials were necessary in order to teach this module effectively, please let us know so that we may include them in the Equipment/Materials list in the Annotated Instructor's Guide.

Write: Product Development and Revision
National Center for Construction Education and Research
3600 NW 43rd St, Bldg G, Gainesville, FL 32606

Fax: 352-334-0932

E-mail: curriculum@nccer.org

Craft _____ Module Name _____

Copyright Date _____ Module Number _____ Page Number(s) _____

Description _____

_____

_____

_____

_____

(Optional) Correction _____

_____

_____

(Optional) Your Name and Address _____

_____

_____

04204-08

Sheet Metal Duct
Fabrication Standards

04204-08
Sheet Metal Duct Fabrication Standards

Topics to be presented in this module include:

Overview

In order to ensure a quality product and promote professionalism in the trade, the sheet metal industry has adopted standards that address the best way to manufacture and install ductwork of different types. In addition to these industry standards, governments at the local, state, and federal level impose a set of building codes that are developed in the interest of public safety. While adherence to industry standards is largely voluntary, contractors must comply with building codes.

As a professional in the sheet metal industry, you will need to know what codes apply to your trade and make sure that your work is in compliance. Additionally, most contract work you perform will require you to conform to certain industry standards. In this module, you will learn about building codes and how to apply industry standards to manufacturing practices in a sheet metal shop.

Objectives

When you have completed this module, you will be able to do the following:

1. Explain the effect of operating pressure on the design of a duct system.
2. Locate standards for selected topics, fittings, or components.
3. Define the difference between standards and codes or ordinances.
4. Determine sealing requirements for a selected ductrun by using reference charts and tables.
5. Determine minimum gauge requirements for a selected ductrun by using reference charts and tables.
6. Determine minimum connector and reinforcing requirements for selected ductruns by using reference charts and tables.
7. Describe the purpose of a tie rod and determine when a tie rod is optional or mandatory by using reference charts and tables.
8. Identify the different types of acceptable longitudinal seams, including applications and any limitations.

Trade Terms

Free area
Inches of water gauge
Pascal
Proprietary connector
Reinforcement

Seal class
Tie rod
Transverse connector
Variable air volume
  (VAV) system

Prerequisites

Before you begin this module, it is recommended that you successfully complete the following: *Core Curriculum; Sheet Metal Level One; Sheet Metal Level Two*, Modules 04201-08 through 04203-08.

This course map shows all of the modules in the second level of the Sheet Metal curriculum. The suggested training order begins at the bottom and proceeds up. Skill levels increase as you advance on the course map. The local Training Program Sponsor may adjust the training order.

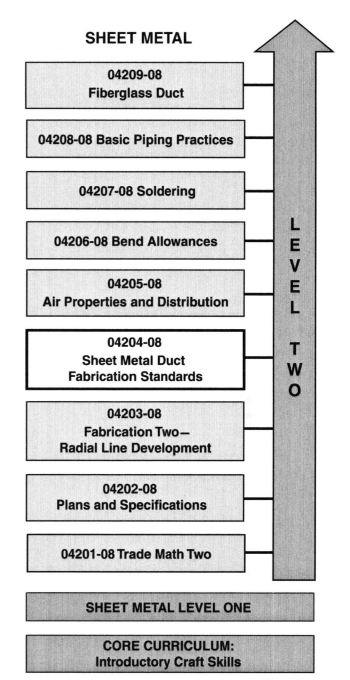

SHEET METAL

04209-08 Fiberglass Duct

04208-08 Basic Piping Practices

04207-08 Soldering

04206-08 Bend Allowances

04205-08 Air Properties and Distribution

04204-08 Sheet Metal Duct Fabrication Standards

04203-08 Fabrication Two— Radial Line Development

04202-08 Plans and Specifications

04201-08 Trade Math Two

LEVEL TWO

SHEET METAL LEVEL ONE

CORE CURRICULUM: Introductory Craft Skills

204CMAP.EPS

1.0.0 ◆ INTRODUCTION

At this point in your training, you have learned how to make, install, support, and test duct and fittings. This module introduces you to building codes and industry standards. You will also learn about the technical manuals that affect the work you do. These manuals contain information developed over time by professionals in your industry. The manuals reflect the best and safest practices in the heating, ventilating, and air conditioning (HVAC) industry.

Many technical manuals are available in the HVAC industry. In this module, you will focus on *HVAC Duct Construction Standards—Metal and Flexible*, which is published by the Sheet Metal and Air Conditioning Contractors' National Association (SMACNA). Once you learn how to work with one manual, you can use your knowledge to work with other technical manuals.

2.0.0 ◆ BUILDING CODES AND INDUSTRY STANDARDS

Your work in the sheet metal industry is governed by building codes and guided by industry standards. Each trade in the construction industry has its own codes and standards. Often, industry standards are made part of building codes. The chief difference between codes and standards is that codes are mandatory and standards are voluntary.

2.1.0 Building Codes

Building codes, also called building ordinances, are designed to protect the health, welfare, and safety of the public. Inspectors use the codes to ensure that completed work is done according to code requirements. Building codes have the force of law, which means that contractors must follow the codes or face legal penalties. A contractor may do more than the code requires, but may never do less. A code describes the minimum acceptable levels of the following:

- Design
- Construction
- Installation
- Materials
- Performance of appliances, equipment, and systems

No single building code governs all aspects of construction in every location. Weather conditions can affect how codes are written in different parts of the country. For example, in Florida, where hurricanes occur, local codes may include

standards that will help a building withstand hurricane-force winds. Local codes in states that experience heavy snowfall and ice storms may reflect those conditions.

Time can have an effect on building codes. It may take many years for a problem to show up in a material that originally was considered safe and appropriate for installation. For example, lead-based paint and asbestos, once widely used, have been found to be unsafe. Therefore, codes have been revised to prohibit their use. Many older buildings still contain these materials. Therefore, updated codes also will include information about safely disposing of or containing these materials. Although there may be general similarities among codes, there are also a number of differences, so you must always follow the local and most current code. The following list includes examples of the more commonly cited building codes and code-writing organizations:

- Building Officials and Code Administrators (BOCA)
- International Association of Plumbing and Mechanical Officials (IAPMO)
- National Electrical Code (NEC)
- Southern Building Code Congress (SBCC)
- Uniform Mechanical Code (UMC)

2.2.0 Industry Standards

A standard is a way of doing work that has been agreed on by professionals in the industry. Contractors voluntarily adopt standards for the following reasons:

- Standards represent the collected knowledge and experience of architects, designers, engineers, contractors, and other professionals in the industry.
- Standards focus on the best and safest work practices.
- Work that is based on standards will be consistent from project to project, which increases efficiency.
- Standards are frequently reviewed and updated to reflect the most current industry practices.
- Apprenticeship training is often based on industry standards. Although local shop or company practices may differ slightly from place to place, basing training on recognized standards helps to ensure a well-trained workforce.

Several associations publish technical manuals for use in the sheet metal and HVAC industries. They include SMACNA and the North American

Insulation Manufacturers Association (NAIMA). Professionals in these industries write and review the standards to ensure that they are accurate and reflect current industry practices.

In addition, many organizations publish standards that relate to their own part of the construction industry. Many of these organizations may affect the work that you do. The following is a sample list of these organizations:

- Air Conditioning and Refrigeration Institute (ARI)
- Air Conditioning Contractors of America (ACCA)
- Air Diffusion Council (ADC)
- Air Distribution Institute (ADI)
- Air Movement and Control Association (AMCA)
- American Conference of Governmental Industrial Hygienists (ACGIH)
- American National Standards Institute (ANSI)
- ASTM International
- American Society of Heating, Refrigeration and Air Conditioning Engineers (ASHRAE)
- American Society of Mechanical Engineers (ASME)
- American Welding Society (AWS)
- Building Officials and Code Administrators (BOCA)
- Fabricators and Manufacturers Association, International (FMA)
- Factory Mutual Engineering Research Corporation (FM)
- Mechanical Contractors Association of America (MCAA)
- National Environmental Balancing Bureau (NEBB)

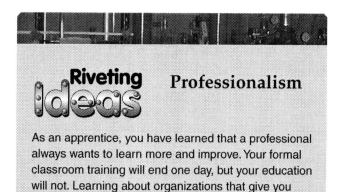

Riveting Ideas — Professionalism

As an apprentice, you have learned that a professional always wants to learn more and improve. Your formal classroom training will end one day, but your education will not. Learning about organizations that give you current information about your industry is one way you can continue to improve your skills.

- National Fire Protection Association (NFPA)
- Thermal Insulation Manufacturers Association (TIMA)
- Underwriters Laboratories (UL)

3.0.0 ◆ SMACNA MANUAL DUCT FABRICATION STANDARDS

SMACNA publishes several technical manuals containing standards that are relevant to your trade. These standards are based on engineering principles, research, and application data supplied by manufacturers, construction industry professionals, testing laboratories, and others with specialized experience in the sheet metal and HVAC industries.

The SMACNA manual containing duct information is titled *HVAC Duct Construction Standards— Metal and Flexible*. The following section gives you an idea of what the SMACNA manual covers.

3.1.0 Section One, Basic Duct Construction

At one time, duct systems were divided into three air pressure classifications: low, medium, and high. Duct systems are now classified according to their pressure class in either **inches of water gauge** or **Pascals** (metric system).

Section One of the SMACNA manual includes tables that show the construction standards for duct by pressure class. It also includes standards for sealing ducts, joints, and **reinforcements** (which strengthen the ducts at certain joints). The figures, text, and tables contain the information you need to properly fabricate, assemble, and install various types of ductruns.

3.2.0 Section Two, Fittings and Other Fabrications

Section Two of the SMACNA manual contains standards for fittings and other sheet metal fabrications. Construction and assembly standards are included for many fittings. Among the fittings covered are elbows, vanes, branches, offsets, transitions, diffuser plenums, connections, access doors, dampers, exhaust hoods, and liners.

A technical manual may refer you to standards established by other associations, testing laboratories, or governmental organizations. Some of these organizations are not primarily concerned with HVAC work, but they may still have an effect on your work. The SMACNA manual directs you to *National Fire Protection Association Standard 90A*.

This standard requires that any duct liner must be interrupted wherever a fire damper is installed in a duct. You must not impair the operation of a fire damper with anything else that you install in a ductrun.

3.3.0 Section Three, Round, Oval, and Flexible Duct

Round, oval, and flexible duct consists of straight sections of duct and fittings. Like rigid rectangular duct, it must be properly supported. The standards affecting this type of duct are covered in Section Three of the SMACNA manual. Like rectangular duct, these types of duct are classified according to their pressure class in either inches of water gauge or Pascals. This section also covers standards for underslab installation and encased cuts.

3.4.0 Section Four, Hangers and Supports

You have covered the basics of properly hanging and supporting duct in earlier modules in the Sheet Metal curriculum. Therefore, much of the material in this section of the SMACNA manual will be familiar to you. Tables in this section show

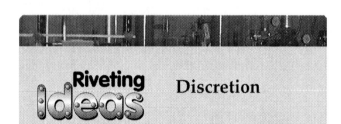

Riveting Ideas   Discretion

Discretion is the ability to make responsible decisions based on knowledge and experience. A professional also relies on the knowledge and experience of others in the industry before making a change to an accepted standard.

Organizations that publish technical manuals know that it is impossible to cover every design and installation situation. Conditions and materials at each job site may vary widely. Therefore, a standard may include a note stating that any change to a duct hanging system will be at the contractor's discretion with the engineer's approval. The contractor and the engineer will determine if a change to an accepted standard will affect the duct system's safety and operating efficiency. Depending on the complexity of the job, other professionals may be consulted before a change is made.

the minimum size and spacing standards for duct hangers. Notice that the duct sizes are stated as half-perimeters. Three tables in this section cover the details of hanger size and spacing.

Section Four also includes several formulas. These are included mainly for engineers to use when designing the duct system.

3.5.0 Section Five, Exterior Components

Section Five of the SMACNA manual covers sheet metal parts installed on a building's exterior. These parts include louvers, rooftop ducts, flashing, exhaust vents, and covers for intake or exhaust ventilation. Standard formulas for calculating such things as **free area** and the percentage of free area are also included in this section. Free area is the minimum area in a louver through which air can pass. This is an important factor, because you do not want to install the wrong size louver. If the louver is too small, the system won't receive enough airflow, and it is wasteful to install a louver that is too large.

3.6.0 Section Six, Equipment and Casings

Section Six of the SMACNA manual covers HVAC equipment and casings such as plenums. Also covered are curbs, drain pans, pipe penetrations of casings, and casing access doors. The standard notes that these parts must be carefully constructed so that they will carry their share of the casing load.

The SMACNA manual advises against using tapered sides and roofs for casings because it is difficult to develop the right strength and air tightness at joints. In theory, tapered casings conserve energy and encourage uniform airflow. In practice, however, there usually is not enough space in most equipment rooms for a tapered design.

3.7.0 Section Seven, Functional Criteria

Section Seven of the SMACNA manual explains how SMACNA conducts tests to establish its standards. It includes the information that engineers and civil authorities need to conduct their own tests. The material in this section is used mainly by engineering designers and building inspectors. However, if you are ever asked to assemble the parts needed for a test, you may have to refer to the figures in this section.

3.8.0 Appendixes

The SMACNA manual concludes with several appendixes. The appendixes include thickness measurements for various metals, a list of metal properties, charts showing duct areas and weights, conversion charts, pressure and gauge tables, models of typical HVAC systems and equipment arrangements, and many useful diagrams.

4.0.0 ◈ OPERATING PRESSURE AND LEAKAGE IN DUCT SYSTEMS

A duct system must stand up to pressure and be properly fabricated and installed to perform at top efficiency. Two factors play an important role in ductwork performance: operating pressure and air leakage. The designer of a duct system relies on industry standards to deal with these two factors.

4.1.0 Operating Pressure

Operating pressure is a measurement of duct pressure in pounds per square foot. It is stated as inches of water gauge. A cubic foot of water in a $1' \times 1' \times 1'$ container weighs 62.4 pounds at sea level. The unit weight of the container is stated as 62.4 pounds per square foot for 12 inches of height. The weight is proportionate to the height. Therefore, a measurement of 1 inch of water gauge is equal to 5.2 pounds per square foot of duct pressure (62.4 pounds ÷ 12 inches). This relationship can be stated more simply as the following equation:

1 inch of water gauge = 5.2 lb/ft^2

The operating pressures commonly used in commercial HVAC work are shown in *Table 1*. The table shows the ranges of duct operating pressure measured in both Pascals and inches of water gauge.

An example of how a duct pressure class designation looks on a schematic drawing of a section of duct is shown in *Figure 1*. The duct pressure class in inches of water gauge is shown inside the triangles. The Pascal (Pa) designation is also shown in the figure.

4.2.0 Air Leakage

An airtight duct system may be a desirable goal, but absolute air tightness is not possible. Even in well-designed systems, a small amount of air will leak along the joints and seams. A certain amount of air leakage is expected and is directly related to the following factors:

- Operating pressure
- Square footage of the duct system
- Duct construction
- Sealing

Table 1 Duct Static Pressure Classes

Duct Static Pressure Class		Operating Pressure
Pascals	Inches of Water Gauge	Pounds per Square Foot
125	0.5	Up to 2.6
250	1.0	2.6 to 5.2
500	2.0	5.2 to 10.4
750	3.0	10.4 to 15.6
1,000	4.0	15.6 to 20.8
1,500	6.0	20.8 to 31.2
2,500	10.0	31.2 to 52.0

To help contain air leakage at acceptable levels, you must properly seal the duct at connections, joints, and seams. Sheet metal workers use approved adhesives, gaskets, or tape systems either alone or in combination to seal the duct. Check the manufacturer's guidelines or follow your shop's standards for choosing and applying the appropriate sealing material.

Duct sealing requirements are placed into categories by pressure class (*Table 2*). The duct system designer is responsible for selecting the pressure class and the **seal class**. This information is usually contained in the job specifications and engineering drawings. If the designer does not specify the seal class, for constant air volume systems use one inch of water gauge pressure class with no seal and for **variable air volume systems** use two inches of water gauge pressure class with a Class C seal.

5.0.0 ◈ DETERMINING GAUGE AND SPACING

Duct panels and connectors will deflect (bend or deform) under air pressure. Air pressure may be exerted from inside or outside the duct. To minimize deflection, it is important to choose the proper sheet metal gauge and connector and reinforcement spacing. Design standards limit duct panel deflection to ¾-inch for widths over 24 inches and limit connector and reinforcement deflection to ¼-inch.

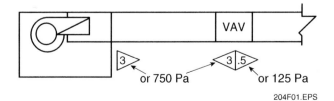

204F01.EPS

Figure 1 ◈ Duct pressure class designation.

Table 2 Pressure Seal Requirements

Class	Requirements
Class **C**	**C**onnector joint sealed for 2 inches of water gauge pressure class and, when specified, for 0.5 inches of water gauge pressure class or for 1 inch of water gauge pressure class.
Class **B**	**B**oth connector joints and longitudinal seams sealed for 3 inches of water gauge pressure class.
Class **A**	**A**ll connector joints, longitudinal seams, and duct wall penetrations sealed for 4 inches of water gauge pressure class and greater.

5.1.0 Determining Duct Gauge

Duct reinforcement tables are available to help you select the minimum sheet metal gauge and reinforcement required (*Table 3*). Take a few moments to study *Table 3* and the examples that follow it.

The letters in *Table 3* stand for the reinforcement grade class, and the numbers represent the sheet metal gauge. Notice that statements such as F + rod or G + rod appear in some of the columns. For example, look at the line in the table for the duct dimensions of 49 to 54 inches. In the column headed 5.0 feet, the following entry appears:

H-18

F + rod

This entry means that the fabricator has the option of using a lower grade class of reinforcement (F) with a **tie rod** equal to reinforcement grade class H.

Example 1:

In this example, you will learn how to use *Table 3* to select the minimum gauge for the ductwork shown in *Figure 2*. The measurements start at 48" × 28" and reduce to 48" × 10". Notice that each section of duct is a different length and that the duct reduces in size in section three (marked with a black triangle). There is also a 90-degree elbow in section five. Pressure is at 3 inches of water gauge. The duct panels are unsupported (not reinforced).

Look at the first column in *Table 3* to find a duct dimension that covers the dimensions in this example. In this case, that would be 43 inches to 48 inches. Read across that line to determine the correct minimum gauge for each of the five sections of duct.

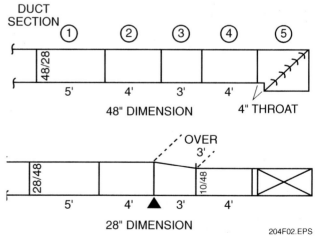

Figure 2 ◆ Ductrun with reducer.

Table 3 Standards for Duct Construction

Duct Dimension (in inches)	Three Inches of Water Gauge Pressure (Positive or Negative) Required Reinforcement Spacing (in feet) Minimum Rigidity Class with Maximum Gauge Duct				
	5.0	4.0	3.0	2.5	2.0
10	A-24	A-24	A-24	A-24	A-24
11 and 12	A-24	A-24	A-24	A-24	A-24
13 and 14	A-24	A-24	A-24	A-24	A-24
15 and 16	A-24	A-24	A-24	A-24	A-24
17 and 18	A-24	A-24	A-24	A-24	A-24
19 and 20	A-24	A-24	A-24	A-24	A-24
21 and 22	B-24	B-24	B-24	B-24	B-24
23 and 24	B-24	B-24	B-24	B-24	B-24
25 and 26	C-24	C-24	C-24	C-24	C-24
27 and 28	C-22	C-24	C-24	C-24	C-24
29 and 30	C-22	C-24	C-24	C-24	C-24
31 to 36	E-20	D-24	D-24	D-24	D-24
37 to 42	E-20	E-22	E-24	E-24	E-24
43 to 48	G-18	F-20	E-22	E-24	E-24
49 to 54	H-18 F + rod	H-18 F + rod	G-22 F + rod	E-24	E-24
55 to 60	H-16 F + rod	H-18 F + rod	G-20 F + rod	G-24 F + rod	G-24 F + rod
61 to 72		I-16 F + rod	H-20 F + rod	H-22 F + rod	H-24 F + rod
73 to 84			J-18 F + rod	J-20 F + rod	I-22 F + rod
85 to 96			L-16 G + rod	K-18 G + rod	J-20 G + rod
97 and higher				H-18t	H-18t

- Section 1 is 5 feet long; the minimum gauge is 18.
- Section 2 is 4 feet long; the minimum gauge is 20.
- Section 3 is where the ductwork reduces in size. This section has a finished length of 3 feet. However, if you take a closer look at the lower drawing in the figure, you will see that the sloping panel on this section is longer than 3 feet. Therefore, you must use 20-gauge duct (which would cover 4-foot spacing). As an alternative, you could reduce the size of the section until the sloping panel measures 3 feet. If you do this, you could use 22-gauge duct.
- Section 4 is 4 feet long; the minimum gauge is 20.
- There is an elbow in section 5 (*Figure 3*). The elbow length and heel length both measure 52 inches. However, the width is still 48 inches. Therefore, the minimum gauge for the elbow is 18.

DID YOU KNOW?

Pascals

In the metric system, the inches of water gauge measurement is stated as Pascals. The unit of measurement is named to honor Blaise Pascal, a 17th-century French mathematician, for his studies of barometric pressure.

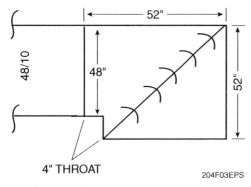

Figure 3 ◆ Ductrun elbow.

204F03EPS

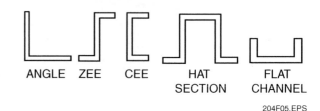

204F05.EPS

Figure 5 ◆ Types of intermediate reinforcement.

Example 2:

In this example, intermediate reinforcement is used on the duct panels. Using reinforcements allows you to reduce the panel distance. It will also allow you to reduce the required sheet metal gauge. However, when installing intermediate reinforcement, you must take both dimensions of the duct into account.

Study *Figure 4*. The large dimension is 48 inches, and the panel spans 5 feet. According to *Table 3*, 18-gauge metal is required. If you install reinforcement in the middle of this duct, you will reduce the span to 2½ feet. It seems logical to say that you could use 24-gauge metal, but you must not forget about the 28-inch side of the duct. That side still has 5 feet between reinforcements, which requires a minimum of 22-gauge metal. If you want to use 24-gauge metal, you must also reinforce the 28-inch side to reduce its span. This reinforcement can be a standard structural angle; a zee, cee, or hat section; or a flat channel (*Figure 5*).

You can use a single piece of intermediate reinforcement for duct up to 3 inches of water gauge pressure. Attach the reinforcement within 2 inches from each duct corner and at maximum intervals of 12 inches on center. You can weld, spot weld, bolt, screw, or rivet to attach the reinforcement.

For duct systems that are more than 3 inches of water gauge pressure, you must fasten the rein-

forcement ends with a ¼-inch tie rod or at least a 1" × 1" × 12"-gauge angle. You must do this even if intermediate reinforcing is not required on the side dimension. Acceptable methods for intermediate reinforcement are shown in *Figure 6*.

5.2.0 Determining Spacing and Rigidity for Connectors

You will also refer to *Table 3* to determine the reinforcement grade class for connectors based on spacing requirements. This is an example of a shop standard. Refer to your local mechanical building code for actual standards used in your jurisdiction. This spacing is figured differently than the spacing for metal gauge. Once you have the correct spacing, you can select the minimum rigidity class from the table.

Example 3:

For this example, refer to *Table 3* and study *Figure 7*. This process is a little more complicated than figuring sheet metal gauge. You may find it somewhat confusing at first, but if you work your way through each section in the figure, you will soon see how the calculations are done. Note that the duct dimensions are 48 inches and 28 inches, so you will refer to those lines in *Table 3*. Notice also that the figure shows two views of the seven sections of duct.

- In section 1 of *Figure 7*, the connector takes up a minimum amount of space; therefore, the minimum gauge in the applicable inches of water gauge table will meet the requirements.

- In section 2 of *Figure 7*, the connector measures 6 inches and the duct measures 60 inches. Take one-half of each measurement and add them: 3 inches + 30 inches = 33 inches. This total is more than 2.5 feet but less than 3 feet. Therefore, refer to the 3-foot spacing column in the table. For the 48-inch dimension, the answer is E. For the 28-inch dimension, the answer is C.

- In section 3 of *Figure 7*, one duct measures 60 inches and the other duct measures 48 inches. Take one-half of each measurement and add them: 30 inches + 24 inches = 54 inches. This total is more than 4 feet but less than 5 feet.

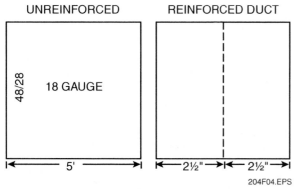

204F04.EPS

Figure 4 ◆ Intermediate reinforcement.

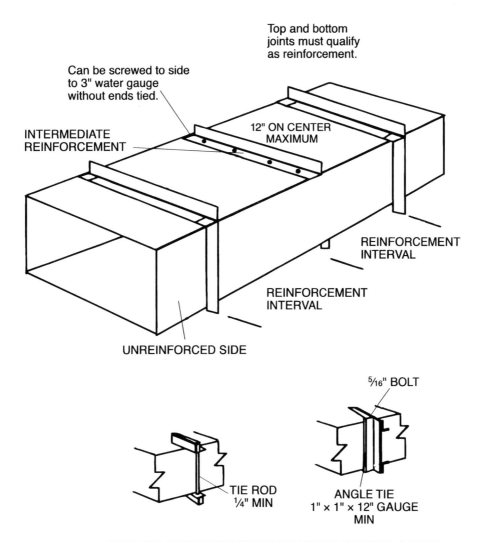

Top and bottom joints must qualify as reinforcement.

Can be screwed to side to 3" water gauge without ends tied.

INTERMEDIATE REINFORCEMENT

12" ON CENTER MAXIMUM

REINFORCEMENT INTERVAL

REINFORCEMENT INTERVAL

UNREINFORCED SIDE

⁵⁄₁₆" BOLT

TIE ROD ¼" MIN

ANGLE TIE 1" × 1" × 12" GAUGE MIN

TIE ENDS OF REINFORCEMENTS ABOVE 3" WATER GAUGE

204F06.EPS

Figure 6 ❖ Intermediate reinforcement on one or both dimensions of duct.

Therefore, refer to the 5-foot spacing column in the table. For the 48-inch dimension, the answer is G. For the 28-inch dimension, the answer is C.

- In section 4 of *Figure 7*, the duct measures 48 inches and the reduced panel measures 36 inches. Take one-half of each measurement and add them: 24 inches + 18 inches = 42 inches. This total is more than 3 feet but less than 4 feet. Therefore, refer to the 4-foot spacing column in the table. For the 48-inch dimension, the answer is F. For the 28-inch dimension, the answer is C.

- In section 5 of *Figure 7*, the reduced panel measures 36 inches and the distance to the intermediate reinforcement measures 30 inches. Take one-half of each measurement and add them: 18 inches + 15 inches = 33 inches. This total is more than 2.5 feet but less than 3 feet. Therefore, refer to the 3-foot spacing column in the table. For the 48-inch dimension, the answer is E. For the 10-inch dimension, the answer is A.

(Notice that in this section, the 28-inch dimension has reduced to 10 inches.)

- In section 6 of *Figure 7*, the distance to the intermediate reinforcement is 30 inches and the distance to the elbow is 30 inches. Take one-half of each measurement and add them: 15 inches + 15 inches = 30 inches. This total is equal to 2.5 feet. Therefore, refer to the 2.5-foot spacing column in the table. The answer for the 48-inch dimension is E-24. There is no side support required for the 28-inch dimension.

- There is an elbow in section 7 of *Figure 7*. The distance to the intermediate reinforcement is 30 inches and the heel of the elbow measures 54 inches. Take one-half of both measurements and add them: 15 inches + 27 inches = 42 inches. This total is more than 3 feet but less than 4 feet. Therefore, refer to the 4-foot spacing column in the table. For the 48-inch dimension, the answer is F. For the 10-inch dimension, the answer is A.

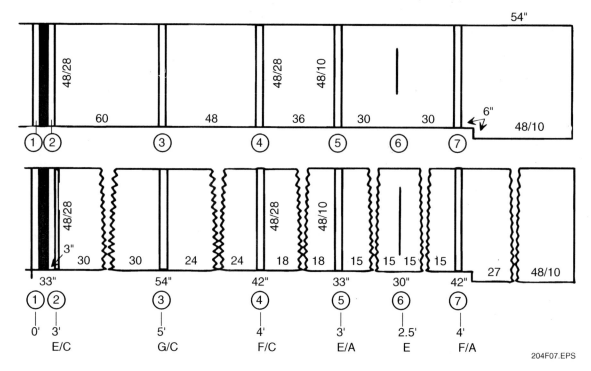

Figure 7 ◆ Connector reinforcement spacing.

5.3.0 Standard Transverse Connectors

Two general types of connectors are used to reinforce and connect duct joints: **transverse connectors** and **proprietary connectors**. Standard transverse connectors include the slip and drive (or slide and drive). These connectors can be reinforced by an additional bar or angle to bring them up to proper rigidity (*Figure 8*).

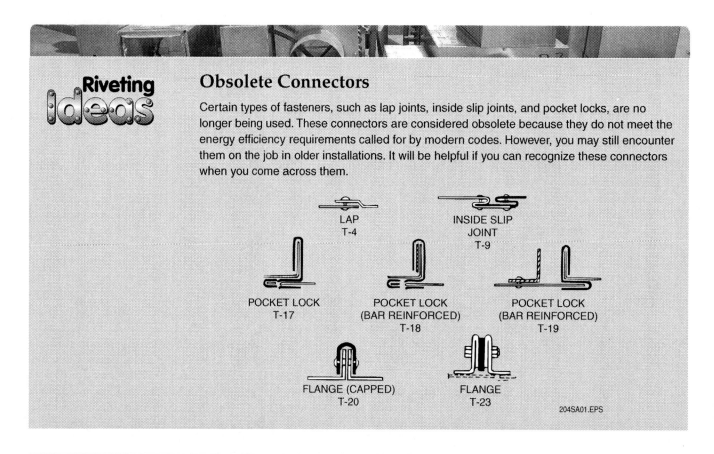

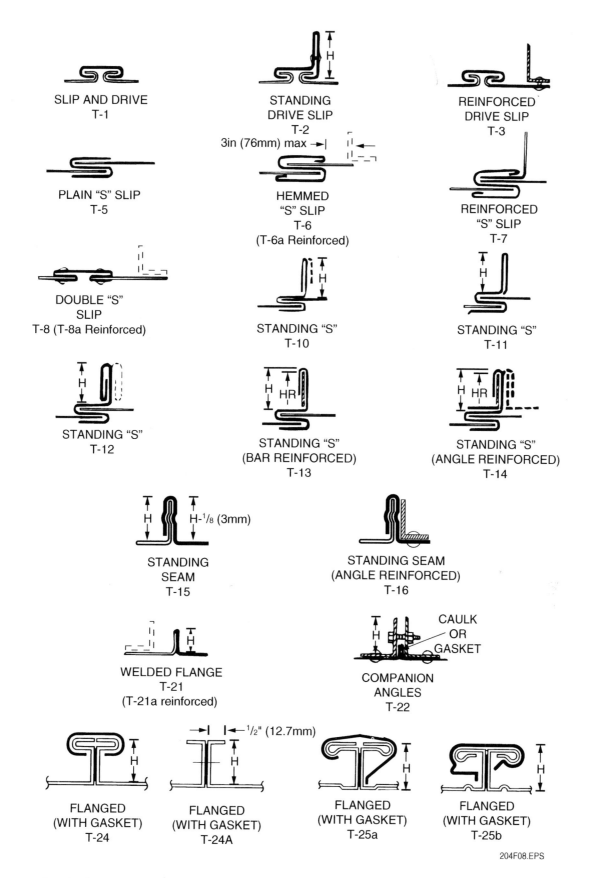

SLIP AND DRIVE
T-1

STANDING
DRIVE SLIP
T-2

REINFORCED
DRIVE SLIP
T-3

PLAIN "S" SLIP
T-5

3in (76mm) max

HEMMED
"S" SLIP
T-6
(T-6a Reinforced)

REINFORCED
"S" SLIP
T-7

DOUBLE "S"
SLIP
T-8 (T-8a Reinforced)

STANDING "S"
T-10

STANDING "S"
T-11

STANDING "S"
T-12

STANDING "S"
(BAR REINFORCED)
T-13

STANDING "S"
(ANGLE REINFORCED)
T-14

STANDING
SEAM
T-15

$H - \frac{1}{8}$ (3mm)

STANDING SEAM
(ANGLE REINFORCED)
T-16

WELDED FLANGE
T-21
(T-21a reinforced)

CAULK
OR
GASKET

COMPANION
ANGLES
T-22

FLANGED
(WITH GASKET)
T-24

$\frac{1}{2}$" (12.7mm)

FLANGED
(WITH GASKET)
T-24A

FLANGED
(WITH GASKET)
T-25a

FLANGED
(WITH GASKET)
T-25b

204F08.EPS

Figure 8 ◈ Standard transverse connectors.

As an option, you may use a tie rod with a lower-rated standard transverse connector in place of a stiffer connector. In some cases, this may be the only acceptable way to reinforce the duct. For example, assume you have a duct at 3 inches of water gauge pressure with 54-inch dimensions and a requirement for 5-foot standard transverse connector spacing. Refer back to *Table 3*. Find the line that shows duct dimensions of 49 to 54 inches. Trace across that line to the column headed 5.0 (for spacing in feet). The entry reads as follows:

H-18

F + rod

As you have already learned, this entry gives the fabricator the option of using a lower-grade class of standard transverse connector (F) with a tie rod equal to standard transverse connector grade class H. You can use a ¼-inch tie rod for duct dimensions up to 36 inches and a ⅜-inch tie rod for duct dimensions of 37 inches up to 97 inches.

For duct lengths greater than 97 inches, the entry in *Table 3* is H-18t. This type of entry indicates that you must use a grade class H standard transverse connector with a tie rod. Several tie rod attachment methods are shown in *Figures 9* and 10. In *Figure 9*, you will also see how to handle intermediate reinforcing, which is done in the same way as connector reinforcing. As you did with the connector, you can use a lower-rated standard transverse connector with a tie rod.

5.4.0 Longitudinal Seams

Longitudinal seams are constructed to withstand 1½ times the system operating pressure without duct failure or metal deformation. The acceptable longitudinal seams and recommended locations are shown in *Figure 11*. The two most commonly used longitudinal seams are the Pittsburgh lock seam and the snap lock seam.

The pocket depth on the Pittsburgh seam varies depending on the roll-forming equipment used. The maximum pocket depth is ⅝-inch. The larger pocket accommodates heavier duct. Pittsburgh lock seams can be used for all pressure classes. The following limitations apply to snap lock seams:

- They can be used only for duct up to a maximum of 4 inches of water gauge pressure.
- They are not recommended for aluminum or other soft metal duct.

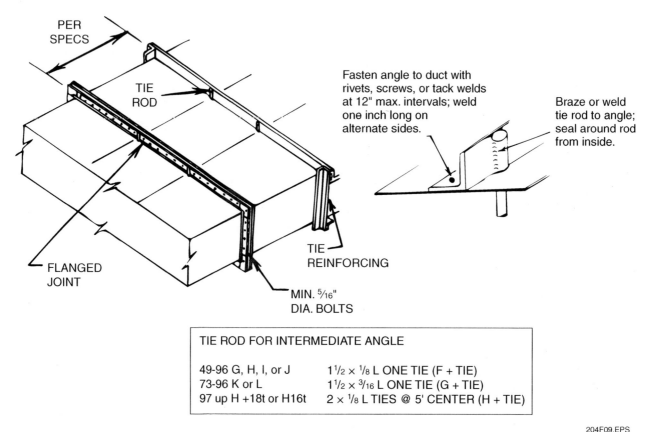

Figure 9 ◆ Connector and intermediate reinforcement with tie rod.

204F09.EPS

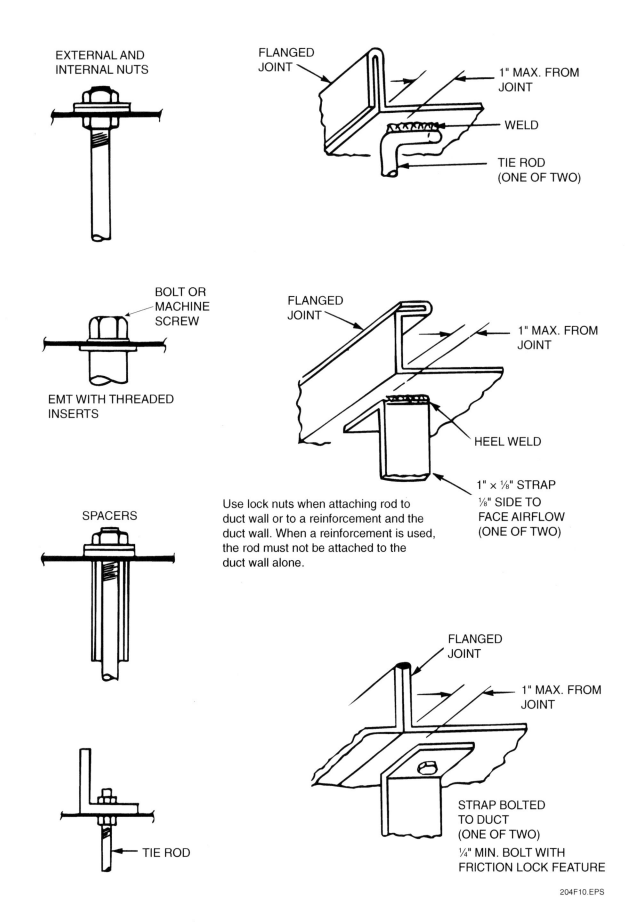

EXTERNAL AND
INTERNAL NUTS

FLANGED
JOINT

1" MAX. FROM
JOINT

WELD

TIE ROD
(ONE OF TWO)

BOLT OR
MACHINE
SCREW

EMT WITH THREADED
INSERTS

FLANGED
JOINT

1" MAX. FROM
JOINT

HEEL WELD

1" × ⅛" STRAP
⅛" SIDE TO
FACE AIRFLOW
(ONE OF TWO)

SPACERS

Use lock nuts when attaching rod to
duct wall or to a reinforcement and the
duct wall. When a reinforcement is used,
the rod must not be attached to the
duct wall alone.

TIE ROD

FLANGED
JOINT

1" MAX. FROM
JOINT

STRAP BOLTED
TO DUCT
(ONE OF TWO)

¼" MIN. BOLT WITH
FRICTION LOCK FEATURE

204F10.EPS

Figure 10 ◆ Tie rod attachments.

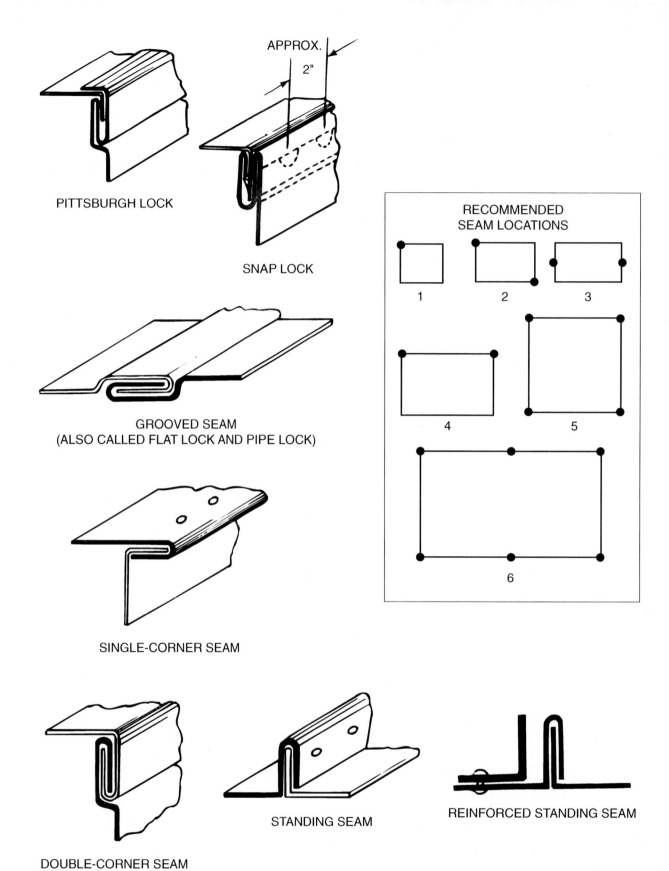

PITTSBURGH LOCK

APPROX.
2"

SNAP LOCK

GROOVED SEAM
(ALSO CALLED FLAT LOCK AND PIPE LOCK)

SINGLE-CORNER SEAM

RECOMMENDED
SEAM LOCATIONS

1

2

3

4

5

6

DOUBLE-CORNER SEAM

STANDING SEAM

REINFORCED STANDING SEAM

204F11.EPS

Figure 11 ◈ Longitudinal seams.

- You must add screws to the duct ends for duct that is over 48 inches wide with 3 inches of water gauge pressure and for all duct with 4 inches of water gauge pressure.

Other longitudinal seams used are the grooved seam (also known as pipe lock or flat lock), single-corner seam, double-corner seam, standing seam, and reinforced standing seam. Of these, the standing seam is most commonly used to splice duct panels. Normally a 1-inch seam is used for ducts with widths of up to 42 inches. A 1½-inch seam is used for wider ducts. The seam is fastened together at the ends and at 8-inch intervals.

You must use cross breaking or beading (*Figure 12*) if the duct is wider than 18 inches, the panel is larger than 10 square feet, and all of the following are true:

- The duct is not insulated.
- The duct pressure is less than 3 inches of water gauge.
- The duct is fabricated from less than 18-gauge metal.

Cross breaking or beading helps to stiffen the metal. It also reduces noise from air rushing inside the duct. This type of noise is often called the oil-can effect. Without the beading or cross breaking, the varying levels of air pressure inside the duct will cause the walls of the duct panel to bend inward and outward, producing a sound like that made when you squeeze the bottom of an oilcan.

5.5.0 Shop Standards

Construction methods and equipment will vary from one shop to another. There are several ways to meet reinforcing requirements. Some shops may have the machinery to form connectors, some

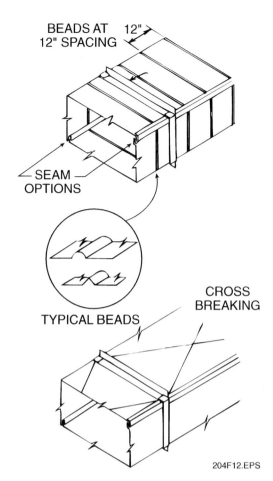

Figure 12 ◆ Cross breaking and beading.

may use preformed connectors such as Duct-mate®, and others may use slip or double-slip joints with angle reinforcing. Information on developing shop standards, along with useful tables and figures, is presented in *Appendixes A* and *B* of this module.

Proprietary Connectors

A contractor may use a proprietary connector. A proprietary connector is one that is developed by a company and registered with a trademark. This type of connector is used when a duct size is beyond the limits of a standard connector or when a duct requires a Class C seal. Proprietary connectors, such as Ductmate® and Lockformer®, are available from several companies. The TDC System, developed by Lockformer®, is used in HVAC ducting. The flange can be rolled through the machine in one pass. Thus, it eliminates the need for a separate flange. This also increases strength and rigidity and prevents leakage. Both TDC and TDF connectors are formed from the duct material. Other types of connectors are attached to the duct as separate components.

Source: Professional Building Services, http://www.hasttiegroup.com/au/PBS/default.apx, June 2008

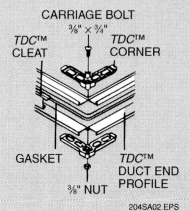

CARRIAGE BOLT
$\frac{3}{8}$" × $\frac{3}{4}$"

TDC™ CLEAT

TDC™ CORNER

GASKET

$\frac{3}{8}$" NUT

TDC™ DUCT END PROFILE

204SA02.EPS

1. Industry standards are _____.
   a. mandatory
   b. voluntary
   c. a compilation of company policies and building codes
   d. used chiefly by engineers

2. Duct systems are classified according to their pressure class in either _____.
   a. inches of mercury gauge or meters
   b. inches of pressure gauge or Pascals
   c. inches of water gauge or Pascals
   d. inches of deflection gauge or air speed gauge

3. The minimum area in a louver through which air can pass is called a _____.
   a. deflection area
   b. casing area
   c. plenum area
   d. free area

4. Operating pressure is a measurement of duct pressure in _____.
   a. cubic feet per minute
   b. pounds per square foot
   c. inches of water gauge
   d. Pascals

5. Duct sealing requirements are placed into categories by _____.
   a. seal class
   b. pressure class
   c. constant air volume systems
   d. variable air volume systems

6. The expression H-18, F + rod in a duct gauge table means that the fabricator _____.
   a. may substitute a lower-grade class of reinforcement with a tie rod equal to reinforcement grade class H
   b. must use reinforcement grade class F with a tie rod equal to reinforcement grade class H and 18-gauge metal
   c. may use reinforcement grade class F and add the tie rod as an option if 18-gauge metal is to be used
   d. may use any reinforcement grade class up to and including F as long as a tie rod is included

7. The two general types of connectors used to reinforce and connect duct joints are _____.
   a. transverse connectors and tie rods
   b. transverse connectors and proprietary connectors
   c. longitudinal seams and proprietary connectors
   d. tie rods and longitudinal seams

8. The two most commonly used longitudinal seams are the _____.
   a. Pittsburgh lock seam and the standing seam
   b. Pittsburgh lock seam and the snap lock seam
   c. snap lock seam and the standing seam
   d. snap lock seam and the button lock seam

9. A _____ is the most commonly used longitudinal seam for splicing duct panels.
   a. single-corner seam
   b. double-corner seam
   c. standing seam
   d. reinforced standing seam

10. Cross breaking or beading helps to _____ the metal.
    a. loosen
    b. secure
    c. stiffen
    d. relax

Summary

In this module, you learned how industry standards and building codes affect duct system design, fabrication, and installation. Building codes are administered by the local governing authority and are mandatory. Industry standards, on the other hand, are voluntary. Many contractors voluntarily adopt industry standards because they represent the best and most current information.

Many organizations have an effect on the work you will do. In this module, you were given a sample list of these organizations, and you practiced using a technical manual published by SMACNA. You also learned how to read and apply tables you will use to select the correct sheet metal gauge and connectors for a duct system.

Notes

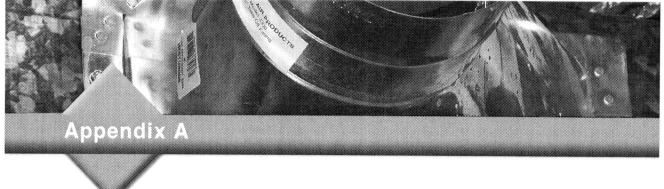

Appendix A

Developing Shop Standards

Establishing a set of shop fabrication rules for complete duct systems can be challenging. This appendix discusses some of the challenges and offers suggestions for developing shop standards.

This appendix includes examples of look-up tables that could be used to determine the correct metal gauge and reinforcement codes for connectors. Tables such as these are built on information found in the *SMACNA Duct Construction Standards* book currently in its third edition. The standards developed for a particular shop may expand upon or reorganize the information from the SMACNA standards to better accommodate the way things work in that shop. As you develop your shop standards, always be careful to check to make sure they comply with local building codes and ordinances.

Basic reinforcement codes are shown in the tables. However, before you can equate these basic codes to an actual connector, you need to look more closely at the diversity in whole ductruns.

Study *Figure A-1* to see how the four primary connector spacing choices were determined for fittings or straight duct lengths. The spacing choices for connectors are 5', 4', 3', 3'-6", 2', and 2'-6".

Study *Figure A-2*. There are two pieces of straight duct with dimensions of 48" and 18". The spacing requirements for the three reinforcement connectors are 5', 4', and 3'. There are also two pieces of straight duct with dimensions of 42" and 12". These two pieces also have three reinforcement connectors with spacing requirements of 4', 3', and 2'-6". If your shop applies the reinforcement spacing rules exactly, these two sections cannot be exchanged. Most shops, however, would want to standardize the sections so that they could be exchanged. To standardize the sections, the shop can adopt two simple rules:

Rule 1 To avoid confusion, evaluate only one end of each piece of duct. It is recommended that a shop use the end toward the air handler.

Rule 2 Assume that a straight piece of duct is the standard length of 4' or 5'. Assume also that the duct piece is joined to another straight piece of duct of the same length.

To standardize duct fittings, follow *Rule 1*. In addition, assume that the fitting is connected to the standard straight duct piece (4' or 5') being used (*Figure A-3*).

Fitting or Straight Duct

		60"	48"	42"	36"	30"	24"
		Use 5' connector column			Use 4' connector column		
5' STR	60"	60	54	51	48	45	42
4' STR	48"	54	48	45	42	39	36
					Use 3' + 2'6" connector column		
5' INTER	30"	45	39	36	33	30	27

204A01.EPS

Figure A-1 ◆ Reinforcement spacing with combinations of straight duct or fittings.

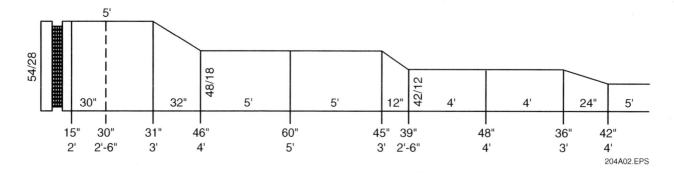

Figure A-2 ◈ Sample ductrun.

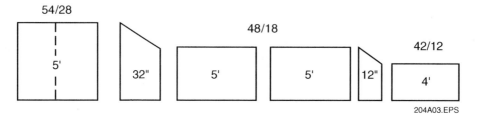

Figure A-3 ◈ Breakdown of a sample ductrun.

These rules are the basis for the basic look-up tables for the sheet metal gauge and connectors (*Tables A-1* through *A-7*). These tables are examples of shop standards. Be sure to consult SMACNA and local standards for work performed in your shop. The tables in this appendix focus on 5-foot shops, but these tables can also be used by 4-foot shops.

The following discussion on how to read the tables will focus on *Table A-1* because the information is organized in the same way in all the tables. Each table is set up for a different duct size (shown in inches of water gauge pressure).

Notice that each table is divided vertically with a double line into two halves. The left half covers all straight duct and fittings to a 5-foot maximum. The right half covers 5-foot intermediate straight duct and 42-inch maximum fittings.

Look at the left-hand side of the table first. You can look up the 5-foot reinforcement connector for each size of duct shown in the ½-inch of water gauge column. The 4-foot reinforcement connectors also are listed. The connectors in this column can be used by 4-foot shops. They also can be used for 36-inch-long fittings with 5-foot straight duct. You will use the same connector requirement for all straight duct and fittings. Choose the gauge from the 5-foot, 4-foot, or 3-foot gauge column.

Look at the right-hand side of the table. The information presented here is used when there is a 2-gauge reduction from 20-gauge metal or when

18-gauge metal is required. Notice the column titled CODE. The same code applies to both the connector and to the intermediate reinforcement except when noted otherwise in the column to the right of the CODE column.

Both sides of the table also provide information you can use to learn when a lighter reinforcement with a tie rod may be substituted for a heavier reinforcement. For example, look at the line in the table that begins with a duct size of 84 and trace across the line until you see the term (Ft) H. This entry means that you can substitute an F-grade connector with a tie rod for an H-grade connector. Sometimes the use of a tie rod is mandatory. In this case, the table entry will read Ht. You must use a tie rod when you see this entry.

The section titled *General Notes* includes information about the type of seam, connectors, intermediate reinforcing, and seal class required. Notice that although some tables allow a choice between Pittsburgh lock seams or snap lock seams, some tables prohibit use of the snap lock seam.

Some of the common standard connector options are shown in *Figure A-4*. Study this figure to see how the basic reinforcement codes in *Tables A-1* through *A-7* (in the column titled CONN.) equate to the connectors shown. For example, a flat S and drive connector conforms to basic code A. For a detailed view of a double-slip joint connection, study *Figure A-5*.

GENERIC CODE	STANDARD CONNECTORS (All Dimensions in Inches)
A	Flat "S" and drive – conforms to A where allowed to 2" wg pressure (T-1 & T-5) 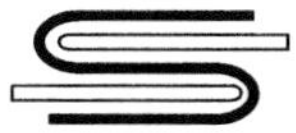 T-5 - PLAIN "S" SLIP  T-1 - SLIP AND DRIVE
A, B, C B C	Hemmed "S" with 1 × 1 × 18 ga angle – A, B, or C, to 2" wg pressure (T-7) Standing Drive (T-2) 1 - 1/8 × 26 ga 1 - 1/8 × 22 ga 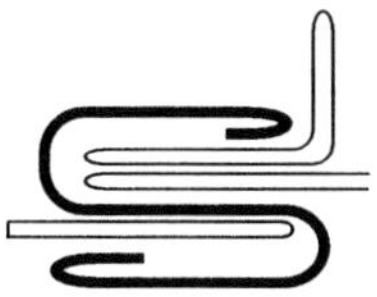 T-7 REINFORCED "S" SLIP 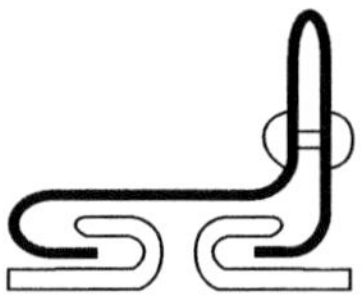 T-2 STANDING DRIVE SLIP
C D	Standing "S" 1 × 26 ga 1 × 24 ga – (T-10 & T-11)  T-10 STANDING "S"  T-11 STANDING "S" (Alt.)
F G	Standing "S" reinforced with bar (T-13) 1-1/2 × 24 ga + 1-1/2 × 1/8 bar 1-1/2 × 22 ga + 1-1/2 × 1/8 bar  T-13 STANDING "S" (Bar Reinforced)
H I J	Standing "S" reinforced with angle (T-14) 1-1/2 × 20 ga + 1-1/2 × 3/16 2 × 20 ga + 2 × 1/8 2 × 20 ga + 2 × 3/16 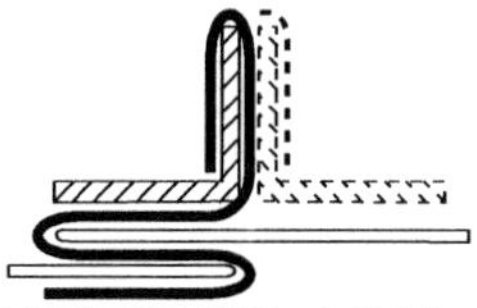 T-14 STANDING "S" (Angle Reinforced)
	Limit Length 3" wg – 36" 4" wg – 30" Not recommended above 4" wg If above 3" wg and 4" wg limits (see above) or 6" wg and 10"wg, use inside slip or double slip with angle reinforcement. The letter "T" is a prefix assigned by SMACNA to identify transverse connectors.

INTERMEDIATE REINFORCEMENT	
C D E F G H I J K L	Angle (inches) 1 × 18 ga 1-1/4 × 18 ga 1-1/2 × 16 ga 1-1/2 × 1/8 – – – – – – – – – –1 (Ft) 1-1/2 × 1/8 with tie rods = G–J 1-1/2 × 3/16 – – – – – – – – – 2 (Gt) 1-1/2 × 3/16 with tie rods = K–L 2 × 1/8 – – – – – – – – – – – – 3 (Ht) 2 × 1/8 with tie rods = Ht 2 × 3/16 2-1/2 × 1/18 2-1/2 × 3/16 2-1/2 × 1/4

204A04.EPS

Figure A-4 ◈ Connectors and intermediate reinforcements.

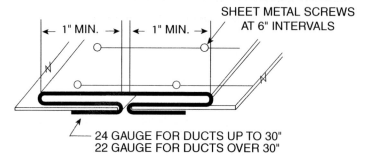

DOUBLE-SLIP JOINT

Above 3" and 4" water gauge limits, use double-slip joint with angle reinforcement.

204A05.EPS

Figure A-5 ◈ Double-slip joint connection.

Table A-1 Duct Connector Selection for ½ Inch of Water Gauge Pressure

ALL STR & FTG to 5' MAX

+ or – 1/2" w.g. SIZE	CONN. GAUGE YOUR SHOP	CONN. (TIE) 5'	GAUGE 5'	GAUGE 4'	GAUGE 3'
18		SLIP	26		
20		DRIVE	26		
26		A	26		
30		B	26		
36		C	26		
42		D	26	C*	
48		D	26		
54		D	26		
60		E	24	26	
72		F	22	24	
84		(Ft) H	22	G*24	
96		(Ft) H	20	24	
97 UP		(Ft) H	18		

5' INTER STR & 42" MAX FTG

CONN. GAUGE YOUR SHOP	CONN. (TIE) 3'	GAUGE 2'6"	GAUGE 3'	GAUGE 4'	INTER CODE

* ALT CONN FOR 4' STR & 4' FTG OR 36" MAX FTG WITH 5' STR

GENERAL NOTES

SEAMS – Pittsburgh or Snap lock
CONNECTOR – Standard connector

Class C seal recommended minimum, but required only when specified

204TA01.EPS

Table A-2 Duct Connector Selection for 1 Inch of Water Gauge Pressure

GENERAL NOTES

SEAMS – Pittsburgh or Snap lock

CONNECTOR – Standard connector

INTERMEDIATE REINFORCING – Need only to be screwed to duct 2" from ends, 12" oc

Class C seal recommended minimum, but required only when specified

ALL STR & FTG to 5' MAX

+ or − 1" w.g. SIZE	CONN. GAUGE YOUR SHOP	CONN. (TIE) 5'	GAUGE 5'	GAUGE 4'	GAUGE 3'
10		SLIP	26		
20		DRIVE	26		
26		A	26		
30		B	26		
36		C	26		
42		D	24	26	
48		E	24	D*26	
54		E	22	24	
60		F	22	24	
72		(Ft) H	18	G*22	24
84		(Ft) I	18	H*20	20
96		(Ft) J	16	I*18	20
97 UP					

5' INTER STR & 42" MAX FTG

CONN. GAUGE YOUR SHOP	CONN. (TIE) 3'	GAUGE 2'6"	GAUGE 3'	GAUGE 4'	INTER CODE
W					
72	(Ft) G	24	24	22	
84	(Ft) H	20	20	20	
96	(Ft) I	20	20	18	
97 UP	Ht	18	NO	NO	

* ALT CONN FOR 4' STR & 4' FTG OR 36" MAX FTG WITH 5' STR

204TA02.EPS

Table A-3 Duct Connector Selection for 2 Inches of Water Gauge Pressure

GENERAL NOTES

SEAMS – Pittsburgh or Snap lock
CONNECTOR – Standard connector

Class C seal recommended minimum, but required only when specified

ALL STR & FTG to 5' MAX

+ or – 1/2" w.g. SIZE	CONN. GAUGE YOUR SHOP	CONN. (TIE) 5'	GAUGE 5'	GAUGE 4'	GAUGE 3'
18		SLIP	26		
20		DRIVE	26		
26		A	26		
30		B	26		
36		C	26		
42		D	26	C*	
48		D	26		
54		D	26		
60		E	24	26	
72		F	22	24	
84		(Ft) H	22	G*24	
96		(Ft) H	20	24	
97 UP		(Ft) H	18		

5' INTER STR & 42" MAX FTG

CONN. GAUGE YOUR SHOP	CONN. (TIE) 3'	GAUGE 2'6"	GAUGE 3'	GAUGE 4'	INTER CODE

* ALT CONN FOR 4' STR & 4' FTG OR 36" MAX FTG WITH 5' STR

© 1992 Michael Bergen. All Rights Reserved

204TA01.EPS

MODULE 04204-08 ◆ SHEET METAL DUCT FABRICATION STANDARDS 4.25

Table A-4 Duct Connector Selection for 3 Inches of Water Gauge Pressure

© 1992 Michael Bergen. All Rights Reserved

GENERAL NOTES

SEAMS – Pittsburgh or Snap lock – (Screw ends of duct over 48")

CONNECTOR – Standard connector not recommended over 36". No slip allowed. Use double slip with angle reinforced over 36".

INTERMEDIATE REINFORCING – Check gauge required when depth is above 26"; screw reinforcement 2" from end, 12" oc.

Class B seal minimum required

ALL STR & FTG to 5' MAX

+ or – 3" w.g. SIZE	CONN. GAUGE YOUR SHOP	CONN. (TIE) 5'	GAUGE 5'	4'	3'
18		DRIVE	24		
20		A	24		
24		B	24		
26		C	24		
30		C	22	24	
36		E	20	D*24	24
42		E	20	22	24
48		(Ft) G	18	F*20	22
54		(Ft) H	18	18	22
60		(Ft) H	16	18	20
72		(Ft) I	NO	16	20

5' INTER STR & 42" MAX FTG TO 72" WIDE

CONN. GAUGE YOUR SHOP	CONN. (TIE) 3'	GAUGE 2'6"	3'	4'	INTER CODE
36	D	24	24	24	
42	E	24	24	22	
48	E	24	22	20	
54	(Ft) G	24	22	18	E
60	(Ft) G	24	20	18	
72	(Ft) H	22	20	16	
84	(Ft) J	20	18	NO	
96	(Gt) L	18	16	NO	K
96 UP	Ht	18	NO	NO	

* ALT CONN FOR 4' STR & 4' FTG OR 36" MAX FTG WITH 5' STR

204TA04.EPS

Table A-5 Duct Connector Selection for 4 Inches of Water Gauge Pressure

GENERAL NOTES

SEAMS – Pittsburgh or Snap lock – (Screw ends of duct)

CONNECTOR – Standard connector not recommended over 30". No pocket lock, no slip allowed. Use double slip with angle reinforced over 30".

INTERMEDIATE REINFORCING – Both W & D or tie ends 1x1x12 ga. min. or 1/4" rod

Class A seal required

+ only 4" w.g.

ALL STR & FTG to 5' MAX

CONN. GAUGE YOUR SHOP	SIZE	CONN. (TIE) 5'	GAUGE 5'	4'	3'
	16	DRIVE	24		
	18	A	24		
	22	B	24		
	26	C	22	24	
	30	D	22	24	24
	36	E	20	22	24
	42	F	18	20	24
	48	(Ft) G	18	18	22
	54	(Ft) H	16	18	20
	60	(Ft) I	16	16	20

5' INTER STR & 42" MAX FTG TO 60" WIDE

CONN. GAUGE YOUR SHOP	SIZE	CONN. (TIE) 3'	GAUGE 2'6"	3'	4'	INTER CODE
	36	D	24	24	22	
	42	E	24	22	20	
	48	F	22	22	18	
	54	(Ft) G	22	20	18	
	60	(Ft) H	22	20	16	
	72	(Ft) I	20	18	NO	
	84	(Gt) K	18	16	NO	J
	96	(Gt) L	18	NO	NO	
	96 UP	Ht	18	NO	NO	

204TA05.EPS

Table A-6 Duct Connector Selection for 6 Inches of Water Gauge Pressure

GENERAL NOTES

SEAMS – Pittsburgh, no Snap lock

CONNECTOR – Standard connector not recommended. Use double slip with angle reinforcement.

INTERMEDIATE REINFORCING – Both W & D. Notice spacing limits above 72".

Class A seal required

5' INTER STR & 42" MAX FTG TO 42" WIDE

CONN. GAUGE YOUR SHOP	CONN. (TIE) 3'	GAUGE 2'	GAUGE 2'6"	GAUGE 3'	INTER CODE
28	C		24	24	
30	D		24	24	
36	E		22	20	E
42	F		22	20	
48	(Ft) H		22	18	G
54	(Ft) H		20	18	
60	(Ft) H		20	18	
72	(Ft) J		18	16	
84	(Gt) L	18	16	NO	
96	(Gt) L	18	NO	NO	
96 UP	Ht	18	NO	NO	

ALL STR & FTG to 5' MAX

+ only 6" w.g. SIZE	CONN. GAUGE YOUR SHOP	CONN. (TIE) 5'	GAUGE 5'	4'	3'
12		DRIVE	24		
16		A	24	24	
20		B	22	24	
22		C	22	24	
24		C	22	22	24
28		D	20	22	24
30		D	18	22	24
36		F	18	E*20	22
42		(Ft) G	16	18	20

* ALT CONN FOR 4' STR & 4' FTG OR 36" MAX FTG WITH 5' STR

204TA06.EPS

Table A-7 Duct Connector Selection for 10 Inches of Water Gauge Pressure

GENERAL NOTES

SEAMS – Pittsburgh, no Snap lock

CONNECTOR – Standard connector not recommended. No slip or drive allowed. Use double slip with angle reinforcement.

INTERMEDIATE REINFORCING – Both W & D. Note – Over 72" same construction

Class A seal required

ALL STR & FTG to 5' MAX

+ only 10" w.g. SIZE	CONN. GAUGE YOUR SHOP	CONN. (TIE) 5'	GAUGE 5'	GAUGE 4'	GAUGE 3'
8		A	24		
12		A	22	24	
14		A	20	22	24
16		B	20	22	24
18		C	20	B*22	24
20		C	18	20	24
22		C	18	20	24
24		D	18	20	24
26		D	18	20	22
28		E	18	D*20	22
30		E	16	18	22
36		F	16	18	20
42		(Ft) H	NO	16	18

5' INTER STR & 42" MAX FTG TO 42" WIDE

CONN. GAUGE YOUR SHOP	CONN. (TIE) 3'	GAUGE 2'	GAUGE 2'6"	GAUGE 3'	INTER CODE
20	B		24	24	
22	C		24	24	
24	C		24	24	
26	D		24	22	C
28	D		24	22	
30	D		24	22	
36	F		22	20	E
42	(Ft) G		20	18	
48	(Ft) H		18	18	
54	(Ft) I		18	16	H
60	(Ft) J		18	16	I
72	(Gt) K		16	NO	
72 UP	Ht	16	NO	NO	

* ALT CONN FOR 4' STR & 4' FTG OR 36" MAX FTG WITH 5' STR

204TA07.EPS

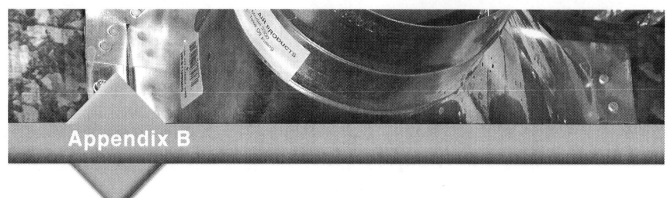

Round Duct

Specifications for longitudinal and spiral seams and round duct joints are shown in *Tables A-8 through A-11*. The tables cover standards in both inches of water gauge and Pascals and in both positive and negative pressures.

Several types of seams are used for round duct. These seams include the grooved seam (also known as pipe lock or flat lock), the butt or lap weld seam, the lock-type spiral seam, the snap lock seam, and the lap and rivet seam (*Figure A-6*).

Round duct transverse joints include the slip joint, the draw band joint, the crimp or lap joint, and the flange joint. Details of these joints are shown in *Figure A-7*.

Table A-8 Round Duct Gauge Seam Standards in Inches of Water Gauge (Positive Pressure)

Maximum Diameter (Inches)	+2 in. wg		+4 in. wg		+10 in. wg	
	Spiral	Long.	Spiral	Long.	Spiral	Long.
6	28	28	28	28	28	28
8	28	28	28	28	28	26
10	28	26	28	26	28	26
12	28	26	28	26	26	24
14	28	26	26	24	26	24
16	26	24	24	24	24	22
18	26	24	24	24	24	22
19 to 26	26	24	24	22	24	22
27 to 36	24	22	22	20	22	20
37 to 50	22	20	20	20	20	20
51 to 60	20	18	18	18	18	18
61 to 84	18	16	18	16	18	16

204TA08.EPS

Table A-9 Round Duct Gauge Seam Standards in Pascals (Positive Pressure)

Maximum Diameter (Millimeters)	+500 Pa		+1,000 Pa		+2,500 Pa	
	Spiral	Long.	Spiral	Long.	Spiral	Long.
150	0.48	0.48	0.48	0.48	0.48	0.48
200	0.48	0.48	0.48	0.48	0.48	0.55
250	0.48	0.55	0.48	0.55	0.48	0.55
300	0.48	0.55	048	0.55	0.55	0.70
360	0.48	0.55	0.55	0.70	0.55	0.70
400	0.55	0.70	0.55	0.70	0.70	0.85
460	0.55	0.70	0.70	0.70	0.70	0.85
660	0.55	0.70	0.70	0.85	0.70	0.85
910	0.70	0.85	0.85	1.00	0.85	1.00
1,270	0.85	1.00	1.00	1.00	1.00	1.00
1,520	1.00	1.31	1.31	1.31	1.31	1.31
2,130	1.31	1.61	1.31	1.61	1.31	1.61

204TA09.EPS

Table A-10 Round Duct Gauge Seam Standards in Inches of Water Gauge (Negative Pressure)

Maximum Diameter (Inches)	−2 in. wg		−4 in. wg		−10 in. wg	
	Spiral	Long.	Spiral	Long.	Spiral	Long.
6	28	28	28	28	26	26
7	28	28	28	28	26	26
8	28	28	28	28	26	26
9	28	28	28	26	26	24
10	28	28	26	26	26	22
11	28	26	26	24	26	22
12	28	26	26	24	24	22
13	28	26	26	24	24	20
14	28	24	24	22	24	20
15	28	24	24	22	22	20
16	26	24	24	22	22	18
17	26	24	24	20	22	18
18	24	22	24	20	22	18
19	24	22	24	20	22	18
20	24	22	22	20	22	18
21	24	20	22	18	22	18
22	24	20	22	18	22	16
23	24	20	22	18.	20	16
24	22	20	22	18	20	16
25 to 26	22	20	20	18	20	18 A4
27 to 29	22	18	20	16	18	16 A4
30	22	18	20	16	18	16 B4
31 to 33	20	18	20	16	18	16 B4
34	20	18	20	20 A6	18	16 B4
35 to 36	20	16	20	20 A6	18	16 B4
37 to 42	20	16	18	18 B6	18 F12	—
43 to 48	20	18 A6	18	18 B5	18 F6	—
49 to 60	18	18 B4	18 F6	16 B4	18 F6	—
61 to 72	16	—	18 F6	—	16 F4	—

An alphabet letter in the table means that reinforcement angles or their equivalent must be used at the foot interval following the letter. The angle sizes are as follows:
A = 1" × 1" × 1/8"; B = 1-1/4" × 1-1/4": × 3/16"; C = 1-1/2" × 1-1/2": × 3/16"; D = 1-1/2" × 1-1/2" × 1/4";
E = 2" × 2" × 3/16"; F = 2" × 2" × 1/4".
If companion flange joints are used as reinforcements, those for 25" to 36" diameter shall be
1-1/2" × 1-1/2" × 3/16"; for 37" to 48" diameter 2" × 2" × 3/16"; for 40" to 60" diameter 2-1/2" × 2-1/2" × 3/16";
for 61" to 72" diameter 3" × 3" × 1/4".

204TA10.EPS

Table A-11 Round Duct Gauge Seam Standards in Pascals (Negative Pressure)

Maximum Diameter (Millimeters)	−500 Pa		−1,000 Pa		−2,500 Pa	
	Spiral	Long.	Spiral	Long.	Spiral	Long.
150	0.48	0.48	0.48	0.48	0.55	0.55
180	0.48	0.48	0.48	0.48	0.55	0.55
200	0.48	0.48	0.48	0.48	0.55	0.55
230	0.48	0.48	0.48	0.55	0.55	0.70
250	0.48	0.48	0.55	0.55	0.55	0.85
280	0.48	0.55	0.55	0.70	0.55	0.85
300	0.48	0.55	0.55	0.70	0.70	0.85
330	0.48	0.55	0.55	0.70	0.70	1.00
360	0.48	0.70	0.70	0.85	0.70	1.00
380	0.48	0.70	0.70	0.85	0.85	1.00
400	0.55	0.70	0.70	0.85	0.85	1.31
430	0.55	0.70	0.70	1.00	0.85	1.31
460	0.70	0.85	0.70	1.00	0.85	1.31
480	0.70	0.85	0.70	1.00	0.85	1.31
500	0.70	0.85	0.85	1.00	0.85	1.31
530	0.70	1.00	0.85	1.31	0.85	1.31
560	0.70	1.00	0.85	1.31	0.85	1.61
580	0.70	1.00	0.85	1.31	1.00	1.61
600	0.85	1.00	0.85	1.31	1.00	1.61
660	0.85	1.00	1.00	1.31	1.00	1.31 A1.2
740	0.85	1.31	1.00	1.6	1.31	1.61 A1.2
760	0.85	1.31	1.00	1.6	1.31	1.61 B1.2
840	1.00	1.31	1.00	1.6	1.31	1.61 B1.2
860	1.00	1.31	1.00	1.00 A1.8	1.31	1.61 B1.2
910	1.00	1.61	1.00	1.00 A1.8	1.31	1.61 B1.2
1,070	1.00	1.61	1.31	1.31 B1.8	1.31 F3.6	—
1,220	1.00	1.31 A1.8	1.31	1.31 B1.8	1.31 F1.8	—
1,520	1.31	1.31 B1.2	1.31 F1.8	—	1.31 F1.2	—
1,830	1.61	—	1.31 F1.8	—	1.61 F1.2	—

An alphabet letter in the table means that reinforcement angles or their equivalent must be used at the foot interval following the letter. The angle sizes are as follows:
A = 25 × 25 × 3.2 mm; B = 32 × 32 × 4.8 mm; C = 38 × 38 × 4.8 mm; D = 38 × 38 × 4.8 mm; E = 51 × 51 × 4.8 mm; F = 51 × 51 × 6.4 mm.

If companion flange joints are used as reinforcements, those for 630 to 910 mm diameter shall be 38 × 38 × 6.4 mm; for 940 to 1,220 mm diameter 51 × 51 × 4.8 mm; for 1,240 to 1,520 mm diameter 64 × 64 × 4.8 mm; for 1,550 to 1,830 mm diameter 76 × 76 × 6.4 mm.

204TA11.EPS

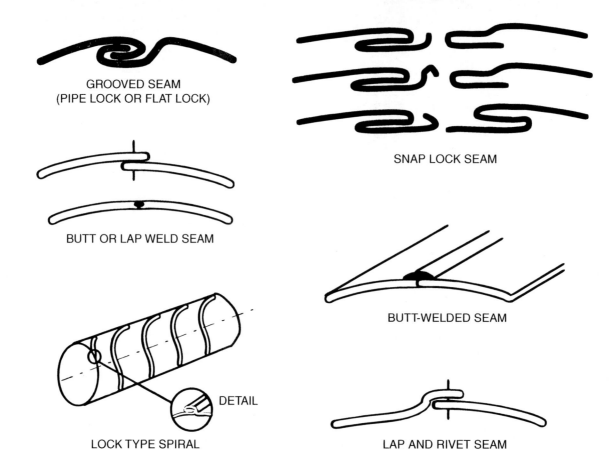

GROOVED SEAM
(PIPE LOCK OR FLAT LOCK)

SNAP LOCK SEAM

BUTT OR LAP WELD SEAM

BUTT-WELDED SEAM

DETAIL

LOCK TYPE SPIRAL

LAP AND RIVET SEAM

204A06.EPS

Figure A-6 ◆ Round duct seams.

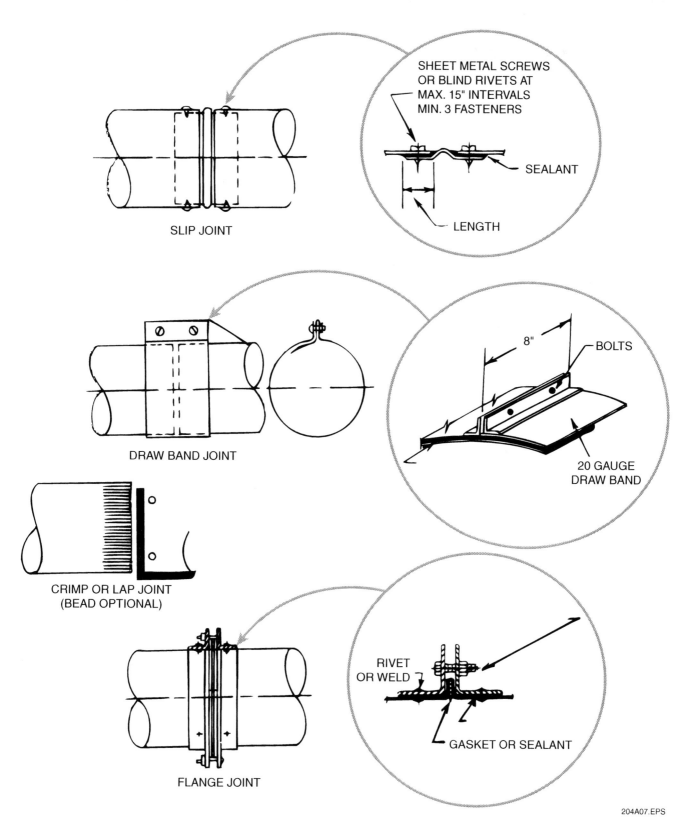

SHEET METAL SCREWS
OR BLIND RIVETS AT
MAX. 15" INTERVALS
MIN. 3 FASTENERS

SEALANT

LENGTH

SLIP JOINT

DRAW BAND JOINT

8"

BOLTS

20 GAUGE
DRAW BAND

CRIMP OR LAP JOINT
(BEAD OPTIONAL)

RIVET
OR WELD

GASKET OR SEALANT

FLANGE JOINT

204A07.EPS

Figure A-7 ◆ Round duct transverse joints.

Free area: The total minimum area of the openings in an air inlet or outlet, such as an air diffuser, grille, or register, through which air can pass. It is usually expressed as a percentage of the total area.

Inches of water gauge: The American system unit of measurement that measures pressure.

Pascal: The metric system unit of measurement that measures pressure.

Proprietary connector: Any connector or connection system developed by a company and registered with a trademark.

Reinforcement: Used at transverse connector joints or at intermediate points on certain sizes and pressure classes of ductrun sections to provide strength and reduce deflection of the duct sheet metal. In some cases, it enables the use of lighter-gauge sheet metal.

Seal class: Any one of three types of sealing methods used to control duct leakage on certain pressure classes of ductruns.

Tie rod: Used with transverse joints and with intermediate reinforcement for certain sizes and pressure classes of ductrun sections to provide strength and reduce deflection of the sheet metal. In some cases, it enables the use of lighter-gauge reinforcement members.

Transverse connector: A type of joint used to connect sections of rectangular ductruns.

Variable air volume (VAV) system: A system in which the temperature of each conditioned space is controlled by the temperature and/or volume of air supplied to that space.

Resources & Acknowledgments

Additional Resources

This module is intended to present thorough resources for task training. The following reference works are suggested for further study. These are optional materials for continued education rather than for task training.

HVAC Duct Construction Standards—Metal and Flexible, Third Edition. 2005. Chantilly, VA: Sheet Metal and Air Conditioning Contractors' National Association.

International Building Code. Falls Church, VA: International Code Council.

International Mechanical Code. Falls Church, VA: International Code Council.

Uniform Building Code. Whittier, CA: International Conference of Building Officials.

Uniform Mechanical Code. Walnut, CA: International Association of Plumbing and Mechanical Officials.

Figure Credits

General Pump, 204F02, 204F03, 204F04

Lockformer Company, 204SA02

Michael Bergen, 204TA01 – 204TA07

NCCER makes every effort to keep these textbooks up-to-date and free of technical errors. We appreciate your help in this process. If you have an idea for improving this textbook, or if you find an error, a typographical mistake, or an inaccuracy in NCCER's Contren® textbooks, please write us, using this form or a photocopy. Be sure to include the exact module number, page number, a detailed description, and the correction, if applicable. Your input will be brought to the attention of the Technical Review Committee. Thank you for your assistance.

Instructors – If you found that additional materials were necessary in order to teach this module effectively, please let us know so that we may include them in the Equipment/Materials list in the Annotated Instructor's Guide.

Write: Product Development and Revision
National Center for Construction Education and Research
3600 NW 43rd St, Bldg G, Gainesville, FL 32606

Fax: 352-334-0932

E-mail: curriculum@nccer.org

Craft _____ Module Name _____

Copyright Date _____ Module Number _____ Page Number(s) _____

Description _____

_____

_____

_____

(Optional) Correction _____

_____

_____

(Optional) Your Name and Address _____

_____

_____

_____

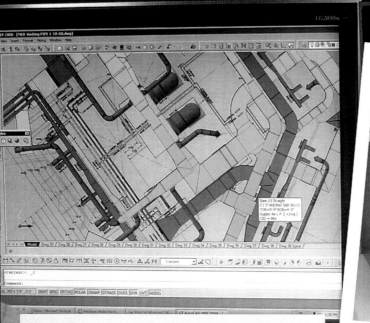

A detailer prepares plans for fabrication of ductwork in coordination with other trades on the job. A detailer's job is one of the most complex in the sheet metal trade.

All work done for a job must conform to construction specifications. Specifications for all trades, including sheet metal work, are listed in the specifications.

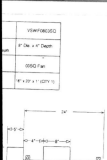

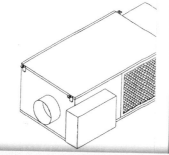

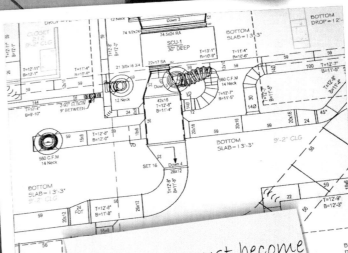

Not all pieces of an HVAC system are fabricated at the shop. Some are purchased from specialty suppliers.

Installers must become adept at reading the shop drawings so they can properly install the system on site.

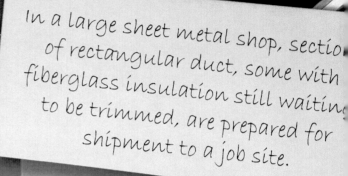

In a large sheet metal shop, section of rectangular duct, some with fiberglass insulation still waiting to be trimmed, are prepared for shipment to a job site.

A Pittsburgh seam on a square elbow is hammered together using an air tool.

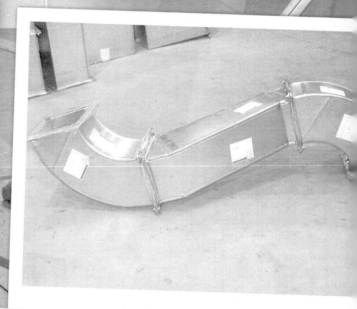

A sheet metal worker prepares a layout for a square-to-round transition using radial line development skills.

Through careful field measurement and coordination with other trades, complicated transitions and offsets are prepared in order to fit around other building features even though the installation location has not been seen by the sheet metal workers.

Sheet metal workers often work with copper pipe and copper tubing.

Copper pipe subassemblies, freshly fabricated, are ready to be staged for shipment to a job site.

A pipe threading machine is used to cut, ream, and thread iron pipe.

Heavy iron pipe for large volume supply and for drainage requirements are frequently handled by mechanical shops where sheet metal work is also done.

Using this punch and die on a power brake, a sheet metal worker creates a bend in a piece. Often when calculating the size of the blank of metal required to fabricate a piece, allowances need to be made for the small length of metal used in the creation of the bend.

A sheet metal worker uses a soldering iron and bar solder to tightly seal a seam on this small fitting.

On the job site, a sheet metal worker studies a shop drawing prepared by the detailer back at the shop. He needs to be able to properly interpret these plans so that the duct is installed the right way the first time.

This ogee offset was prepared using radial line layout techniques.

04205-08

Air Properties
and Distribution

04205-08
Air Properties and Distribution

Topics to be presented in this module include:

Overview

As a sheet metal professional or HVAC mechanic, you will often be called upon to measure certain properties of the air in a building and make adjustments to the system to optimize its performance. To do this effectively, you need a working knowledge of the properties of air and how they affect, and are affected by, the HVAC system. If an area in a building is not receiving sufficient duct static pressure, how can the output of the system's fan be adjusted to compensate? What effect does the air's relative humidity have on the ability of a system to cool or heat the air? What instruments should be used to measure different properties of the air, and what is the proper way to use them? In this module, you'll learn the answers to questions like these that you may encounter as an HVAC mechanic.

Objectives

When you have completed this module, you will be able to do the following:

1. Explain the gas laws (Dalton's, Boyle's, and Charles's) used when dealing with air and its properties and explain how the properties of air relate to one another.
2. Use a psychrometric chart to evaluate air properties and changes in air properties.
3. Explain the differences between propeller and centrifugal fans and blowers.
4. Recognize the instruments used to make measurements in air systems and explain the use of each instrument.
5. Make basic temperature, air pressure, and velocity measurements in an air distribution system.
6. Explain the three fan laws.

Trade Terms

Absolute pressure
Air-handling unit
Atmospheric pressure
British thermal unit (Btu)
Constant air volume (CAV) system
Cubic feet per minute (cfm)
Dew point
Dry-bulb thermometer
Enthalpy
Evaporator
Free air delivery
Gauge pressure
Latent heat
Pitot tube
Plenum
Propeller fan

Psychrometric chart
Psychrometry
Relative humidity
Sensible heat
Specific density of air
Specific heat of air
Specific volume of air
Static pressure
Static pressure tip
Thermistor
Thermocouple
Total heat
Total pressure
Velocity pressure
Venturi
Volume
Wet-bulb thermometer

Required Trainee Materials

1. Calculator
2. Pencil and paper

Prerequisites

Before you begin this module, it is recommended that you successfully complete the following: *Core Curriculum; Sheet Metal Level One; Sheet Metal Level Two*, Modules 04201-08 through 04204-08.

This course map shows all of the modules in the second level of the Sheet Metal curriculum. The suggested training order begins at the bottom and proceeds up. Skill levels increase as you advance on the course map. The local Training Program Sponsor may adjust the training order.

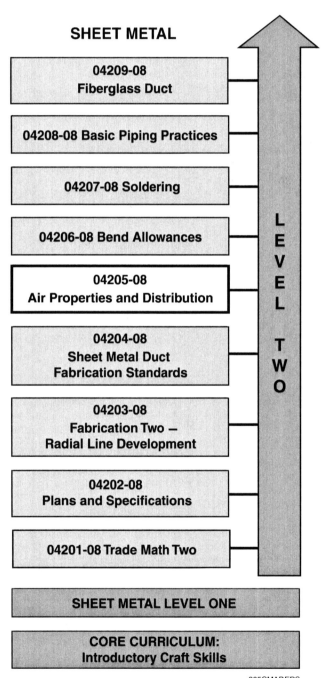

SHEET METAL

04209-08
Fiberglass Duct

04208-08 Basic Piping Practices

04207-08 Soldering

04206-08 Bend Allowances

04205-08
Air Properties and Distribution

04204-08
**Sheet Metal Duct
Fabrication Standards**

04203-08
**Fabrication Two —
Radial Line Development**

04202-08
Plans and Specifications

04201-08 Trade Math Two

SHEET METAL LEVEL ONE

**CORE CURRICULUM:
Introductory Craft Skills**

L E V E L   T W O

205CMAP.EPS

1.0.0 ◆ INTRODUCTION

Efficient air distribution systems are designed to deliver the right amount of conditioned air to occupied spaces. These systems make our working and living spaces comfortable, and they are essential in many industries.

In this module, you will be introduced to the science of **psychrometry**, which is the study of air properties, and you will see how those properties react to changes in pressure, temperature, and **volume**. Blowers and fans are an important part of any air distribution system, and this module highlights the blowers and fans commonly used in heating, ventilating, and air conditioning (HVAC) work. Finally, you will learn about the various instruments used to measure air properties in an air distribution system.

2.0.0 ◆ AIR PRESSURE

Before looking at how an air distribution system works, you need a basic understanding of how air reacts to changes in volume, temperature, and pressure. This section gives you an overview of this topic. For more in-depth information, refer to the books listed in the *Additional Resources* section of this module.

Our atmosphere, which extends about 600 miles above the earth, is made up of a mixture of gases: 78% nitrogen, 21% oxygen, and a 1% mix of other gases. The total weight of these gases is measured in pounds per square inch (psi). At sea level, a square-inch column of air extending 600 miles up pushes down on the earth with a pressure of 14.7 psi (*Figure 1*). This air pressure is called **atmospheric pressure**.

Atmospheric pressure is measured with a barometer, which is a mercury-filled tube that is open on one end and closed on the other (like a test tube). The tube is inverted into a cup that also contains mercury. A scientist named Evangelista Torricelli invented the barometer in 1643. He used a 30-inch-long glass tube in his experiments. He noticed that, at sea level at 70°F, the mercury in the tube fell until its top was 29.92 inches above the surface of the mercury in the cup. The atmospheric pressure on the liquid in the cup kept the mercury in the tube from falling farther. Another scientist named Blaise Pascal found that the air pressure at higher altitudes was different from that found at sea level. Thus, the barometer also can be used to measure altitude. In addition, it was found that temperature had an effect on the mercury in the barometer. In HVAC work, you will often use the values of 14.7 psi and 29.92 inches of mercury (inches Hg) at sea level at 70°F.

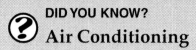

DID YOU KNOW?

Air Conditioning

The term *air conditioning* is commonly associated with air cooling. However, in HVAC work, the term has a much larger meaning. Conditioned air is air that has been treated to control its temperature (heated or cooled), humidity, cleanliness, and distribution within an interior space.

2.1.0 Air Pressure Scales and Gauges

The pressures of fluids such as air, water, and gas are important factors in HVAC work. A number of pressures, including atmospheric pressure and the air pressure created by a fan or a blower, are at work in an air distribution system. If you could remove all of the air from an air distribution system, you would also remove all of the pressure. This creates a vacuum with zero pressure.

The pressure exerted by a fluid above zero pressure is called its **absolute pressure,** which is measured by an absolute pressure scale. On this scale, air pressure is expressed in psi or in pounds per square inch absolute (psia).

Another scale, called **gauge pressure,** is frequently used to define air pressure levels. Gauge pressure scales use atmospheric pressure as their zero starting point. Positive gauge pressures, those above zero, are expressed in pounds per square inch gauge (psig). Negative pressures,

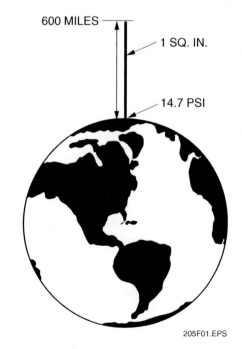

Figure 1 ◆ Atmospheric pressure.

those below 0 psig, are expressed in inches of mercury vacuum (inches Hg vacuum). You can easily convert gauge pressure to absolute pressure by adding 14.7 to the gauge pressure value. For example, a gauge pressure of 10 psig equals an absolute pressure of 24.7 psi (10 + 14.7). A comparison of the gauge and absolute pressure scales is shown in *Figure 2*. Converting one scale to another is often required when you make calculations involving air pressure relationships.

2.2.0 Pressure, Temperature, and Volume Relationships

Air is a mixture of gases. No matter what their chemical makeup, gases are governed by gas laws. The gas laws are the result of experiments made by scientists who have studied how gases react in response to three variables: pressure, temperature, and volume. Because gases have no fixed shape, the gas laws describe how gases react in enclosed spaces such as an air distribution system. Two English scientists, John Dalton and Robert Boyle, and one French scientist, Jacques Charles, developed three of the basic laws that govern gases.

2.2.1 Dalton's Law

Dalton's law is used to calculate the total air pressure in a container. This law states the following:

> The total air pressure in a container is the sum of the individual pressures of all the gases in the mixture. In addition, each gas behaves as if it occupies the space alone.

For example, to find the total air pressure of the cylinder shown in *Figure 3*, you would find the individual pressure for each gas (oxygen, nitrogen, carbon dioxide, and water vapor) and then add them up. Note that even though the gases are mixed together in the cylinder, the pressure of each, as Dalton noted, can be measured separately.

2.2.2 Boyle's Law

Air pressure also has an effect on the volume of a gas. This effect is known as Boyle's law, which states the following:

> At a constant temperature, the pressure exerted by a gas varies inversely with its volume.

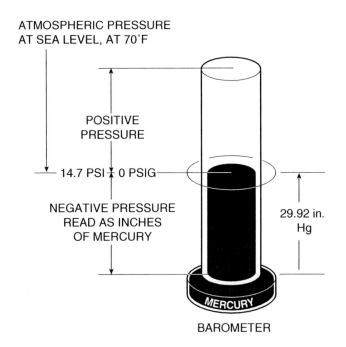

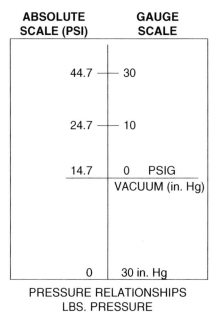

205F02.EPS

Figure 2 ◆ Atmospheric, absolute, and gauge pressure relationships.

DALTON'S LAW

- EACH GAS IN A MIXTURE OF GASES ACTS INDEPENDENTLY

- THE TOTAL PRESSURE CREATED IN A CYLINDER IS EQUAL TO THE SUM OF THE PRESSURES OF EACH GAS IN THE MIXTURE

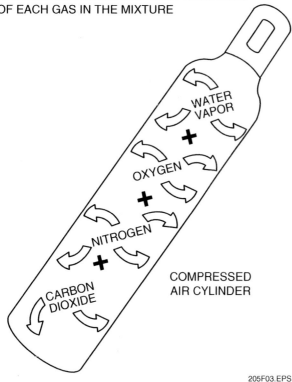

205F03.EPS

Figure 3 ◈ An illustration of Dalton's law.

For example, if the pressure on a quantity of gas is doubled, its volume is halved. If the volume is doubled, the pressure is halved. This relationship can be expressed mathematically in the three formulas that follow:

Formula 1 $(Po)(Vo) = (Pn)(Vn)$

Formula 2 $Vn = \dfrac{(Vo)(Po)}{Pn}$

Formula 3 $Pn = \dfrac{(Po)(Vo)}{Vn}$

Where:

Po = Old absolute pressure (psia)

Pn = New absolute pressure (psia)

Vo = Old volume (cubic feet)

Vn = New volume (cubic feet)

psia = psig + 14.7

Example

What is the new volume of 5 cubic feet of gas at 20 psig if it is compressed to 60 psig, providing the temperature remains constant? Formula 2 is used to solve this problem.

$$Vn = \frac{(Vo)(Po)}{Pn}$$

$$\frac{(5 \text{ cu. ft.})(20 \text{ psig} + 14.7)}{60 \text{ psig} + 14.7}$$

$$\frac{(5)(34.7)}{74.7}$$

Vn = 2.32 cubic feet

2.2.3 Charles's Law

Charles's law deals with how the volume and pressure of a gas react to changes in temperature. This law is stated as follows:

> At a constant pressure, the volume of a confined gas varies directly with the temperature. At a constant volume, the pressure of a confined gas varies directly with the temperature.

In other words, at a constant pressure if the temperature rises, the volume of a gas will increase. At a constant volume, if the pressure increases, so will the temperature.

Formulas 4 through 6 apply to the constant pressure aspect of Charles's law. Formulas 7 through 9 apply to the constant volume aspect of this law.

With constant pressure:

Formula 4 $(Vo)(Tn) = (Vn)(To)$

Formula 5 $Vn = \dfrac{(Vo)(Tn)}{To}$

Formula 6 $Tn = \dfrac{(Vn)(To)}{Vo}$

Where:

To = Old absolute temperature

Tn = New absolute temperature

Vo = Old volume (cubic feet)

Vn = New volume (cubic feet)

Absolute temperature = °F + 460

With constant volume:

Formula 7 $(Po)(Tn) = (Pn)(To)$

Formula 8 $Tn = \dfrac{(Pn)(To)}{Po}$

Formula 9 $Pn = \dfrac{(Po)(Tn)}{To}$

Where:

To = Old absolute temperature

Tn = New absolute temperature

Po = Old absolute pressure (psia)

Pn = New absolute pressure (psia)

Absolute temperature = °F + 460

psia = psig + 14.7

Examples

1. What is the new volume of 5 cubic feet of gas at 37°F if the temperature is raised to 90°F at a constant pressure? Formula 5 is used to solve this problem.

$$Vn = \frac{(Vo)(Tn)}{To}$$

$$\frac{(5 \text{ cu. ft.})(90°F + 460)}{37°F + 460}$$

$$\frac{(5)(550)}{497}$$

Vn = 5.53 cubic feet

2. What is the new pressure (in psig) of a quantity of gas at 40°F and 35 psig if its temperature is raised to 60°F, at a constant volume? Formula 9 is used to solve this problem.

$$Pn = \frac{(Po)(Tn)}{To}$$

$$\frac{(35 \text{ psig} + 14.7)(60°F + 460)}{40°F + 460}$$

$$\frac{(49.7)(520)}{500}$$

Pn = 51.69 psia

51.69 psia − 14.7 = 36.99 psig

3.0.0 ◈ PSYCHROMETRY

Psychrometry is a branch of physics. It is used to measure air properties, especially the amount of moisture in the air. This might seem like a simple matter. However, the amount of moisture in the air is affected by many factors, and understanding them can be complicated. As an apprentice, you will not have to master psychrometry, but you should have a basic understanding of this science. HVAC professionals use psychrometric measurements to install the right air conditioning equipment and to troubleshoot existing systems. Some of the air properties you should become familiar with are discussed in the following sections.

3.1.0 Dry Air

Dry air is a mixture of gases composed of approximately 78% nitrogen, 21% oxygen, and 1% other gases. The characteristics of dry air (volume, density, and heat) are shown in *Figure 4.*

Specific volume of air is the amount of space one pound of dry air occupies. At sea level at 70°F, 1 pound of dry air occupies a volume of 13.33 cubic feet. If air is heated and maintained at a constant pressure, it will expand and weigh less per cubic foot of volume.

Specific density of air is how much one pound of dry air weighs. At sea level at 70°F, 1 pound of dry air weighs 0.075 pounds per cubic foot. You can find the specific density per cubic foot of dry air by dividing the volume of air into one pound. The formula is as follows:

Specific density = 1 lb ÷ 13.33 ft³ = 0.075 lb/ft³

Specific heat of air is the amount of heat needed to raise one pound of dry air one degree Fahrenheit. Specific heat is measured as **British thermal units (Btus)** per pound per degrees Fahrenheit. Air at sea level has a specific heat of 0.24 Btu/lb/°F.

To calculate how many Btus are needed to raise air temperature, use the **sensible heat** formula. Sensible heat is the amount of heat that, when added to the air, causes a change in temperature with no change in the amount of moisture. It is measured with a standard thermometer. The sensible heat formula, which involves using a measurement of **cubic feet per minute (cfm),** is as follows:

Sensible heat (Btu/hr) =
(specific heat)(specific density)(60 min/hr)(cfm)(ΔT)

Where:

cfm = velocity of airflow in cubic feet per minute

ΔT = change in temperature (°F)

Example

Btu/hr = (0.24)(0.075)(60)(cfm)(ΔT)

Btu/hr = (1.08)(cfm)(ΔT)

3.2.0 Humidity and Relative Humidity

A certain amount of water vapor (humidity) is almost always present in the air. The amount of moisture that air will hold depends on the temperature. Warm air holds more moisture than cold air, but as warm air becomes more humid, it cannot pick up as much moisture as cold air can. That is why you feel so much warmer on a hot, humid day than on a hot, dry day. Perspiration, which helps you to stay cool, cannot evaporate as easily on a humid day, so it stays on your skin. That's what makes you feel both hot and sticky.

The amount of moisture in the air is usually stated as grains of moisture per pound of dry air. It also can be stated as pounds of moisture per pound of dry air, but the grains-of-moisture measurement results in larger numbers that are easier to work with. There is usually very little water vapor in a pound of air. At sea level at 70°F, one pound of water contains 7,000 grains of moisture (*Figure 5*). Under these same conditions, one

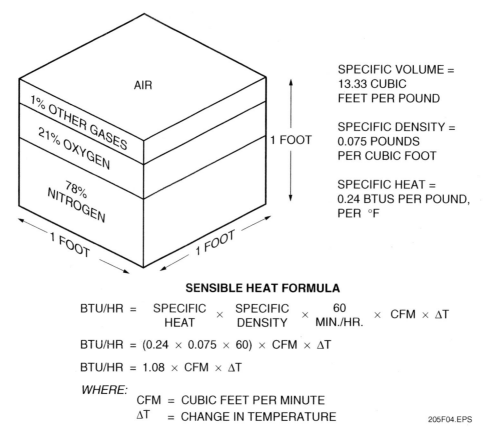

SPECIFIC VOLUME =
13.33 CUBIC
FEET PER POUND

SPECIFIC DENSITY =
0.075 POUNDS
PER CUBIC FOOT

SPECIFIC HEAT =
0.24 BTUS PER POUND,
PER °F

SENSIBLE HEAT FORMULA

$$\text{BTU/HR} = \frac{\text{SPECIFIC}}{\text{HEAT}} \times \frac{\text{SPECIFIC}}{\text{DENSITY}} \times \frac{60}{\text{MIN./HR.}} \times \text{CFM} \times \Delta T$$

$$\text{BTU/HR} = (0.24 \times 0.075 \times 60) \times \text{CFM} \times \Delta T$$

$$\text{BTU/HR} = 1.08 \times \text{CFM} \times \Delta T$$

WHERE:
CFM = CUBIC FEET PER MINUTE
ΔT = CHANGE IN TEMPERATURE

205F04.EPS

Figure 4 ◈ Dry air at sea level at 70°F.

pound of saturated air contains only 110.5 grains of moisture. If this amount of moisture were stated in pounds of moisture per pound of dry air, it would weigh only 0.01579 pound (110.5 ÷ 7,000).

Relative humidity is the ratio of the amount of moisture already in the air to the total amount it can hold. If air is saturated, its relative humidity is 100 percent. At this point, the air can hold no more moisture. If air contains only half the moisture it can hold, its relative humidity is 50 percent (*Figure 6*).

Technicians use relative humidity measurements to provide indoor comfort. Most people feel best when indoor temperature and humidity fall within certain ranges called comfort zones. In winter, the comfort zone is between 67°F and 76°F with a relative humidity of about 30 percent. In summer, the comfort zone is between 72°F and 81°F with a relative humidity of about 40 percent. However, human comfort is only part of the story. Controlling temperature and humidity results in improved energy efficiency. For many businesses, controlling the amount of humidity in the air is essential.

Technicians use several instruments to measure relative humidity. These instruments are discussed later in this module. For this part of the module, you need to know about the two thermometers found in some of these instruments.

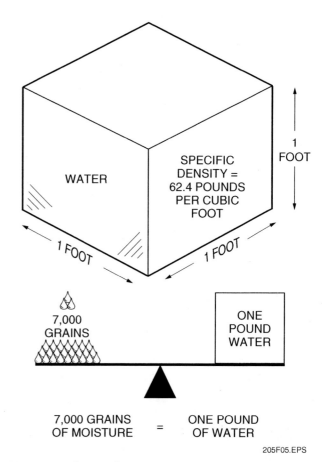

7,000 GRAINS OF MOISTURE = ONE POUND OF WATER

205F05.EPS

Figure 5 ◈ Grains of moisture in one pound of water.

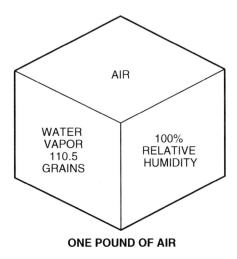

ONE POUND OF AIR

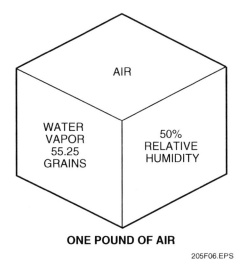

ONE POUND OF AIR

205F06.EPS

Figure 6 ◆ Relative humidity.

The **dry-bulb thermometer** is the standard, every-day thermometer. It measures the amount of sensible heat in the air. The **wet-bulb thermometer** measures the moisture content of the air and looks like the dry-bulb thermometer, but its sensing bulb is wrapped with a water-soaked wick. Evaporation takes place at the wick, so the wet-bulb temperature usually is lower than the dry-bulb temperature.

3.3.0 Dew Point

Dew point is the temperature at which air becomes saturated with moisture. At this point, the air condenses into water droplets. At 100 percent relative humidity, the dew point temperature, wet-bulb temperature, and dry-bulb temperature are all the same. Normally, the wet-bulb temperature is lower than the dry-bulb temperature. However, saturated air cannot hold any more moisture, so no evaporation can take place at the wick of the wet-bulb thermometer.

3.4.0 Enthalpy

Moisture vapor in the air has its own heat content called **latent heat.** Latent heat added to the sensible heat of dry air equals the **total heat** content of the air. **Enthalpy** is the total heat content in the air at 100 percent saturation. Enthalpy is measured using a wet-bulb thermometer and is stated in Btu/lb of air. Technicians use a value called *change in enthalpy* to calculate how many Btus per hour the air temperature has increased or decreased. For example, change in enthalpy can be used to find the difference in heat content between the dry- and wet-bulb temperatures of an air stream entering and leaving an **air-handling unit.**

4.0.0 ◆ THE PSYCHROMETRIC CHART

The **psychrometric chart** is a graph used to determine how air properties vary as the amount of moisture in the air changes (*Figure 7*). Manufacturers of air conditioning equipment provide these charts, and there may be some slight differences from one manufacturer to another. The temperatures on the chart used in this module range from 20°F to 110°F at sea level.

Other types of psychrometric charts are available that graph air properties in temperatures ranging from –20°F to 250°F. Some charts graph

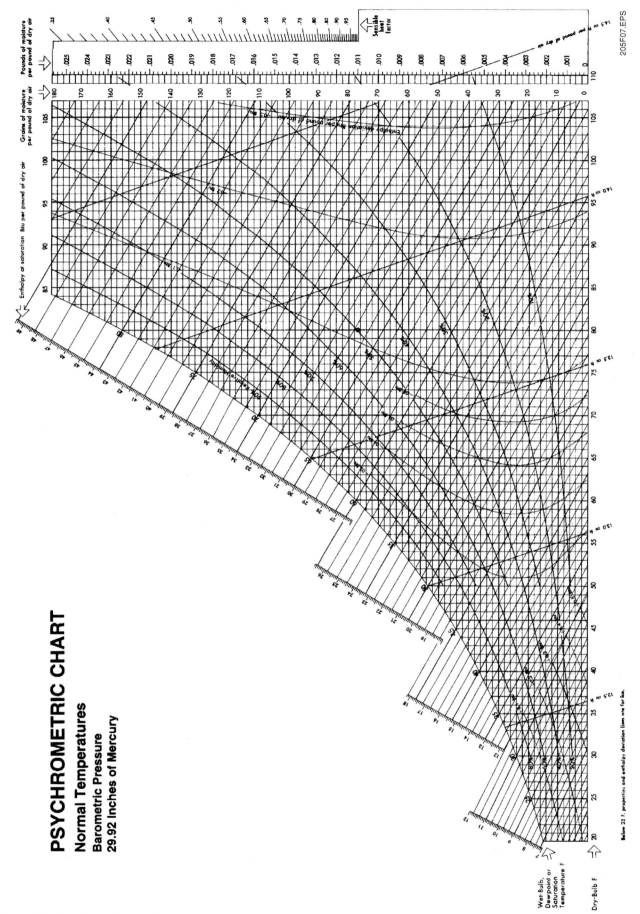

Figure 7 ◆ Psychrometric chart.

air properties at elevations other than sea level to correct for changes in barometric pressure. Psychrometric charts that use metric measurements also are available.

4.1.0 Scales on the Psychrometric Chart

In this section, the scales found on the psychrometric chart are shown and discussed as individual figures so you can easily see how to read each scale. Study each figure to understand the scale and then compare each scale to its position on the psychrometric chart.

4.1.1 Dry-Bulb Temperature Scale

The dry-bulb temperature scale is laid out horizontally (*Figure 8A*). The temperatures, which range from 20°F to 110°F, are read on the vertical lines. The temperature markings are read from left (lowest temperature) to right (highest temperature).

4.1.2 Grains of Moisture Scale (Specific Humidity)

The grains of moisture scale, shown in *Figure 8B*, is also called the specific humidity scale. It is used to find the amount of water vapor mixed with each pound of air. This scale ranges from 0 to 180 grains of moisture per pound of dry air. Refer again to the psychrometric chart. Located to the right of the grains of moisture scale is another scale. This scale, which ranges from 0 to 0.025 pounds, is used to find the pounds of moisture per pound of dry air.

4.1.3 Relative Humidity Scale

The relative humidity scale uses curved lines to represent relative humidity (*Figure 8C*). This scale ranges from 0 to 100 percent relative humidity in increments of 10 percent.

4.1.4 Dew Point and Wet-Bulb Temperature Scales

The dew point temperature scale (*Figure 8D*) and the wet-bulb temperature scale (*Figure 8E*) appear in the same spot on the psychrometric chart. Both scales range from 20°F to 85°F. The difference is in how the lines run. The lines on the dew point temperature scale run horizontally. The lines on the wet-bulb temperature scale slant at about a 45-degree angle. Both scales are also used for saturation (100 percent relative humidity) temperatures.

4.1.5 Enthalpy Scale

The lines on the enthalpy scale extend to the left beyond the psychrometric chart (*Figure 8F*). You can find an approximate value for enthalpy by extending a wet-bulb temperature line past the 100 percent saturation line on the psychrometric chart. The enthalpy scale ranges from 7 Btu/lb to 49 Btu/lb.

Recall that enthalpy is the total heat in the air at 100 percent saturation. Therefore, if the air is not completely saturated, a slight error is present in the enthalpy reading. When complete accuracy is needed, the enthalpy reading can be corrected using the enthalpy deviation lines on the psychrometric chart. The enthalpy deviation lines are used chiefly for design work, where precise measurement is required. Technicians generally do not use them for field service work.

4.1.6 Specific Volume Scale

The lines on the specific volume scale range from 12.5 to 14.5 cubic feet per pound (*Figure 9A*). These lines represent the space occupied in cubic feet per pound of dry air. Refer to the psychrometric chart. Run your finger along the dry-bulb temperature line at the bottom of the chart. You will see that the specific volume of 1 pound of dry air at 75°F displaces a volume of 13.5 cubic feet. If the air is heated to 95°F, it expands and takes up 14 cubic feet. If the air is cooled to 35°F, it occupies only 12.5 cubic feet. This effect of temperature on volume is an example of Charles's law: the higher the temperature, the greater the volume of air; the lower the temperature, the smaller the volume of air. Technicians use specific volume mainly to determine performance and select motor sizes for fans and blowers.

4.1.7 Sensible Heat Factor Scale

The lines on the sensible heat factor scale represent the percentage of sensible heat in the air (*Figure 9B*). The scale, which is located to the far right of the psychrometric chart, ranges from 0.35 to 1.00 (35 percent to 100 percent). The percentages are read from top (lowest percentage) to bottom (highest percentage).

Technicians use this scale to plot such processes as cooling or dehumidification. Once the process is plotted, the ratio of sensible heat to total heat can be found. Refer again to the psychrometric chart. Run your finger along the dry-bulb temperature line at the bottom of the chart until you find 80°F. Now run your finger up the chart until you find the 50 percent relative humidity line. You will see a dot. You can use this dot, together with the sensible heat factor scale, to find the dew point when you know the sensible heat factor.

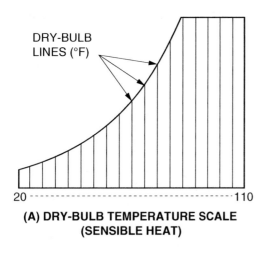

(A) DRY-BULB TEMPERATURE SCALE (SENSIBLE HEAT)

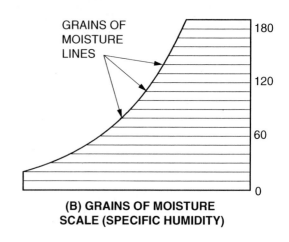

(B) GRAINS OF MOISTURE SCALE (SPECIFIC HUMIDITY)

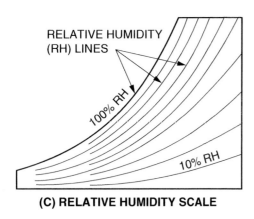

(C) RELATIVE HUMIDITY SCALE

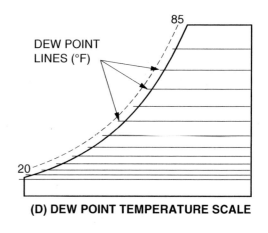

(D) DEW POINT TEMPERATURE SCALE

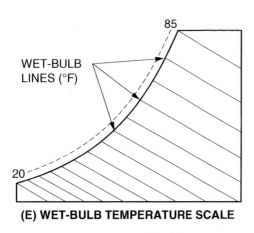

(E) WET-BULB TEMPERATURE SCALE

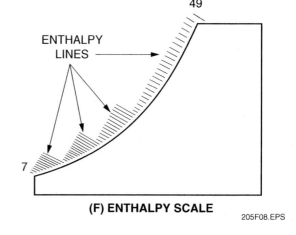

(F) ENTHALPY SCALE

205F08.EPS

Figure 8 ◈ Scales on the psychrometric chart.

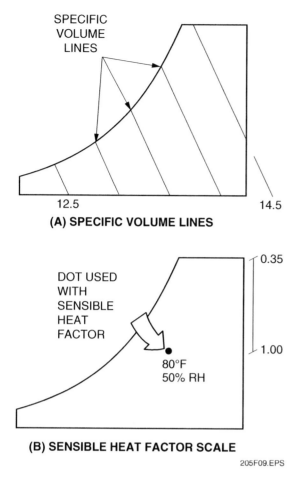

(A) SPECIFIC VOLUME LINES

(B) SENSIBLE HEAT FACTOR SCALE

205F09.EPS

Figure 9 ◆ Specific volume and sensible heat factor scales.

4.2.0 Using the Psychrometric Chart

The psychrometric chart can be used to show what is happening during a specific air conditioning process. Engineers use it to design air distribution systems, and technicians use it to service those systems. You will use the main psychrometric chart together with the plot charts included with the examples in this section.

4.2.1 Using Known Values to Find Unknown Values

Figure 10 is a simplified psychrometric chart with lines plotted for the following air properties: dry- and wet-bulb temperatures, dew point temperature, specific humidity, and relative humidity. If the values for any two of these air properties is known, you can find the values for all of the other properties on the main psychrometric chart.

For example, assume that for a given sample of air the dry-bulb temperature is 75°F and the wet-bulb temperature is 65°F. Knowing these two values, you can find the relative humidity, the specific humidity, and the dew point temperature. Refer to the main psychrometric chart to plot the dry- and wet-bulb temperatures. Remember that dry-bulb

temperature lines run vertically and wet-bulb temperature lines slant at about a 45-degree angle. You will find that the relative humidity is about 58 percent, specific humidity is 76 grains, and the dew point temperature is 59.5°F. A plot of this example is shown in *Figure 11*. Note that the known values are shown in boldface type.

4.2.2 Finding Changes in Sensible Heat

In an air distribution system, an increase in sensible heat occurs when air passes over a heating coil. A decrease in sensible heat occurs when air passes over a cooling coil. You can use the psychrometric chart to plot the changes in air properties as a result of an increase or decrease in heat.

Study the plot for a sensible heat increase shown in *Figure 12*. In this example, air at 30°F dry-bulb temperature and 80 percent humidity is heated to 75°F. You know two values: dry-bulb temperature (before and after heating) and existing relative humidity. Therefore, you can use the main psychrometric chart to find the values for

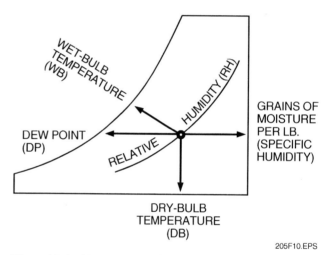

Figure 10 ◆ Air properties plotted on a simplified psychrometric chart.

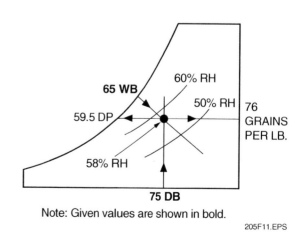

Note: Given values are shown in bold.

205F11.EPS

Figure 11 ◆ Using known values to find unknown values.

wet-bulb temperature, new relative humidity, and dew point temperature. Notice that the relative humidity decreases from 80 percent to 15 percent. Recall that as air becomes warmer it can hold more moisture; therefore, the ratio of the moisture in the air to the total amount it can hold decreases. Notice, also, that the dew point temperature does not change because no water vapor has been added. Therefore, the constant temperature of the dew point is plotted in a straight line from point 1 to point 2.

Study the plot for a sensible heat decrease shown in *Figure 13*. In this example, air at 95°F dry-bulb temperature and 75°F wet-bulb temper-

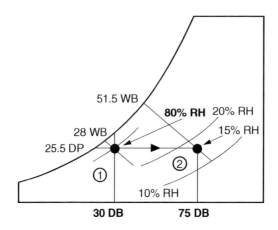

	DRY-BULB (DB) TEMPERATURE °F	WET-BULB (WB) TEMPERATURE °F	RELATIVE HUMIDITY (RH) %	DEW POINT (DP) °F
AIR AT	**30**	28	**80**	25.5
IS HEATED TO	**75**	51.5	15	25.5

Note: Given values are shown in bold.

205F12.EPS

Figure 12 ◈ Plotting a sensible heat increase.

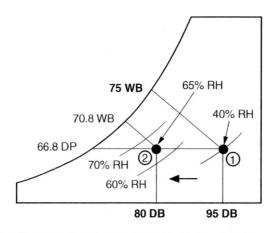

	DRY-BULB (DB) TEMPERATURE °F	WET-BULB (WB) TEMPERATURE °F	RELATIVE HUMIDITY (RH) %	DEW POINT (DP) °F
AIR AT	**95**	**75**	40	66.8
IS COOLED TO	**80**	70.8	65	66.8

Note: Given values are shown in bold.

205F13.EPS

Figure 13 ◈ Plotting a sensible heat decrease.

ature is cooled to 80°F dry-bulb temperature. You know two values: dry-bulb temperature (before and after cooling) and wet-bulb temperature. Therefore, you can use the main psychrometric chart to find the values for the changes in wet-bulb temperature, relative humidity, and dew point temperature. Notice that when the air temperature drops, its relative humidity increases from 40 percent to 65 percent. Notice, also, that the dew point temperature remains the same because no water vapor has been condensed. Therefore, the constant temperature of the dew point is plotted in a straight line from point 1 to point 2.

4.2.3 Finding Change in Enthalpy (Total Heat)

Recall that technicians use a value called *change in enthalpy* to calculate how many Btus per hour the air temperature has increased or decreased. In this example, you will see how to calculate the capacity of an **evaporator.** A plot of the dry- and wet-bulb temperatures for air entering and leaving the evaporator is shown in *Figure 14.* Notice that the wet-bulb temperatures of 51°F and 61.5°F are extended out to the enthalpy scale. These temperatures correspond to enthalpy scale readings of 20.8 Btu/lb and 27.5 Btu/lb, respectively. The dif-

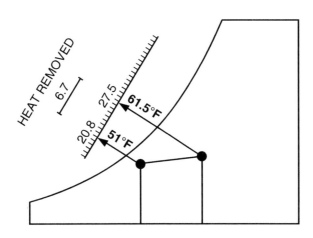

	WET-BULB (WB) TEMPERATURE °F	ENTHALPY BTU/LB	AIR FLOW VOLUME (CFM)
AIR ENTERING EVAPORATOR	**61.5**	27.5	**1,000**
AIR LEAVING EVAPORATOR	**51**	20.8	**1,000**

Note: Given values are shown in bold.

CAPACITY (TOTAL HEAT) BTU/HR = $4.5 \times$ CFM $\times \Delta$ H

WHERE:
4.5 = CONSTANT
CFM = VELOCITY OF AIR FLOW IN CUBIC FEET PER MINUTE
ΔH = CHANGE IN ENTHALPY IN BTU/LB

CAPACITY = $4.5 \times 1,000 \times 6.7 = 30,150$ BTU/HR

CALCULATION FOR CONSTANT:
POUNDS OF AIR PER CUBIC FOOT $\times$ 60 MINUTES PER HOUR
0.075×60 MIN./HR. = 4.5

205F14.EPS

Figure 14 ◆ Using enthalpy to calculate evaporator capacity.

ference between the enthalpy scale readings is the amount of heat removed (27.5 – 20.8 = 6.7 Btu/lb). Once you have the value of enthalpy, you can calculate the capacity of the evaporator in Btus per hour by using the following formula:

Capacity (Btu/hr) =
(specific density)(60 min./hr.)(cfm)(ΔH)

Where:

(specific density)(60 min./hr.) = (0.075)(60) = 4.5

cfm = velocity of airflow in cubic feet per minute

ΔH = change in enthalpy (Btu/lb)

(Note: 4.5 is a constant that is used when determining change in enthalpy.)

Example

How many Btu/hr (total heat) are removed if 1,000 cfm of air is circulated over the evaporator?

Capacity (Btu/hr) =
(specific density)(60 min./hr.)(cfm)(ΔH)

Capacity (Btu/hr) = (4.5)(1,000)(6.7)

Capacity = 30,150 Btu/hr

5.0.0 ◆ THE AIR DISTRIBUTION SYSTEM

The efficiency of an air conditioning system depends on the efficiency of the air distribution system. An efficient air distribution system has the following characteristics:

- Supplies the right amount of air to all conditioned spaces
- Minimizes drafts
- Maintains good comfort zones for occupants or maintains the proper air conditions required for a business or for a manufacturing process
- Provides for the return of air from conditioned areas to the air-handling unit
- Is energy efficient and quiet
- Requires minimum maintenance

Many air distribution systems are **constant air volume (CAV) systems.** A CAV system provides a constant airflow while varying the temperature to meet heating and cooling needs.

In a typical air distribution system, outside (or makeup) air is mixed with conditioned air in the air supply system. Grilles, registers, and diffusers allow air to flow into occupied spaces and out through the return air system.

In many commercial applications, the air-handling unit, which contains the cooling and heating coils, fan or blower, filters, air mixing section, and dampers, is suspended from the roof. These over-head air distribution systems are typically used in buildings on concrete slabs. Ducts branch out from the air-handling unit to deliver conditioned air to occupied spaces and return air to the unit. Attic **plenum** duct systems are installed in buildings with attic areas. Attic installations eliminate the need for drop ceilings or soffit enclosures, resulting in more usable space in the building. In the system shown in *Figure 15*, each room air outlet is connected to a central supply air plenum on the air-handling unit through ducts called runouts. Flexible, pre-insulated, round ductwork is most often used for the runouts, but fiberglass duct board and round sheet metal ducts with insulating vapor barrier sleeves also are used.

5.1.0 Airflow in Ductwork

To understand how air moves in an air distribution system, let's compare it to a plumbing system. In a plumbing system, water will not move unless there is a difference in pressure. For example, a pump moves water by creating a vacuum in the pipe running from a well to the surface of the ground. Atmospheric pressure pushing down on the well water forces the water up the pipe. Here is a simplified example of this concept at work. Place a straw in a soft drink can and take a sip. You may think that you are drawing the soda up by sucking on the straw, but you are actually changing the air pressure inside the straw. When you suck on the straw, you remove the air inside it, creating a vacuum. Air pressing down on the liquid in the can forces it up the straw to fill the vacuum. The same principle is at work in an air distribution system. Air in the system can be moved only by creating a difference in air pressure. This means that a fan or blower operates like a pump, but instead of water, air is moved.

Air moves through the air distribution system in a circular fashion. Study *Figure 16* for a simple example of how changes in air pressure affect air-

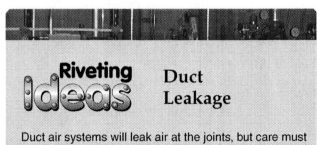

Riveting Ideas Duct Leakage

Duct air systems will leak air at the joints, but care must be taken to limit the amount of air leakage to acceptable levels. Too much air leakage will result in a poorly performing system. A careful installation will limit duct leakage to five percent or less of the total cfm.

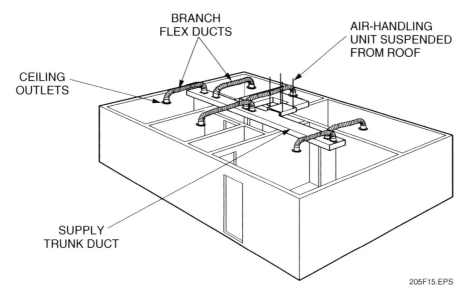

Figure 15 ◈ Attic plenum duct system.

flow. At the return air grille, the air pressure is slightly lower than the atmospheric pressure in the room; therefore, air flows into the duct. The air pressure decreases to its lowest point at the intake of the blower. The action of the blower increases air pressure, pulling air through the blower and into the duct. Air pressure is at its highest level at the blower discharge. The air pressure causes the air to flow away from the blower into the duct, which delivers it to the conditioned space. Then the cycle begins again.

5.2.0 Air Velocity and Volume in Ductwork

Designers of HVAC systems must calculate the amount of pressure needed to move air through the duct system. The amount of air pressure required depends on the air velocity and air volume.

Air velocity within a duct is not uniform. It varies from zero at the duct walls to a maximum speed in the center of the duct. This variation in

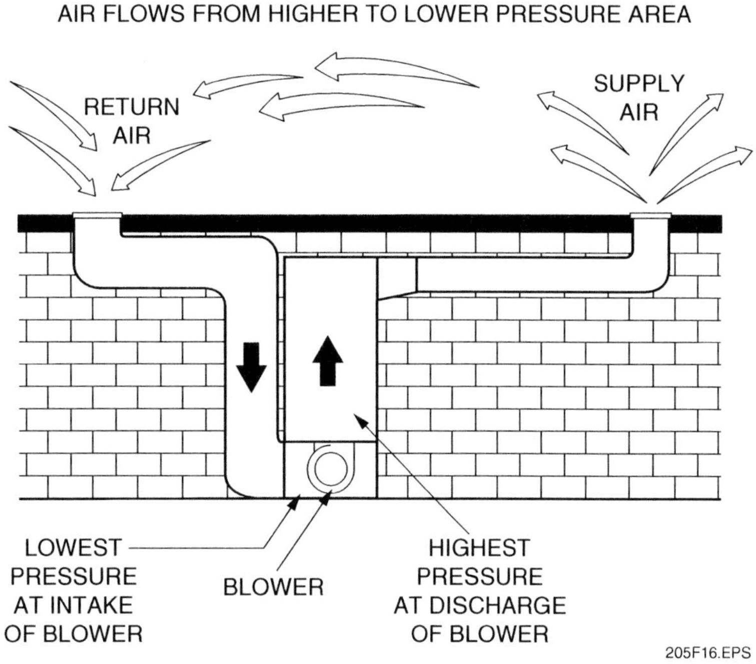

Figure 16 ◈ Airflow pressure in an air distribution system.

speed is caused by friction, which occurs as the air molecules are dragged over the duct surfaces. Velocity is also affected by turbulence, which occurs where there are changes in the size or direction of the duct (*Figure 17*). Velocity is measured in feet per minute (fpm).

Air volume is a measure of the amount of air in cubic feet that flows past a point in one minute. You can calculate the volume of air in cfm by multiplying the air velocity in fpm by the area through which it moves. The formula for this calculation is:

cfm = (area)(velocity)

If you are solving for velocity, the formula is:

$$velocity = \frac{cfm}{area}$$

If you are solving for area, the formula is:

$$area = \frac{cfm}{velocity}$$

Example

Determine the cfm for a 12" × 12" duct with a velocity of 1,000 fpm.

cfm = (1')(1')(1,000 fpm)

cfm = 1,000

5.3.0 Pressures in Ductwork

Three pressures exist in ductwork. They are as follows:

- **Static pressure** – This pressure is exerted uniformly in all directions inside the ductwork. In a supply air duct, static pressure is the bursting or exploding pressure that acts on all internal surfaces of the duct.

Riveting Ideas

Internal and External Static Pressure Loss

The air-handling unit is subject to both internal static pressure loss and external static pressure loss. The manufacturer of the air-handling unit accounts for internal static pressure loss. Therefore, in the field, you will be concerned only with external static pressure loss. External static pressure loss is the total pressure loss of an air distribution system, excluding the air-handling unit. Technicians usually must size the duct to match the external static pressure of the selected equipment.

- **Velocity pressure** – The movement of air in the duct system creates velocity pressure. Velocity pressure acts only in the direction of the airflow.
- **Total pressure** – The fan or blower produces total pressure. Total pressure is the sum of static pressure and velocity pressure.

You can see the mathematical relationship among these pressures in *Figure 18*.

The levels of the static, velocity, and total pressures in ductwork are quite small. Therefore, to be accurate, the scale used to measure them must be numerically large. To measure static, velocity, and total pressure, technicians use a measurement known as inches of water column (inches wc). This is the height, in inches, to which the pressure will lift a column of water. Recall that the atmosphere exerts a pressure of approximately 14.7 psi

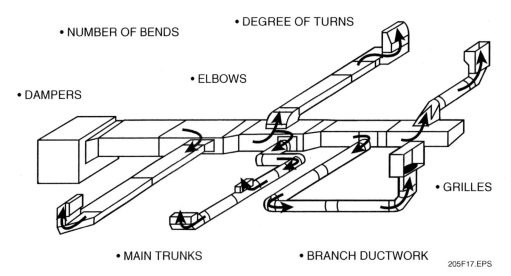

• NUMBER OF BENDS • DEGREE OF TURNS
• ELBOWS
• DAMPERS
• GRILLES
• MAIN TRUNKS • BRANCH DUCTWORK

205F17.EPS

Figure 17 ◆ Turbulence occurs at changes in duct size or direction.

(14.696) at sea level at 70°F. This amount of pressure will support a column of water 33.9 feet or 406.8 inches high (*Figure 19*). Therefore, for every 1 pound per square inch of pressure, a column of water will rise to a height of 27.68 inches (406.8 ÷ 14.696) or about 2.3 feet.

5.4.0 Air Distribution in a Cooling System

You have seen a simplified example of how air flows through an air distribution system. This section contains a more detailed example that includes measurements showing air pressure and airflow in cfm (*Figure 20*).

Generally, more airflow is needed to support the cooling mode than is needed for the heating mode. This example assumes a requirement of 3 tons of cooling. A rule of thumb used in HVAC work is that cooling requires about 400 cfm ±50 cfm of air per ton of cooling. Therefore, the air-handling unit in this example must supply air at a minimum of 1,200 cfm (3 × 400 cfm). For this example, we will make the following assumptions:

- The static pressure of the blower is 0.4 inches wc.
- The static pressure of the supply duct is 0.2 inches wc.
- The static pressure of the two return ducts is –0.2 inches wc.
- The system has six air outlets, each requiring 200 cfm.

Because the building itself supplies the necessary atmospheric pressure to force air into the return grilles, it is considered part of the air distribution system. The air pressure at the grilles (–0.03 inches wc) is lower than the atmospheric pressure of the building (0.03 inches wc), so air is forced into the return grilles. As the air flows toward the blower, the air pressure continues to decrease as a result of friction losses in the duct. At the blower inlet, the air pressure is at –0.2 inches wc, the lowest pressure in the system. At the blower outlet, the air pressure increases to its highest level (0.2 inches wc), which is well above the building's pressure. The difference in static pressure between the inlet and outlet of the blower is 0.4 inches wc.

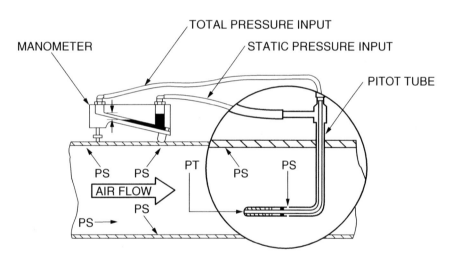

THE PITOT TUBE SENSES TOTAL AND STATIC PRESSURES. THE MANOMETER MEASURES VELOCITY PRESSURE (DIFFERENCE BETWEEN TOTAL AND STATIC PRESSURES)

PT = TOTAL PRESSURE
PS = STATIC PRESSURE
PV = VELOCITY PRESSURE

- STATIC PRESSURE = TOTAL PRESSURE MINUS VELOCITY PRESSURE
  PS = PT – PV

- TOTAL PRESSURE = STATIC PRESSURE PLUS VELOCITY PRESSURE
  PT = PS + PV

- VELOCITY PRESSURE = TOTAL PRESSURE MINUS STATIC PRESSURE
  PV = PT – PS

205F18.EPS

Figure 18 ◈ Static, velocity, and total pressures.

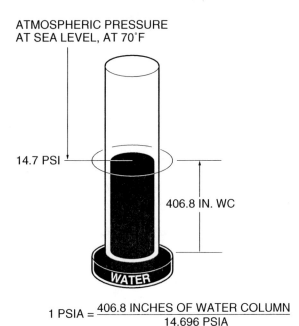

ATMOSPHERIC PRESSURE
AT SEA LEVEL, AT 70°F

14.7 PSI

406.8 IN. WC

WATER

$$1 \text{ PSIA} = \frac{406.8 \text{ INCHES OF WATER COLUMN}}{14.696 \text{ PSIA}}$$

1 PSIA = 27.68 INCHES OF WATER COLUMN

205F19.EPS

Figure 19 ◆ Atmospheric pressure compared to inches of water column.

The air at the blower outlet is pushed through the cooling coil, where it encounters a pressure drop of 0.1 inches wc. After the air enters the supply duct, the air pressure reduces slightly at the tee where the duct splits into two reducing trunks that supply air to opposite ends of the building. The trunks are called reducing trunks because they reduce in size as they move away from the tee. Reducing the duct size as air is distributed off the main trunk keeps the pressure in the duct system at the desired level throughout the duct.

In *Figure 20*, both trunks reduce in size three times, and there are branch outlets (one on each side) in each section, for a total of six outlets. The outlets take part of the volume of air supplied in each section. Study one side of the trunk to see how each 200 cfm outlet uses part of the volume of air as follows:

- *First section* – The first 200 cfm outlet reduces the 600 cfm supplied to one side of the trunk to 400 cfm.
- *Second section* – The second 200 cfm outlet reduces the remaining 400 cfm to 200 cfm.

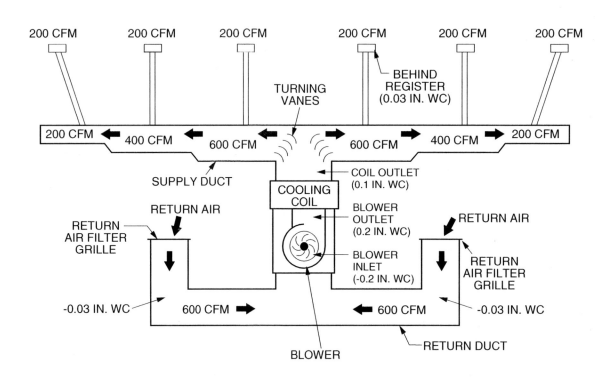

SYSTEM CAPACITY 3 TONS
CFM REQUIREMENT 400 CFM PER TON = 400 × 3 = 1,200 CFM
BLOWER STATIC PRESSURE (0.4 IN. WC)
SUPPLY DUCT STATIC PRESSURE (0.2 IN. WC)
RETURN DUCT STATIC PRESSURE (-0.2 IN. WC)

205F20.EPS

Figure 20 ◆ Measuring airflow and pressure in an air distribution system.

- *Third section* – The third 200 cfm outlet uses the remaining 200 cfm.

The same thing happens on the other side of the trunk. Therefore, the total 1,200 cfm is distributed evenly throughout the system.

In an actual system, technicians would install dampers in each branch to adjust the quantity of air supplied to each room. The system in our example will furnish 200 cfm to each outlet, but if a room does not need that much air, the dampers can be adjusted to reduce the quantity of air supplied.

6.0.0 ◈ BLOWERS AND FANS

In HVAC work, the terms *blower* and *fan* are often used interchangeably. Generally speaking, though, these words are used to describe specific applications. For example, blower usually describes applications where the air-moving device must overcome the resistance caused by friction in a duct system and fan usually describes applications where high quantities of air are needed.

6.1.0 Belt-Drive and Direct-Drive Blowers

In a belt-drive blower, the motor is connected to the blower by a belt and pulley. The speed is adjusted mechanically by a change in the pulleys. In a direct-drive blower, the blower wheel is mounted directly on the motor shaft. The speed is adjusted electrically by changing the motor terminals or by changing the settings of motor speed selection switches on a motor control board. A belt-drive and a direct-drive blower are shown in *Figure 21*.

DID YOU KNOW?

Friction and Turbulence

Friction and turbulence are naturally occurring forces. Each has an effect on air velocity inside an air distribution system. Friction occurs when two surfaces rub against one another. It slows the motion of an object or a gas and, in some cases, prevents motion entirely. Air moving against the surface of a duct creates friction. Turbulence breaks up the forward motion of air into small currents that then move in a circular fashion. Air flowing in a straight line through a duct is affected by turbulence when the duct changes size or direction.

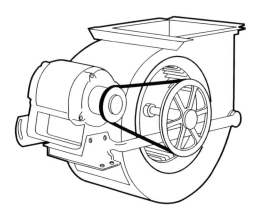

BELT-DRIVE BLOWER

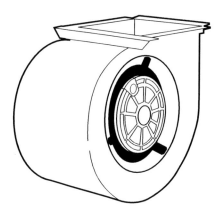

DIRECT-DRIVE BLOWER

205F21.EPS

Figure 21 ◈ Belt-drive and direct-drive blowers.

6.2.0 Centrifugal Blowers

Centrifugal blowers are designed to overcome the resistance caused by friction in the duct system (*Figure 22*). They can be used in large high-pressure systems with pressures of one inch wc or greater. The airflow is at right angles to the shaft on which the wheel is mounted. The wheel is mounted in a scroll-type housing that allows the blower to develop the needed pressures. Centrifugal blowers are classified by the wheel blade position with respect to the direction of rotation. Types of centrifugal blowers include the following:

- Forward-curved
- Backward-curved
- Radial

6.2.1 Forward-Curved Blower

The forward-curved blower is used mostly in residential heating and air conditioning systems. It also is used in light-duty exhaust systems where maximum air delivery and low noise are required.

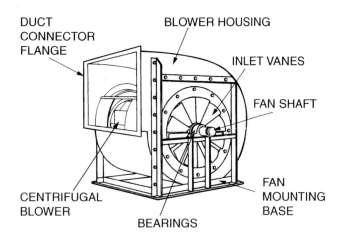

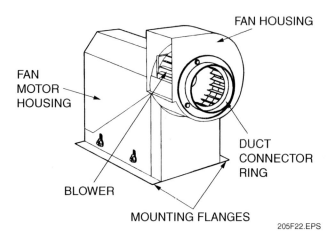

Figure 22 ◆ Centrifugal blowers.

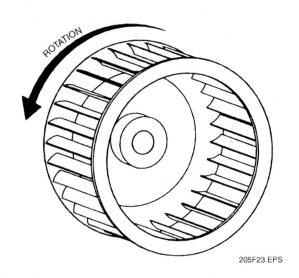

Figure 23 ◆ Forward-curved centrifugal blower blades.

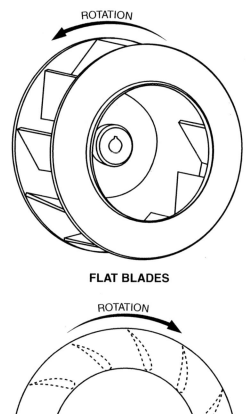

FLAT BLADES

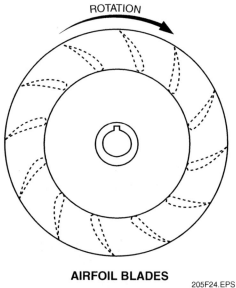

AIRFOIL BLADES

Figure 24 ◆ Flat and airfoil blades.

Typically, this type of blower can handle pressures up to 1½ inches wc. The tips of the blades in a forward-curved blower slant in the direction of rotation (*Figure 23*).

6.2.2 Backward-Curved Blower

The blades of the backward-curved blower slant away from the direction of rotation. This type of blower is used in commercial and industrial heating and cooling systems, which require heavy-duty construction and stable air delivery. It is also used as a ventilator. Backward-curved blowers operate at higher efficiencies, higher speeds, and tend to be noisier than forward-curved blowers. Typically, these blowers can handle pressures up to 3 inches wc.

The wheel on this type of blower may have flat blades or airfoil blades (*Figure 24*). Usually, small wheels have flat blades and large wheels have airfoil blades. Blowers with airfoil blades generally run more quietly than other types of centrifugal blowers. Also, air flows through the airfoil blade with less turbulence.

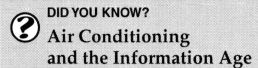
6.2.3 Radial Blower

The wheel on a radial blower has straight blades that are mostly self-cleaning. This feature makes radial blowers suitable for use in air systems with large amounts of particles or grease in the air, such as commercial kitchens. The wheel is simple in construction with narrow blades. It can withstand the high speeds needed to operate at higher static pressures (up to 12 inches wc). Examples of radial blower wheels are shown in *Figure 25.*

6.3.0 Propeller Fans

Propeller fans produce an airflow that is parallel to the shaft on which the propeller is mounted (*Figure 26*). These fans are efficient with nearly **free air delivery.** This means that there is no effective restriction to airflow. Propeller fans usually are mounted in a **venturi** to cause the air to flow in a straight line from one side of the fan to the other. A venturi is a ring or panel surrounding the blades on a propeller fan. To achieve the best performance from a propeller fan, the blade must be properly set in the venturi opening. The setting must be as specified by the manufacturer or performance will drop off. Propeller fans make more noise than centrifugal blowers, and are therefore installed in applications where noise is not an important consideration.

6.4.0 Duct Fans

In duct fans, airflow is parallel to the shaft on which the wheel is mounted. The propeller is housed in a cylindrical duct or tube. This design allows duct fans to operate at higher static pres-

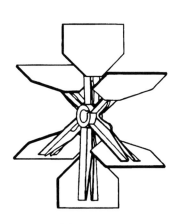

OPEN-TYPE

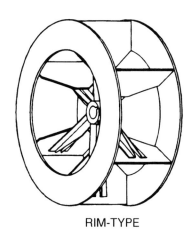

RIM-TYPE

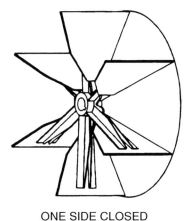

ONE SIDE CLOSED

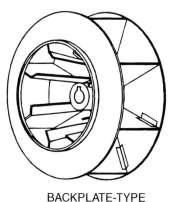

BACKPLATE-TYPE

205F25.EPS

Figure 25 ◆ Examples of radial blower wheels.

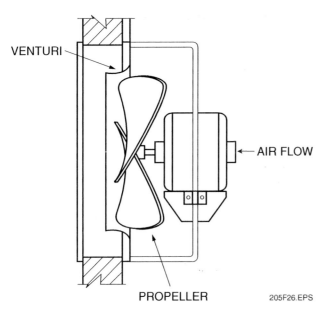

Figure 26 ◆ Propeller fan.

sures than propeller fans. Duct fans are used in spray booths and ducted exhaust systems. A spray booth is an enclosed room, containing at least three sides and an exhaust fan, used for painting. A fan is considered ducted if the duct length is more than the distance between the inlet to and the outlet from the fan blades.

Two types of ducted fans are used in HVAC work: tube-axial fans and vane-axial fans. A tube-axial fan discharges air with a screw-like motion. A vane-axial fan discharges air in a straight line. Both the tube-axial and vane-axial types of duct fans are shown in *Figure 27*.

7.0.0 ◆ FAN LAWS

The performance of fans and blowers is governed by three rules commonly known as the fan laws. The forces generated by fans and blowers are all related. Therefore, when one force changes, so

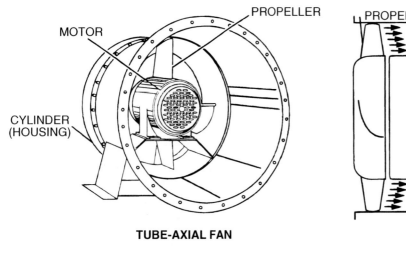

TUBE-AXIAL FAN

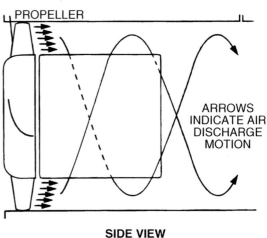

SIDE VIEW

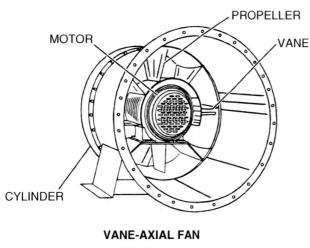

VANE-AXIAL FAN

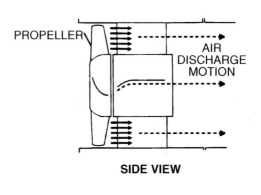

SIDE VIEW

205F27.EPS

Figure 27 ◆ Tube-axial and vane-axial duct fans.

will the others. The forces generated by fans and blowers are as follows:

- Cubic feet per minute (cfm)
- Revolutions per minute (rpm)
- Static pressure (sp)
- Horsepower (hp)

The fan laws can be stated in words or mathematically. Both a word definition and a mathematical definition are provided.

Law 1 – The amount of air delivered by a fan varies directly with the speed of the fan.

$$\text{New cfm} = \frac{(\text{new rpm} \times \text{existing cfm})}{\text{existing rpm}}$$

or

$$\text{New rpm} = \frac{(\text{new cfm} \times \text{existing rpm})}{\text{existing cfm}}$$

Law 2 – The static pressure of a system varies directly with the square of the ratio of the fan speed.

$$\text{New sp} = (\text{existing sp})\left(\frac{\text{new rpm}}{\text{existing rpm}}\right)^2$$

Law 3 – The horsepower of a fan varies directly with the cube of the ratio of the fan speed.

$$\text{New hp} = (\text{existing hp})\left(\frac{\text{new rpm}}{\text{existing rpm}}\right)^3$$

Example

Assume the following existing system conditions:

- 5,000 cfm
- 1,000 rpm
- sp of 0.5 inches wc
- hp of 0.5

With an increase in airflow to 6,000 cfm, what are the new rpm, sp, and hp?

Use Law 1 to calculate the new rpm:

$$\text{New rpm} = \frac{(\text{new cfm} \times \text{existing rpm})}{\text{existing cfm}}$$

$$\text{New rpm} = \left(\frac{6,000 \times 1,000}{5,000}\right)$$

$$\text{New rpm} = 1,200$$

Use Law 2 to calculate the new sp:

$$\text{New sp} = (\text{existing sp})\left(\frac{\text{new rpm}}{\text{existing rpm}}\right)^2$$

$$\text{New sp} = (0.5)\left(\frac{1,200}{1,000}\right)^2$$

$$\text{New sp} = 0.72 \text{ inches wc}$$

Use Law 3 to calculate the new hp:

$$\text{New hp} = (\text{existing hp})\left(\frac{\text{new rpm}}{\text{existing rpm}}\right)^3$$

$$\text{New sp} = (0.5)\left(\frac{1,200}{1,000}\right)^3$$

$$\text{New hp} = 0.864$$

7.1.0 Fan Curve Charts

You can also use manufacturers' fan curve charts to determine values for the forces generated by fans and blowers (cfm, rpm, sp, and hp). A typical fan curve chart is shown in *Figure 28*. If you know the values for any two of the forces shown on the chart, you can easily find an unknown value. For example, assume that the sp is 1.4 inches wc and the fan is running at 900 rpm. To find the cfm, locate the spot where the sp line and the rpm curve meet. From this point, drop down vertically to the cfm scale, and read the value of 7,500 cfm.

8.0.0 ◆ MEASUREMENT INSTRUMENTS

To properly install, balance, and service HVAC equipment, technicians rely on a variety of measurement instruments. You learned about dry- and wet-bulb thermometers earlier in this module. In this section, you will learn about the instruments in which they are used. Many of the measurement instruments covered in this section are available in several styles. Some have built-in features such as charts, calculators, digital readouts, and printers. The instructions for setting up, operating, and caring for these instruments will be different for each manufacturer. Always read and follow the manufacturer's instructions.

DID YOU KNOW?

A 15th-Century Ventilator

In addition to being an artist, Leonardo da Vinci was an inventor. He designed a ventilator for the palace of the Duke of Milan. The ventilator was a huge wheel turned by waterpower. Valves that opened and closed automatically drew air into the drum of the wheel, where it was cooled by water. Air pressure generated by the wheel (which acted like a giant fan) forced the cooled air into a hollow shaft, which was connected to the palace by pipes. This giant ventilator, a full story in height, cooled a single room in the palace: the bedroom of the Duke's wife.

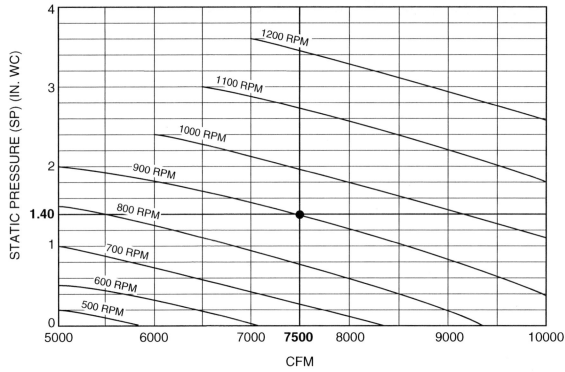

Figure 28 ◆ Typical fan curve chart.

Technicians take the following types of measurements in an air distribution system:

- Temperature and humidity
- Air pressure
- Air velocity

8.1.0 Temperature and Humidity

The correct temperature and humidity are an important part of indoor comfort, and they play a key role in many businesses and manufacturing processes. The instruments commonly used by technicians to measure temperature and humidity include the electronic thermometer, psychrometer, and hygrometer.

8.1.1 Electronic Thermometer

The electronic thermometer measures dry- and wet-bulb temperatures in air distribution systems and HVAC equipment. It includes either a **thermocouple** or a **thermistor** temperature probe to generate the temperature reading. Often, several different probes are used with one instrument to allow measurement of a wide range of temperatures. These probes also allow you to make measurements at several locations at the same time.

With an electronic thermometer you can calculate the difference in temperature between the locations. Electronic thermometers are available with either an analog meter or digital display (*Figure 29*).

8.1.2 Psychrometer

Technicians use the psychrometer to measure relative humidity. This instrument includes two thermometers: dry-bulb and wet-bulb. As you learned earlier in this module, the dry-bulb thermometer is an ordinary thermometer. The wet-bulb ther-

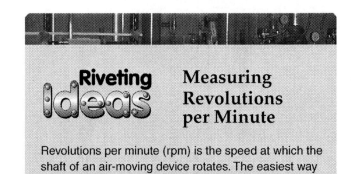

Riveting Ideas

Measuring Revolutions per Minute

Revolutions per minute (rpm) is the speed at which the shaft of an air-moving device rotates. The easiest way to determine rpm is to measure it directly with a tachometer.

Measurement Instruments Versus the Psychrometric Chart

Measurement instruments give a direct read of atmospheric conditions. They are fast and highly accurate. One advantage in using a psychrometric chart instead of a direct-read instrument is that you can use the chart to estimate the changes in value for various properties when adjustments are made to an HVAC system.

mometer is similar to the dry-bulb thermometer except that its sensing bulb is wrapped with a water-soaked wick. Evaporation from the wick causes the wet-bulb temperature to be lower than the dry-bulb temperature. To determine relative humidity, readings taken from a psychrometer are compared to a psychrometric chart. Some psychrometers have a built-in chart.

There are several types of psychrometers. The sling psychrometer is shown in *Figure 30.* To take a reading, the operator spins the tool rapidly. The spinning causes the water on the wick to evapo-

rate. The rate of evaporation depends on the amount of moisture already in the air when a reading is taken.

Squeeze-bulb and battery-operated fan psychrometers are also available. These psychrometers are used when there is not enough room to swing the sling psychrometer. The squeeze-bulb psychrometer is equipped with a hollow rubber bulb. To take a reading, the operator rapidly squeezes the bulb to draw air over the wick. In the battery-operated model, a fan draws air over the wick.

8.1.3 Hygrometer

The hygrometer also measures relative humidity. For field service work, the electronic hygrometer is commonly used (*Figure 31*). Electronic hygrometers measure the temperature and humidity of conditioned air with sensing probes and display the readings digitally. Many types of hygrometers are available. Some also measure and give a direct reading of the dew point. Others can display temperature and relative humidity readings simultaneously.

8.2.0 Air Pressure

Technicians must measure three types of air pressure in an air distribution system: static, velocity, and total. The instruments most commonly used to measure these pressures are the manometer, differential pressure gauge, and the **pitot tube** and **static pressure tips.** Pitot tubes and static pressure tips are frequently used together with manometers and differential pressure gauges.

8.2.1 Manometer

The simplest type of manometer consists of a U-shaped tube filled with either water or mercury. Generally, water is used to measure small pressures and mercury is used to measure large pressures. One end of the manometer is connected to

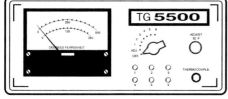

DIGITAL

ANALOG

205F29.EPS

Figure 29 ◆ Digital and analog electronic thermometers.

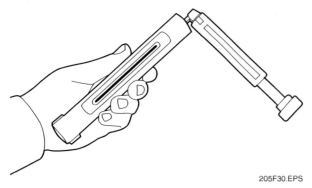

Figure 30 ◆ Sling psychrometer.

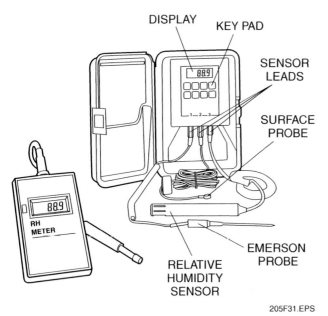

DISPLAY   KEY PAD

SENSOR
LEADS

SURFACE
PROBE

RH
METER

RELATIVE
HUMIDITY
SENSOR

EMERSON
PROBE

205F31.EPS

Figure 31 ◆ Electronic hygrometer.

the location where the pressure reading is to be taken. The other end is left open to atmospheric pressure. The reading is the difference between the pressure being measured and the atmospheric pressure. Like the other measuring instruments used by technicians, manometers come in mechanical and electronic versions. Three types of manometers are shown in *Figure 32.*

8.2.2 Differential Pressure Gauge

The differential pressure gauge provides a direct reading of pressure. This gauge is used to measure fan and blower pressures, filter resistance, air velocity, and furnace draft. Some gauges can measure only pressure; others can measure both pressure and air velocity. Single-scale gauges are calibrated as either inches wc or psi. Dual-scale gauges are calibrated in inches wc for pressure and in fpm for air velocity. Several models are available covering pressures from 0 to 10 inches wc and air velocities from 300 to 12,500 fpm. Usu-

ally these gauges are installed in the air conditioning equipment, but portable gauges also are available. A portable differential pressure gauge is shown in *Figure 33.*

8.2.3 Pitot Tube and Static Pressure Tips

The pitot tube and static pressure tips (*Figure 34*) are probes that you use with manometers and pressure gauges to take measurements inside an air distribution system. Pitot tubes come in lengths ranging from 6 to 60 inches, with graduation marks at every inch to show the depth of insertion in the duct.

The standard pitot tube used for measurements in ducts 8 inches and larger has a $\frac{1}{16}$-inch outer tube with eight equally spaced 0.04-inch diameter holes used to sense static pressure. For measurements in ducts smaller than 8 inches, a pocket-size pitot tube with a $\frac{1}{8}$-inch outer tube and four equally spaced 0.04-inch diameter holes is used.

The pitot tube consists of an impact tube, which receives the total pressure input, fastened inside a larger tube that receives static pressure input from sensing holes. The air space between the inner and outer tubes permits transfer of pressure from the sensing holes to the static pressure connection at the opposite end of the pitot tube. From there, the pressure is transferred through the connecting

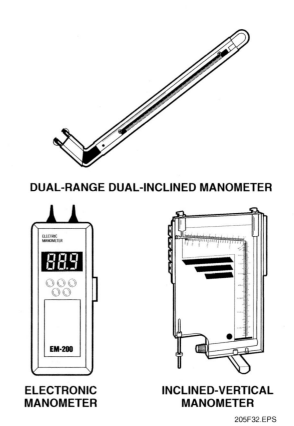

DUAL-RANGE DUAL-INCLINED MANOMETER

ELECTRIC
MANOMETER

EM-200

**ELECTRONIC
MANOMETER**

**INCLINED-VERTICAL
MANOMETER**

205F32.EPS

Figure 32 ◆ Types of manometers.

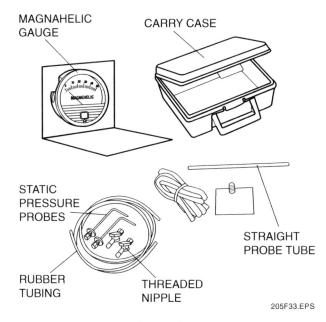

Figure 33 ◈ Portable differential pressure gauge.

tubing to the low- or negative-pressure side of a manometer. When the total pressure tube is connected to the high-pressure side of the manometer, velocity pressure is indicated directly. To ensure accurate velocity pressure readings, the pitot tube tip must be pointed directly into the duct airstream.

Static pressure tips, like pitot tubes, are used with manometers and differential pressure gauges to measure static pressure in a duct system. They are typically L-shaped with four 0.04-inch sensing holes.

8.3.0 Air Velocity

Technicians measure airflow velocity to check the operation of an air distribution system. They also use measurements of air velocity to balance system airflow. The instruments used to measure air velocity include the velometer and the anemometer.

8.3.1 Velometer

A velometer provides direct readings of air velocity in fpm or cfm. Some electronic velometers contain a microprocessor, which enables them to quickly take up to 250 individual readings across a duct area. Some velometers also include an optional microprinter to record the readings.

There are several types of velometer. Swing-vane, rotating vane, flow hood, and air volume balancer velometers are shown in *Figure 35*.

The vane on a rotating vane velometer revolves at different speeds depending on the speed of the airflow. The rate of rotation is converted into an equivalent velocity reading and is displayed on the velometer.

The vane on a swing-vane velometer tilts at different angles depending on the speed of the airflow. The angle of the vane is converted into an equivalent velocity reading and is displayed on the velometer.

The flow hood velometer is used to measure the output of large air diffusers. This type of velometer is used mainly in large commercial air distribution systems.

Special velometers, known as air volume balancers, are used to balance air distribution systems. This type of velometer is held directly against the diffuser or register to get a direct reading of air velocity.

8.3.2 Anemometer

The anemometer gives direct readings of air velocity in fpm. It uses a sensing probe that contains a resistance heater element. The temperature of the element changes in response to changes in air velocity, which triggers a change in the amount of current in the anemometer. The anemometer converts the change in current into an equivalent air velocity and displays it on a meter.

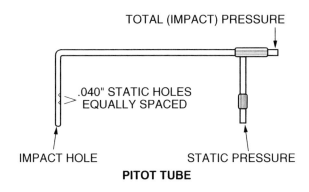

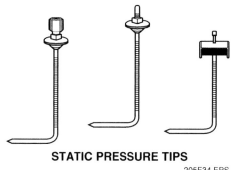

Figure 34 ◈ Pitot tube and static pressure tips.

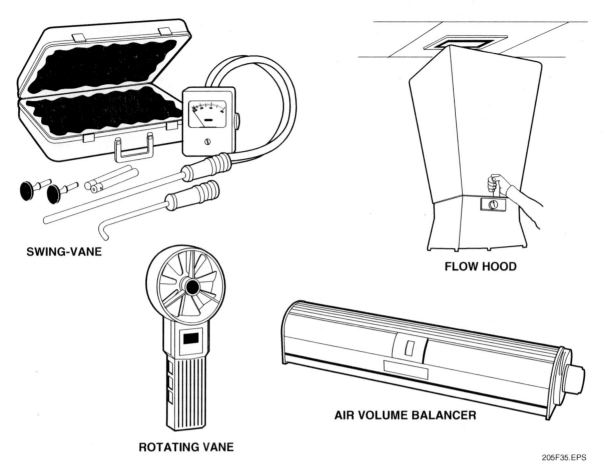

SWING-VANE

FLOW HOOD

ROTATING VANE

AIR VOLUME BALANCER

205F35.EPS

Figure 35 ◈ Examples of velometers.

1. An atmospheric pressure of 14.7 psi at sea level will result in a barometer reading of _____.
   a. 29.92 inches Hg
   b. 24.7 inches Hg
   c. 406.8 inches wc
   d. 14.7 inches Hg

2. A reading of 0 psi on an absolute pressure scale is equal to a gauge pressure scale reading of _____.
   a. 0 psig
   b. 30 inches Hg vacuum
   c. 30 inches Hg
   d. 0 inches Hg vacuum

3. According to Dalton's law, the total air pressure in a container is equal to _____.
   a. the sum of the pressures of the gases in dry air minus water vapor
   b. the barometric pressure at sea level at 70°F
   c. the atmospheric pressure at sea level at 70°F
   d. the sum of the pressures of each gas in the mixture

4. Dry air consists of _____.
   a. 78% nitrogen, 21% oxygen, and 1% water vapor
   b. 78% oxygen, 21% nitrogen, and 1% other gases
   c. 78% oxygen, 21% nitrogen, and 1% water vapor
   d. 78% nitrogen, 21% oxygen, and 1% other gases

5. At sea level at 70°F, the specific density of air is _____.
   a. 0.075 pounds per cubic foot
   b. 13.33 pounds per cubic foot
   c. 0.24 pounds per cubic foot
   d. 0.75 pounds per cubic foot

6. When the dry-bulb, wet-bulb, and dew point temperatures are all the same, _____.
   a. specific humidity is 50 grains of moisture per pound of dry air
   b. relative humidity is 50 percent
   c. specific humidity is 100 grains of moisture per pound of dry air
   d. relative humidity is 100 percent

7. Plotting a 95°F dry-bulb and a 75°F wet-bulb temperature on a psychrometric chart shows that the relative humidity is _____.
   a. 30 percent
   b. 40 percent
   c. 50 percent
   d. 60 percent

8. Plotting a 95°F dry-bulb and a 75°F wet-bulb temperature on a psychrometric chart shows that the dew point temperature is _____.
   a. 75°F
   b. 65°F
   c. 67°F
   d. 62°F

9. Plotting a 95°F dry-bulb and a 75°F wet-bulb temperature on a psychrometric chart shows that the specific humidity is _____.
   a. 70 grains of moisture per pound of dry air
   b. 100 grains of moisture per pound of dry air
   c. 120 grains of moisture per pound of dry air
   d. 65 grains of moisture per pound of dry air

10. Plotting a 95°F dry-bulb and a 75°F wet-bulb temperature on a psychrometric chart shows that the enthalpy of the air is _____.
    a. 24.7 Btu/lb
    b. 38.6 Btu/lb
    c. 40.6 Btu/lb
    d. 38.0 Btu/lb

11. In an air distribution system, the highest air pressure level is found _____.
    a. at the return air grille
    b. between the return air grille and the blower
    c. at the intake to the blower
    d. at the discharge of the blower

12. The static pressure, velocity pressure, and the total pressure measured in a duct system are measured in _____.

    a. psi
    b. inches of mercury
    c. inches of water column
    d. inches of mercury vacuum

13. _____ can be used in large high-pressure systems with pressures of one inch wc or greater.

    a. Centrifugal blowers
    b. Propeller fans
    c. Duct fans
    d. Belt-drive blowers

14. According to Fan Law 2, the static pressure of a fan's output varies _____.

    a. as the ratio of the square of the fan's speed
    b. directly as the fan motor horsepower
    c. as the ratio of the cube of the fan's speed
    d. directly as the speed of the fan

15. Of the following measurement instruments, the one used to determine relative humidity is the _____.

    a. psychrometer
    b. velometer
    c. pressure gauge
    d. manometer

Summary

Because of air distribution systems, we live and work in comfortable homes and businesses. You would find it very hard to study if your classroom had a temperature of 85°F and 100 percent humidity. Almost every business and industry must have properly conditioned air. Without it, materials break, adhesives won't stick, food spoils more quickly, and computers go down.

As long as the air distribution system works, most of us don't think twice about it. However, there is a quite a bit of science involved in the study of air properties. In this module, you were introduced to the science of psychrometry. You saw how this science applies to HVAC work and you learned how to read a psychrometric chart.

Although the chart probably looked confusing at first, you learned how to figure it out and plot changes in air properties.

You also learned about the operation of blowers and fans and the three fan laws governing how adjustments can be made to these devices. You learned about the instruments used to measure temperature, humidity, air pressure, and air velocity. The science of psychrometry also figures in the operation of these HVAC tools.

You will not have to master the science of psychrometry, but a basic understanding of its principles will make you a more effective craftworker.

Notes

Trade Terms Introduced in This Module

Absolute pressure: The sum of gauge pressure plus atmospheric pressure.

Air-handling unit: In an air distribution system, a device that consists of the cooling and heating coils, fan or blower, filters, air mixing section, and dampers enclosed in a sheet metal casing.

Atmospheric pressure: The pressure exerted on the earth by the weight of the atmosphere.

British thermal unit (Btu): The amount of heat required to raise the temperature of 1 pound of water by 1°F.

Constant air volume (CAV) system: An air distribution system that provides a constant airflow while varying the temperature to meet demand for heated or cooled air.

Cubic feet per minute (cfm): A measure of the amount or volume of air in cubic feet flowing past a point in one minute.

Dew point: The temperature at which water vapor in the air becomes saturated and starts to condense into water droplets.

Dry-bulb thermometer: A standard thermometer used to measure the amount of sensible heat in the air.

Enthalpy: The total heat content of the mixture of air and water vapor measured from a predetermined base.

Evaporator: The part of a refrigeration system in which the refrigerant is evaporated, thereby taking up outside heat and producing cool air.

Free air delivery: The condition that exists when there are no effective restrictions to airflow (no static pressure) at the inlet or outlet of an air-moving device.

Gauge pressure: The pressure measured on a gauge, expressed as pounds per square inch gauge or as inches in mercury vacuum.

Latent heat: The heat content of moisture vapor in the air at a constant temperature. Latent heat added to sensible heat equals total heat.

Pitot tube: A device used with a manometer or other pressure-reading instrument to measure the velocity of air in a duct or water in a pipe.

Plenum: In an air conditioning system, a sealed chamber at the inlet or outlet of an air-handling unit to which the duct is attached. In construction, the space between a ceiling and the structure above it.

Propeller fan: A fan that uses curved blades to direct the airflow in a parallel path to the propeller shaft. Propeller fans are often mounted in venturis.

Psychrometric chart: A chart used to determine how air properties vary based on changes in any of the air properties included on the chart.

Psychrometry: The study of air and its properties.

Relative humidity: The ratio of the amount of moisture present in a given sample of air to the amount it can hold at saturation, expressed as a percentage.

Sensible heat: An amount of heat that, when added to the air, causes a change in temperature with no change in the amount of moisture.

Specific density of air: The weight of 1 pound of air. At sea level at 70°F, 1 pound of dry air weighs 0.075 pounds per cubic foot.

Specific heat of air: The amount of heat required to raise 1 pound of air 1°F. Expressed as Btu/lb/°F. At sea level, air has a specific heat of 0.24 Btus per pound, per degree Fahrenheit.

Specific volume of air: The space 1 pound of dry air occupies. At sea level at 70°F, 1 pound of dry air occupies a volume of 13.33 cubic feet.

Static pressure: The pressure that is exerted uniformly in all directions within a duct system.

Static pressure tip: A probe used with a manometer or other pressure-reading instrument to measure air pressure in a duct system.

Thermistor: An electrical resistor that varies sharply in a predictable manner with the temperature.

Thermocouple: A device used for measuring temperatures; it uses the voltage generated by the junction of two dissimilar metals at different temperatures to produce a reading.

Total heat: The sum of sensible heat plus latent heat.

Total pressure: The sum of static pressure plus velocity pressure in an air duct.

Velocity pressure: The difference between total pressure and static pressure in an air duct.

Venturi: A ring or panel surrounding the blades on a propeller fan that is used to improve fan performance.

Volume: The amount of air in cubic feet flowing past a point in one minute.

Wet-bulb thermometer: A standard thermometer with a saturated wick wrapped around a sensing bulb that measures the moisture content of the air.

Resources & Acknowledgments

Additional Resources

This module is intended to present thorough resources for task training. The following reference works are suggested for further study. These are optional materials for continued education rather than for task training.

Air Conditioning Principles and System, Fourth Edition. 2002. Edward G. Pita. Upper Saddle River, NJ: Prentice Hall.

Audel Air Conditioning Home and Commercial, All New 5th Edition. 2004. Rex Miller et al. Hoboken, NJ: Wiley.

NCCER makes every effort to keep these textbooks up-to-date and free of technical errors. We appreciate your help in this process. If you have an idea for improving this textbook, or if you find an error, a typographical mistake, or an inaccuracy in NCCER's Contren® textbooks, please write us, using this form or a photocopy. Be sure to include the exact module number, page number, a detailed description, and the correction, if applicable. Your input will be brought to the attention of the Technical Review Committee. Thank you for your assistance.

Instructors – If you found that additional materials were necessary in order to teach this module effectively, please let us know so that we may include them in the Equipment/Materials list in the Annotated Instructor's Guide.

Write: Product Development and Revision
National Center for Construction Education and Research
3600 NW 43rd St, Bldg G, Gainesville, FL 32606

Fax: 352-334-0932

E-mail: curriculum@nccer.org

Craft _____ Module Name _____

Copyright Date _____ Module Number _____ Page Number(s) _____

Description _____

_____

_____

_____

(Optional) Correction _____

_____

_____

(Optional) Your Name and Address _____

_____

_____

_____

04206-08

Bend Allowances

04206-08
Bend Allowances

Topics to be presented in this module include:

Overview

When metal is bent, a certain amount of the length of the flat piece is used to make the bend. Metal on the inside of the bend is compressed, and the metal on the outside is stretched. To make sure the final piece has the proper dimensions after the metal is bent, an allowance must be added to the dimensions of the stretchout to compensate for the bending. This module reviews the interplay of different factors affecting the amount of bend allowance you need and the methods for calculating that allowance.

Objectives

When you have completed this module, you will be able to do the following:

1. Describe the factors that influence bend allowances on sheet metal blanks.
2. Perform the calculations necessary for determining proper bend allowances on selected sheet metal problems.
3. Apply bend allowance calculations to lay out and fabricate a mating hat channel.
4. Determine bend allowances on selected sheet metal problems.

Trade Terms

Air bend
Empirical method
Forming punch
Geometric Method
Mild steel
Neutral axis
Setback

Sheet metal blank
Sight line
Springback
Swinging
Tangent line
Yield strength

Required Trainee Materials

1. Appropriate personal protective equipment
2. Calculator
3. Pencil and Paper

Prerequisites

Before you begin this module, it is recommended that you successfully complete the following: *Core Curriculum; Sheet Metal Level One; Sheet Metal Level Two*, Modules 04201-08 through 04205-08.

This course map shows all of the modules in the second level of the Sheet Metal curriculum. The suggested training order begins at the bottom and proceeds up. Skill levels increase as you advance on the course map. The local Training Program Sponsor may adjust the training order.

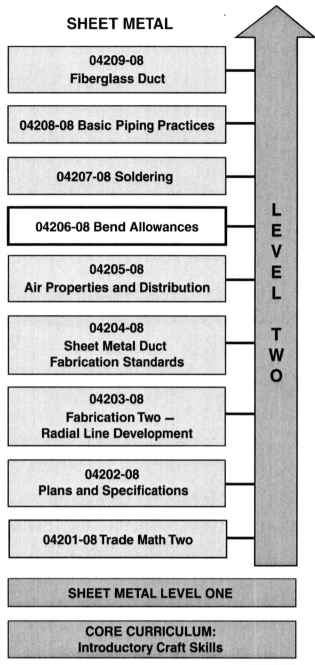

SHEET METAL

04209-08
Fiberglass Duct

04208-08 Basic Piping Practices

04207-08 Soldering

04206-08 Bend Allowances

04205-08
Air Properties and Distribution

04204-08
Sheet Metal Duct
Fabrication Standards

04203-08
Fabrication Two —
Radial Line Development

04202-08
Plans and Specifications

04201-08 Trade Math Two

LEVEL TWO

SHEET METAL LEVEL ONE

CORE CURRICULUM:
Introductory Craft Skills

206CMAP.EPS

1.0.0 ◆ INTRODUCTION

Bending is a forming operation used to make a number of sheet metal parts such as ducts and column guards. The bending process gives the metal shape and rigidity. Metal can be bent using bending machines, such as the cornice brake, or by using hand tools, such as stakes. Although the actual bending process may seem quick and easy, a number of factors affect how the metal will react when it is bent.

When you lay out sheet metal, you must be able to calculate how much metal you will need to accommodate the bend. This amount is called the bend allowance, and this module will introduce you to three methods for calculating that allowance.

This module also includes tasks that will give you an opportunity to put what you learn to practical use. As always, when performing any sheet metal tasks, you must put safety first. Be sure to wear the appropriate personal protective equipment and know and observe all shop safety rules.

2.0.0 ◆ BEND SHAPES

Some of the more common bend shapes are shown in *Figure 1*. The sharp break corner (A) is most often used on thin ductile (easily shaped) metal sheets. This kind of bend is not usually recommended, because the bend line along the corner is weak. If the bend is formed in tempered metal, fracture lines can develop, and the bent section can separate. The remaining four bend shapes shown (corner radius, bevel corner, trough section, and hat section) are much stronger and also look better. These bends include a radius that forms the bend

DID YOU KNOW?

A Brief History of Metal Bending

People have been bending metal since ancient times. The earliest metalworkers bent and shaped metal after it was heated to form tools, weapons, and everyday objects such as cups and bowls. As advances in metalworking took place, metalworkers produced thinner sheets of metal that could be bent and formed without heat. Artisans used the thinner metals to create works of art as well as for decoration on suits of armor. Today, metalworkers use computers for layout and machines to quickly bend metal, but the fundamental principles of bending metal remain the same.

shape. You must take this radius into account when laying out the **sheet metal blank**.

2.1.0 Bending Factors

Two physical forces occur when sheet metal is bent: tension (or stress) and compression. These forces reshape the metal at the bend radius without causing the metal to fracture or fail. When a piece of sheet metal is bent, the metal is stressed at the bend radius (*Figure 2*). The metal on the outside of the radius is under tensile stress (stretched and elongated). The metal on the inside of the radius is under compression stress (squeezed or compressed). If a fracture occurs during bending, it will appear on the outside surface of the bend. If any wrinkling occurs, it will appear on the inside

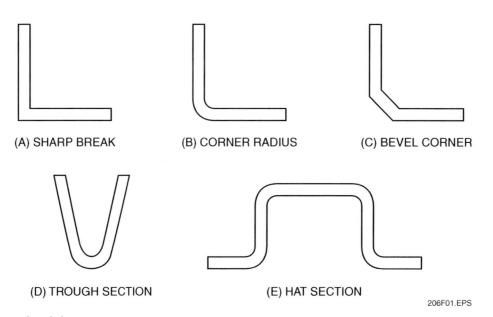

(A) SHARP BREAK   (B) CORNER RADIUS   (C) BEVEL CORNER

(D) TROUGH SECTION   (E) HAT SECTION

206F01.EPS

Figure 1 ◆ Common bend shapes.

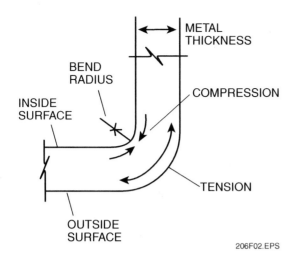

Figure 2 ✧ Bend forces.

surface of the bend. Here is an easy way to remember these principles: make a fist. Observe that the skin over your knuckles (the outside surface of your bent hand) stretches over your bones and that the skin on the inside surface wrinkles.

At some point in the bend, the tensile and compression stresses disappear. This point is called the **neutral axis**, or zero-stress line (*Figure 3*). When you begin bending a metal sheet, the neutral axis is near the center, but as the bending progresses, the neutral axis shifts toward the inside or compression side of the bend. The neutral axis remains the same length after it is bent as it was before it was bent. Its thickness is about 44.5 percent of the thickness of the metal, measuring from the inside of the bend (*Figure 4*). The position of the neutral axis depends on two variables: the radius of the bend and the thickness of the metal.

Therefore, the neutral axis will shift for each variable, and precise blank size calculations may be difficult. As a result, you may need to alter the blank dimensions after making a few trial bends.

2.1.1 Metal Movement

During the bending process, one area of the sheet metal blank is stationary and the remaining area is bent. The **forming punch** forces the bent metal up or down (*Figure 5*). This movement is called **swinging**, and you must ensure that no obstacles are in the way of the moving metal.

Bending tonnage tables show the approximate pressure, in tons per linear foot, required to make a 90-degree **air bend** on **mild steel** with various width die openings (*Table 1*). The boldface figures in the table show the pressures required using a punch with a radius equal to the metal thickness

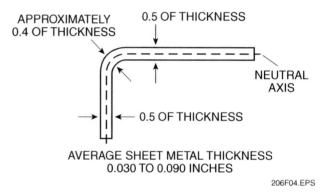

Figure 4 ✧ Thickness of the neutral axis.

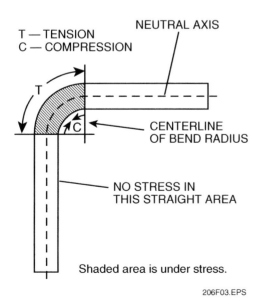

Figure 3 ✧ The neutral axis.

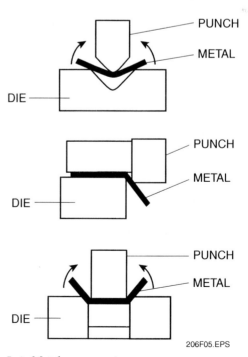

Figure 5 ✧ Metal movement.

Table 1 Bending Tonnage Chart

Thickness of Metal

> With these thicknesses, it is usual practice to have die opening at least 10 times metal thickness.

Width of V-Die Opening	26 G.	24 G.	22 G.	20 G.	18 G.	16 G.	14 G.	13 G.	12 G.	11 G.	10 G.	9 G.	7 G.	1/4"	5/16"	3/8"	7/16"	1/2"	5/8"	3/4"	7/8"	1'
	.018"	.024"	.030"	.036"	.048"	.064"	.075"	.090"	.105"	.120"	.135"	.149"	.187"	.250"	.313"	.375"	.437"	.500"	.625"	.750"	.875"	1.00"
1/8"	1.2	2.1	3.6																			
3/16"	0.8	1.4	2.5	4.1																		
1/4"	0.5	1.1	1.8	2.9	5.4																	
5/16"		0.7	1.4	2.2	4.0	7.0																
3/8"			1.0	1.7	2.9	5.6	8.8															
1/2"				1.2	2.2	3.6	6.0	10.0														
5/8"					1.6	2.7	4.5	6.8	10.1													
3/4"					1.3	.2	3.4	5.4	7.4	10.5												
7/8"						1.7	3.0	4.3	6.3	8.8	11.3											
1"						1.4	2.5	3.7	5.4	7.2	9.6	13.1										
1 1/8"							2.1	3.3	4.4	6.2	8.4	11.9	16.4									
1 1/4"							1.7	2.9	4.0	5.4	7.0	9.0	14.0	28.8								
1 1/2"									3.2	4.3	5.6	6.7	11.2	22.0	38.0							
2"										3.2	4.1	5.2	7.6	15.3	26.0	41.0						
2 1/2"											2.4	3.5	5.8	11.5	19.2	29.9	45.2					
3"												2.2	4.5	9.1	16.0	24.0	35.0	47.9				
3 1/2"														7.5	12.5	19.4	28.0	39.0	69.5			
4"														6.2	10.6	16.0	24.0	33.1	58.0	92.0		
5"															7.6	12.3	17.0	24.0	42.2	69.0	104.0	
6"																9.3	14.6	19.0	32.4	52.2	80.0	112.2
7"																	11.1	15.6	26.0	42.2	63.0	90.2
8"																		12.7	23.0	36.0	52.5	76.0
10"																			16.5	27.0	39.4	56.2
12"																			21.0	31.2	44.0	

and a die opening about eight times the metal thickness. This combination will produce an air bend with an inside radius nearly equal to the metal thickness. Any punch and die that is controlled by air pressure is an air bend.

2.1.2 Springback

The stress on the metal creates a narrow band of elastic, or flexible, metal on both sides of the neutral axis. When the stress is removed, this band will reshape slightly in an attempt to return the metal to its original flat shape. This is called **springback** (*Figure 6*). To get a practical demonstration of how springback works, conduct this simple experiment: Get a metal coat hanger and compress or bend the sides inward. When you release the sides, they will spring back slightly; but, as you will see, the hanger cannot return completely to its original shape on its own. Springback must be dealt with so that the finished bend maintains the proper degree of angle. There are three methods of dealing with springback:

- Overbending
- Bottoming or setting
- Stretch bending

In the overbending method, you plan for the springback and overbend the metal to allow for it. This method produces the desired degree of bend after springback occurs (*Figure 7*).

In the bottoming or setting method, you use a punch to strike the metal very hard at the radius area of the flexible band (*Figure 8*). This method compresses the metal and sets it at the desired degree of bend. During this method, clearance (air space) occurs between the punch and the die. You must leave enough clearance to avoid bottoming out the die. If you bottom out the die, you could cause a stress fracture of the metal die or the punch.

DID YOU KNOW?

Plastically Deformed Metal

When metal is reshaped in the bending process, it is plastically deformed. This means that the metal, while it is bent, loses some of its rigidity and becomes plastic. In this sense, the term *plastic* means flexible. The term *deform* means that the metal shape has changed from one form (straight) to another form (bent).

Stretch bending is generally not used in the sheet metal trade. In this method, the metal is stressed beyond its **yield strength** before bending so that very little springback occurs (*Figure 9*). This method is restricted to bending fairly large radii and is used mostly in milling operations.

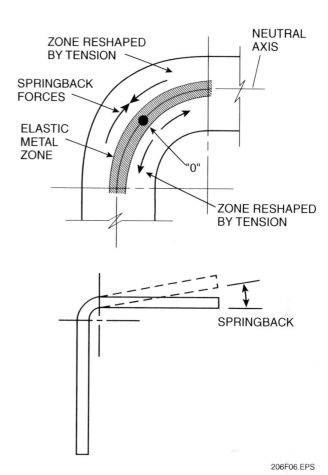

206F06.EPS

Figure 6 ◆ Springback.

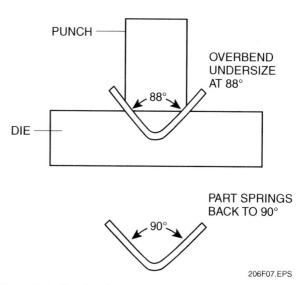

206F07.EPS

Figure 7 ◆ Overbending.

3.0.0 ◆ APPROXIMATE BEND ALLOWANCE

There are several methods for calculating bend allowances. Some of these methods are quite complicated and must be used when precision bends are required. When precise bend allowances are not required, a simple method for calculating approximate bend allowances is used. To calculate approximate bend allowances for sharp bends in thin metal, add a percentage of the metal thickness to the inside bend dimension and subtract a percentage of the metal thickness from the outside bend dimension.

In the following example, calculate the amount of metal required for the stretchout for a 90-degree bend (*Figure 10*). Add one-half the metal thickness (t) of the inside dimensions to the sum of the lengths of the legs (L). The formula follows:

Stretchout = $L_1 + L_2 + \frac{1}{2}t$

Use the following values:

L₁ = 1.250 inches

L_1 = 1.250 inches

L_2 = 2.50 inches

t = 0.050 inches (18-gauge metal)

Stretchout = $1.25 + 2.50 + \frac{(0.050)}{2} = 3.75 + 0.025$

Stretchout = 3.78 inches

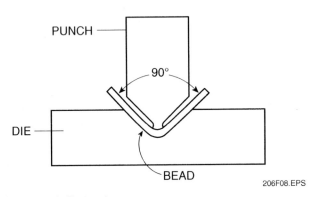

Figure 8 ◆ Bottoming.

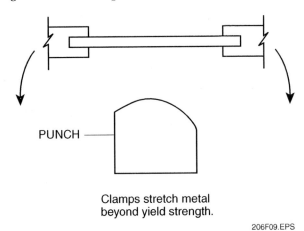

Clamps stretch metal
beyond yield strength.

206F09.EPS

Figure 9 ◆ Stretch bending.

To calculate the stretchout using the outside dimensions, subtract one-half the metal thickness from the sum of the legs. The formula follows:

Stretchout = $L_1 + L_2 - \frac{1}{2}t$

Use the following values:

L_1 = 1.5 inches

L_2 = 3 inches

t = 0.031 inches (22-gauge metal)

$1.5 + 3 - \frac{0.031}{2} = 4.5 - 0.0155$

Stretchout = 4.5 − 0.0155

Stretchout = 4.48 inches

4.0.0 ◆ PRECISION BEND ALLOWANCE

Some of the terms associated with bend allowance are shown in *Figure 11*. Flat 1 and Flat 2, represent two sides of a single piece of sheet metal. You have already learned about compression, tension, and the neutral axis. Study the figure to understand the following terms better:

- Bend **tangent lines** are where the bend begins and ends.
- The bend allowance is the distance between the bend tangent lines measured on the neutral axis.
- **Setback** cannot be shown in the figure. It is the amount of metal saved when metal is bent around a given radius. Less metal is needed to form a radius than a square bend; therefore, you can subtract the amount of metal saved from the layout dimensions to determine the final blank size.

 NOTE

Actual shop process calls for subtracting the setback of the bend allowances from the sum of the outside dimensions. Calculating setbacks will be covered later in your training.

The variables that affect the calculation of precise bend allowances are the thickness of the metal, type of metal, and the radius of the bend. Metal thickness charts are provided in *Appendixes A–D*. You can calculate more precise bend allowances using several methods: bend allowance tables, which are available from manufacturers; the **empirical method**; or the **geometric method**. In this module, you will learn about the empirical method and the geometric method, both of which will give you the same answer.

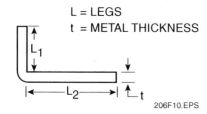

Figure 10 ◆ Approximate bend allowance.

4.1.0 The Empirical Method

Manufacturing engineers have developed several formulas to calculate precise bend allowances. These formulas have been developed over time as a result of observation and experimentation. A constant of 0.01743 is used as the multiplier to solve for bend allowance. The following is the formula for the empirical method:

B = A[(0.01743R) + (kt)]

Where:

B = the bend allowance

A = the angle of the bend in degrees

R = the bend radius

k = a constant (0.0078) related to the neutral axis

t = the thickness of the metal

The two constants in the formula (0.01743 and 0.0078) are calculated using another mathematical formula that will not be covered at this stage in your training. The k factor is related to the location of the neutral axis, which will shift based on the angle of the bend. As the metal is bent, the neutral axis moves from the centerline toward the inside radius of the bend. The sharper the bend, the more the neutral axis moves toward the inside radius.

4.2.0 The Geometric Method

The geometric method can also be used to calculate the bend allowance. In this method, the length of metal in the bend is represented as if it were part of a circle. This means that you can calculate the bend allowance by finding the circumference of a circle with a radius equal to the radius of the neutral axis. In the formulas that follow, the letters and symbols represent the following:

C = 1 degree of the circumference of a circle

π = 3.14

R = radius

k = 0.333, 0.4, 0.455, or 0.5

t = thickness of the metal

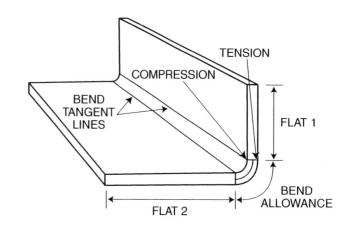

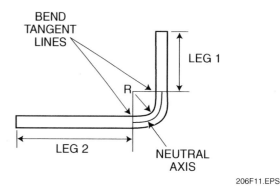

Figure 11 ◆ Bending terms.

Note that in the geometric method, k is not shown as a constant, but rather as a range of values. As you saw in the empirical method, the value of k can be calculated. However, as a result of experimental evaluation, k has been found to range from values of 0.333 to 0.5, depending on the type of metal being used. The value of 0.5 for k places the neutral axis exactly in the center of the metal. This value is often used where the bend radius is greater than twice the thickness of the metal.

In the geometric method, you will solve for three values: the radius of the neutral axis, the length on the circumference for each degree of circumference, and the bend allowance.

To calculate the radius of the neutral axis, use this formula:

R = R + kt

To calculate the length on the circumference for each degree of circumference, use this formula:

$$C = \frac{2\pi R}{360}$$

To calculate the bend allowance, use this formula:

B = AC

4.3.0 Examples Using the Empirical Method and the Geometric Method

In the following examples, you will see how both the empirical method and the geometric method are used to solve two bend allowance problems. Look at the figure associated with each example and observe how the numbers are plugged into each formula. Work through the steps in each example so that you can understand how to apply both methods. Finally, compare the answers obtained with each method. For Examples 1 and 2, use *Figure 12*.

Example 1

Solve for the bend allowance using the empirical method. Remember to use the constant (0.0078) for k.

Step 1 Place the values in the formula.

$$B = A[(0.01743R) + (kt)]$$
$$B = 110[(0.01743 \times 0.5) + (0.0078 \times 0.1875)]$$

Step 2 Solve for B. Round up to the nearest hundredth.

$$B = 110[0.008715 + 0.0014625]$$
$$B = 110[0.0101775]$$
$$B = 1.119525 \text{ inches}$$
$$B = 1.12 \text{ inches (approximately)}$$

Example 2

Solve for the bend allowance using the geometric method. Remember to use the amount given for k in the figure.

Step 1 Solve for R. Place the values in the formula.

$$R = R + kt$$
$$R = 0.5 + (0.445 \times 0.1875)$$
$$R = 0.5 + 0.0853125$$
$$R = 0.585$$

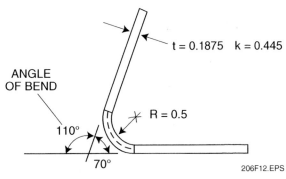

ANGLE OF BEND

t = 0.1875   k = 0.445

R = 0.5

110°

70°

206F12.EPS

Figure 12 ◆ Bend allowance for Examples 1 and 2.

Step 2 Solve for C. Place the values in the formula.

$$C = \frac{2\pi R}{360}$$
$$C = \frac{(2 \times 3.14) \times 0.5853125}{360}$$
$$C = \frac{6.28 \times 0.5853125}{360}$$
$$C = \frac{3.6757625}{360}$$
$$C = 0.0102 \text{ inches for 1 degree of circumference}$$

Step 3 Solve for B. Place the values in the formula.

$$B = AC$$
$$B = 110 \times 0.0102$$
$$B = 1.122 \text{ inches}$$
$$B = 1.12 \text{ inches (approximately)}$$

As you can see, both the empirical method and the geometric method produced the same answer. For Examples 3 and 4, use *Figure 13*.

Example 3

Solve for the bend allowance using the empirical method. Be sure to use the constant (0.0078) for k.

$$B = A[(0.01743R) + (kt)]$$
$$B = 135[(0.01743 \times 0.125) + (0.0078 \times 0.1875)]$$
$$B = 135[0.0021787 + 0.0014625]$$
$$B = 135[0.0036412]$$
$$B = 0.491562$$
$$B = 0.49 \text{ inches (approximately)}$$

Example 4

Solve for the bend allowance using the geometric method. Be sure to use the value given for k in the figure.

Step 1 Solve for R.

$$R = R + kt$$
$$R = 0.125 + (0.445 \times 0.1875)$$
$$R = 0.125 + 0.0834$$
$$R = 0.2084$$

Step 2 Solve for C.

$$C = \frac{2\pi R}{360}$$
$$C = \frac{(2 \times 3.14) \times 0.2084}{360}$$
$$C = \frac{6.28 \times 0.2084}{360}$$

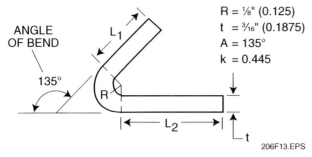

$R = \frac{1}{8}"$ (0.125)
$t = \frac{3}{16}"$ (0.1875)
$A = 135°$
$k = 0.445$

206F13.EPS

Figure 13 ◆ Bend allowance for Examples 3 and 4.

$$C = \frac{1.308752}{360}$$

C = 0.0036354 inches for 1 degree of circumference

Step 3 Solve for B.

B = AC

B = 135 × 0.0036372

B = 0.491022

B = 0.49 inches (approximately)

5.0.0 ◆ STRETCHOUT DEVELOPMENT

A stretchout is a flat pattern for sheet metal fittings. Developing a stretchout is the process of figuring out what size metal (length and width) will produce a correctly sized fitting with minimum waste.

When the length of the flat portion of the metal to be bent is known, developing the stretchout is fairly simple. Just add the bend allowance to that length to get the dimensions for the bend layout.

If the given dimensions are outside dimensions, first locate the bend tangent lines (*Figure 14*). To locate the bend tangent lines, find the length of the unbent portion of each leg of the bend and add the bend allowance between the flat sections. The result is the final dimension for the stretchout of the form.

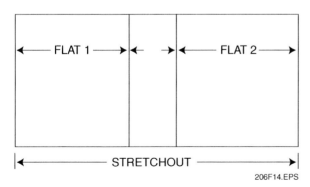

206F14.EPS

Figure 14 ◆ Stretchout development.

What if more than one bend is to be made in the metal? In this case, if the given dimensions are outside dimensions, locate the bend tangent lines and then follow these steps to calculate the stretchout:

Step 1 Note the number of bends and the radius of the angle of the bend (*Figure 15*), the channel (*Figure 16*), and the channel and flange (*Figure 17*).

Step 2 Determine the thickness of the metal.

Step 3 Calculate the bend allowance for each bend. You can use either the empirical method or the geometric method.

Step 4 Add all of the outside dimensions and note the final sum.

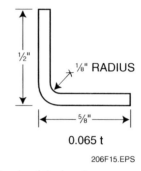

$\frac{1}{2}"$ $\frac{1}{8}"$ RADIUS $\frac{5}{8}"$

0.065 t

206F15.EPS

Figure 15 ◆ Angle of the bend.

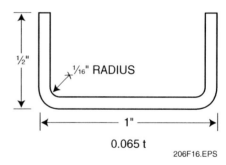

$\frac{1}{2}"$ $\frac{1}{16}"$ RADIUS 1"

0.065 t

206F16.EPS

Figure 16 ◆ The channel.

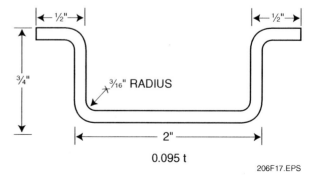

$\frac{1}{2}"$ $\frac{1}{2}"$ $\frac{3}{4}"$ $\frac{3}{16}"$ RADIUS 2"

0.095 t

206F17.EPS

Figure 17 ◆ The channel and flange.

6.0.0 ◆ REFERENCE LINES

When a piece of sheet metal is bent, the bend must begin at the tangent line. On most bending machines, the tangent line is hidden from view by part of the machine. For example, when using a cornice brake, the tangent line lies directly below the radius bar, so you won't be able to see it (*Figure 18*). To solve this problem, establish another reference line, called the **sight line**. Locate the sight line at a distance equal to the radius of the bend from the tangent line. When the metal is properly placed in the brake, the sight line is directly below the nose of the radius bar, so it can easily be seen. The sight line is the only line required on a flat layout. A sight line can be used when bending metal with a cornice brake, bar folder, or a box and pan brake because they all bend metal in a similar way.

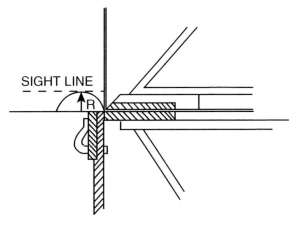

206F18.EPS

Figure 18 ◆ Brake setup showing the straight line.

Riveting Ideas   **Accurate Bends**

Because the actual process of bending metal goes quickly, it is important to set up the bending process with care to produce accurate bends. The bending process involves close tolerances, so there is little room for error. To produce accurate bends, place the metal on the machine carefully and check its placement. Then set the machine for the type and number of bends required. Check the setup before you make the bend. As an apprentice, you should have your supervisor check your setup before you proceed. Pay attention to these details and your bends will be smooth and accurate every time.

7.0.0 ◆ TASK 1: CALCULATING BEND ALLOWANCES

In this task, calculate the bend allowance for selected sections of metal using the approximate method, empirical method, and geometric method. You will need pencils, paper, and a calculator. To complete this task, follow these steps:

Part 1 –

Step 1   Review the material in this module that describes the method for calculating approximate bend allowances.

Step 2   Use this method to calculate the stretch-out for each of the three sketches in *Figure 19*. Assume the metal to be either 24-gauge galvanized steel or 22-gauge aluminum.

Step 3   Check your calculations for accuracy, and show them to your instructor for approval.

Step 4   If you have made an error, do the calculations again. Ask your instructor for help if you need it. When your calculations are correct, proceed to Part 2.

Part 2 –

Step 1   Review the material in this module that describes how to use the empirical method and the geometric method for calculating bend allowances.

Step 2   Use these methods (one at a time) to calculate the stretchout for each of the corner guard sections shown in *Figure 20*. Assume the metal to be either 20- or 22-gauge galvanized steel or 24-gauge or lighter stainless steel.

Step 3   Check your calculations for accuracy and show them to your instructor for approval.

Step 4   If you have made an error, do the calculations again. Ask your instructor for help if you need it. When your calculations are correct, proceed to Part 3.

Part 3 –

Step 1   Using any of the three methods (approximate, empirical, or geometric), calculate the stretchout for the section shown in *Figure 21*. This figure shows an inside mating hat channel. (Note: Your instructor may choose one method or have you calculate all three.)

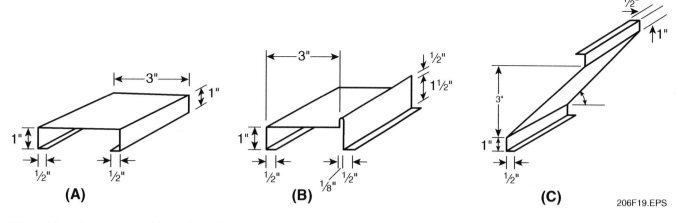

Figure 19 ◆ Stretchout problems for Task 1, Part 1.

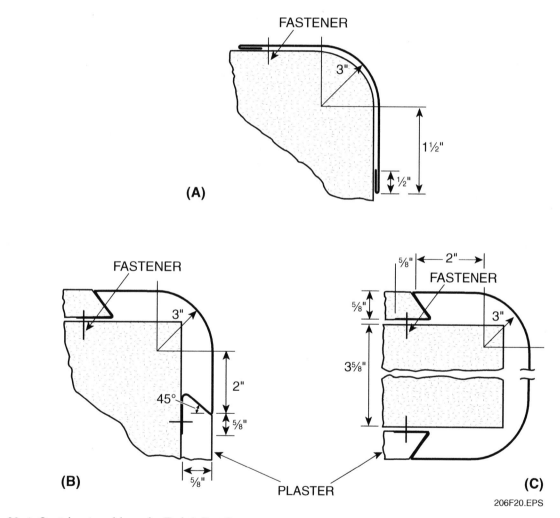

Figure 20 ◆ Stretchout problems for Task 1, Part 2.

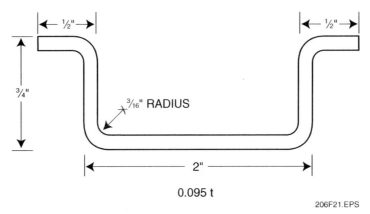

Figure 21 ◆ Stretchout problem for Task 1, Part 3.

Step 2   Check your calculations for accuracy, and show them to your instructor for approval.

Step 3   If you have made an error, do the calculations again. Ask your instructor for help if you need it.

8.0.0 ◆ TASK 2: MATING HAT CHANNEL

In this task, calculate bend allowances and lay out and develop a hat section mold and a mating hat channel. The hat section mold and mating hat channel are used for gutters and downspouts. Your instructor will determine the gauge and type of metal to be used. To complete this task, you will need the following tools and equipment:

- Appropriate personal protective equipment
- Calculator
- Scriber
- Straightedge

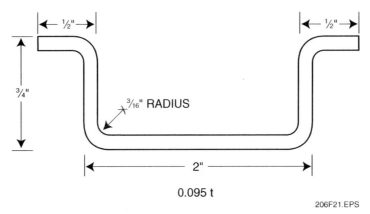

Riveting Ideas — **Checking Your Own Work**

As an apprentice, you will initially rely on your instructor to check your work for accuracy while you learn and perfect your skills. True professionals check the work themselves to ensure that it is done accurately. Get into the habit now of checking and double-checking your own work. Set high standards for yourself and you'll never need to feel nervous when others check your work.

- Combination square
- Pencil and felt-tipped marker
- Hand punch and prick punch
- Sheet metal hammer
- Mallet
- Screwdriver and pliers
- Drill and drill bits
- Reamer
- ¼-inch tape measure
- Circumference rule
- Sheet metal gauge
- Sheet metal snips
- Shearing, bending, and forming machines

Step 1   Study *Figure 22* to visualize the size and final shape of the molded hat section.

Step 2   Calculate the bend allowance for the mating hat channel (*Figure 23*).

Step 3   Begin the layout procedure on the metal. Locate and mark the bend tangent lines. Recheck your work before cutting the stretchout pattern from the sheet metal blank. Consult your instructor for tolerance limits.

Step 4   Check your layout for accuracy, locate the bend lines, and then bend and form the molded hat section.

Step 5   Present the molded hat section to your instructor, if necessary, before proceeding to the next step.

Step 6   Consult your instructor for tolerance limits required for the mating hat channel.

Step 7   Make a sketch of the molded hat section, and record measurements taken from the hat section on the sketch. Double-check your measurements for accuracy.

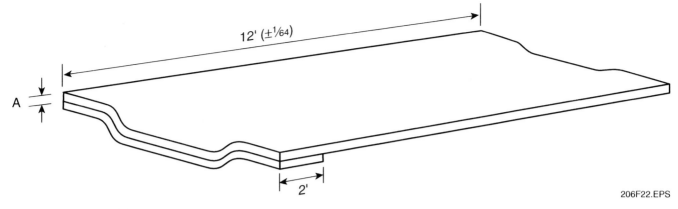

Figure 22 ◆ Molded hat section.

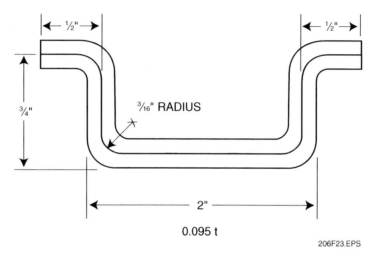

Figure 23 ◆ Mating hat channel.

Step 8 Using the sketch for reference, calculate the bend allowances necessary for the mating hat channel and place the proper dimensions on a sheet metal blank of the appropriate size. Be sure to recheck your calculations before scribing any lines. (Note: You may need to use a lighter gauge metal for this step.)

Step 9 Scribe the reference lines on the stretch-out and recheck them before forming the channel.

Step 10 Your instructor may ask you to drill and ream a hole in the mold section. If so, locate and lay out the hole in the mating hat channel. Locate this dimension properly so that it mates precisely with the hole in the mold section.

Step 11 Bend and form the mating hat channel.

Step 12 Check the mating hat channel for fit in the molded hat section, and present both components to your instructor for evaluation.

Step 13 Clean up your work area.

1. If used on tempered metal, the _____ shape is likely to cause fracture lines and separation of the bent sections.
   a. corner radius
   b. sharp break corner
   c. bevel corner
   d. trough section

2. Metal on the outside of a bend radius is under _____ stress.
   a. tensile
   b. compression
   c. positive
   d. neutral

3. If the metal wrinkles when bent, the wrinkles will appear on the _____.
   a. outside surface of the bend
   b. inside surface of the bend
   c. straight sections on either side of the bend
   d. inside and outside surfaces of the bend

4. The three methods of dealing with springback are _____.
   a. crimping, forming, or stretching
   b. prebending, rolling, or bottoming
   c. underbending, bottoming, or stretch bending
   d. overbending, bottoming, or stretch bending

5. To calculate the approximate bend allowances for sharp bends in thin metal, _____ a percentage of the metal thickness to/from the inside bend dimension and _____ a percentage of the metal thickness to/from the outside bend dimension.
   a. add; subtract
   b. subtract; add
   c. add; add
   d. subtract; subtract

6. The amount of metal saved when it is bent around a given radius is called the _____.
   a. yield
   b. setback
   c. springback
   d. fitting allowance

7. In the empirical method of calculating bend allowances, the constant used as the multiplier to solve for bend allowances is _____.
   a. 0.0078
   b. 0.01743
   c. 0.333
   d. 0.455

8. To find the stretchout of a flat piece of metal, _____ the bend allowance and each of the flat sections.
   a. divide
   b. multiply
   c. add
   d. subtract

9. When a piece of sheet metal is bent, the bend must begin at the _____.
   a. sight line
   b. reference line
   c. inside line
   d. tangent line

10. Locate sight lines at a distance _____ from the tangent line.
    a. equal to the length of the bend allowance
    b. equal to the radius of the bend
    c. ½ the sum of the outside dimensions
    d. adjacent to the tangent line

Summary

Metal is bent to form ducts and other fittings. Although bending metal may seem fairly simple at first, several factors influence how the metal reacts to the bending process. As a sheet metal professional, you will determine bend allowances, taking into account the thickness of the metal, type of metal, and the radius of the bend. In addition, you must be aware of how metal reacts when it is stretched and compressed during the bending process. Learning how metal reacts to the bending process will help you produce smooth, accurate bends.

This module introduced three methods for calculating bend allowances: the approximate method, the empirical method, and the geometric method. The approximate method is perhaps the easiest to learn, but it can be used only when precise bends are not a requirement of the task. More precise calculations can be obtained with the empirical and geometric methods. After you practice these methods and learn how to apply the formulas, you will find them fairly easy to do as well.

Notes

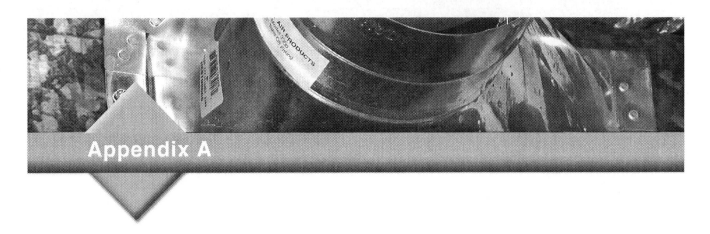

Thickness of Galvanized Steel

Gauge (U.S. Standards)	Approximate Thickness Galvanized (Inches)
7	0.1793
8	0.1664
9	0.1494
10	0.1345
11	0.1196
12	0.1046
13	0.0897
14	0.0747
15	0.0673
16	0.0598
17	0.0538
18	0.0478
19	0.0418
20	0.0359
21	0.0329
22	0.0299
23	0.0269
24	0.0239
25	0.0209
26	0.0179
27	0.0164
28	0.0149
29	0.0135
30	0.0120

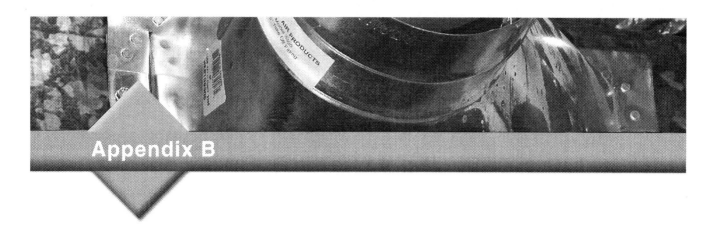

Thickness of Stainless Steel

Gauge (U.S. Standards)	Approximate Thickness (Inches)	
8		0.1719
9	5/32	0.1563
10	9/64	0.1406
11	1/8	0.1250
12	7/64	0.1094
13	3/32	0.0938
14	5/64	0.0781
15		0.0703
16	1/16	0.0625
17		0.0563
18		0.0500
19		0.0438
20		0.0375
21		0.0344
22	1/32	0.0313
23		0.0281
24		0.0250
25		0.0219
26		0.0188
27		0.0172
28	1/64	0.0156
29		0.0141
30		0.0125

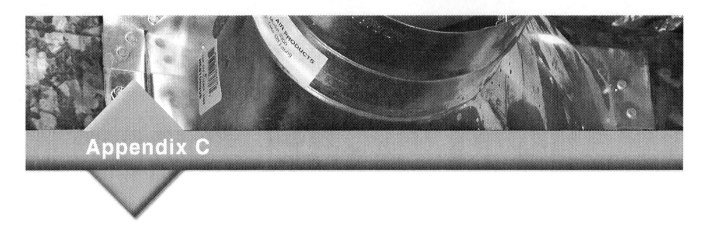

Thickness of Aluminum

Gauge (B & S)	Approximate Thickness (Inches)	
—		0.0403
9		0.0359
—		0.0320
10	1/32	0.0313
—		0.0285
11		0.0253
12		0.0226
—		0.0201
13		0.0179
14		0.0159
—	1/64	0.0156
15		0.0142
16		0.0126
—		0.0113
17		0.0100

Thickness of Copper

Ounces (Sq. Ft.)		Approximate Thickness
96	1/8	0.1290
88		0.1190
80	7/64	0.1080
72	3/32	0.0972
64		0.0863
56	5/64	0.0755
48	1/16	0.0647
44		0.0593
40		0.0539
36	3/64	0.0485
32		0.0431
28		0.0377
24	1/32	0.0323
20		0.0270
18		0.0243
16		0.0216
15		0.0202
14		0.0189
13		0.0175
12	1/64	0.0162
11	1/64	0.0148
10		0.0135
9		0.0121
8		0.0108
7		0.0094
6		0.0081
4		0.0054
2		0.0027

Air bend: Any punch and die that is pneumatically controlled.

Empirical method: A method that relies on observation supplemented by experiments to develop a working practice, procedure, or formula.

Forming punch: Any punch and die set used for forming air bends.

Geometric method: A method of calculating bend allowance where the length of metal in the bend is represented as if it were part of a circle. The bend allowance is found by calculating the circumference of a circle with a radius equal to the radius of the neutral axis.

Mild steel: Ductile (flexible) steel that is nearly pure iron with very low carbon content.

Neutral axis: In a bend, the point where the tensile and compression stresses disappear. The position of the neutral axis depends on the radius of the bend and the thickness of the metal. The neutral axis is also called the zero-stress line.

Setback: The amount of metal that is saved when metal is bent around a radius (as opposed to bending it around a square).

Sheet metal blank: The measured piece of sheet metal from which a fitting is made.

Sight line: A reference line established to enable the operator of a bending machine to position the metal accurately so that it may be bent on the tangent line, which is hidden from direct view by machine parts.

Springback: The slight reshaping of metal toward its original shape after the release of stress.

Swinging: The movement of a metal blank, either up or down, as it is bent in a forming punch.

Tangent line: On a sheet metal blank, the line that establishes where a bend is to begin and end.

Yield strength: The point at which metal no longer maintains its proportion of stress to strain.

Resources & Acknowledgments

Additional Resources

This module is intended to present thorough resources for task training. The following reference works are suggested for further study. These are optional materials for continued education rather than for task training.

Machinery's Handbook, 25th Edition, 1996. New York, NY: Industrial Press, Inc., pp 1246-1251.

Press Brake Technology: A Guide to Precision Sheet Metal Bending, 1997. Steve D. Benson. Dearborn, MI: Society of Manufacturing Engineers.

www.solidworks.com

Figure Credits

Chicago Dreis and Krump Manufacturing Co., 206F18, Table 1

NCCER makes every effort to keep these textbooks up-to-date and free of technical errors. We appreciate your help in this process. If you have an idea for improving this textbook, or if you find an error, a typographical mistake, or an inaccuracy in NCCER's Contren® textbooks, please write us, using this form or a photocopy. Be sure to include the exact module number, page number, a detailed description, and the correction, if applicable. Your input will be brought to the attention of the Technical Review Committee. Thank you for your assistance.

Instructors – If you found that additional materials were necessary in order to teach this module effectively, please let us know so that we may include them in the Equipment/Materials list in the Annotated Instructor's Guide.

Write: Product Development and Revision
National Center for Construction Education and Research
3600 NW 43rd St, Bldg G, Gainesville, FL 32606

Fax: 352-334-0932

E-mail: curriculum@nccer.org

Craft

Module Name

Copyright Date

Module Number

Page Number(s)

Description

(Optional) Correction

(Optional) Your Name and Address

04207-08

Soldering

04207-08
Soldering

Topics to be presented in this module include:

Overview

Soldering is a process in which a heated metal alloy—solder—is used as a filler metal to fasten sheet metal parts, to seal metal seams, or to seal electrical connections. To do a professional soldering job, you must know how to handle the material being soldered, how to choose the materials best suited to the job, and how to select and use the right soldering tools. Because the soldering process involves using chemicals and high heat, it is potentially dangerous, so you must learn how to solder carefully and safely. In this module, you will learn about these topics, and you will also practice soldering.

Objectives

When you have completed this module, you will be able to do the following:

1. Identify soldering tools and materials.
2. Safely and properly use soldering tools and materials in selected tasks.

Trade Terms

Flux
KEVLAR®
Lap joint
Liquidus
Muriatic acid
Oxidation
Pointing up

Sal ammoniac block
Solidus
Sweat solder
Tacking
Tinning
Transite
Wetting

Required Trainee Materials

1. Appropriate personal protective equipment
2. Hammers (ball peen and sheet metal)
3. Hand seamer
4. Metal scribe

Prerequisites

Before you begin this module, it is recommended that you successfully complete the following: *Core Curriculum; Sheet Metal Level One; Sheet Metal Level Two*, Modules 04201-08 through 04206-08.

This course map shows all of the modules in the second level of the Sheet Metal curriculum. The suggested training order begins at the bottom and proceeds up. Skill levels increase as you advance on the course map. The local Training Program Sponsor may adjust the training order.

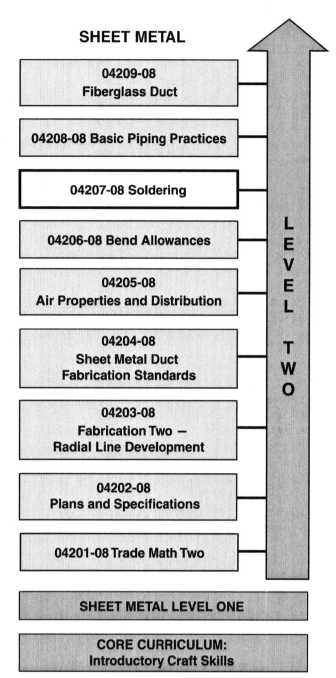

SHEET METAL

04209-08
Fiberglass Duct

04208-08 Basic Piping Practices

04207-08 Soldering

04206-08 Bend Allowances

04205-08
Air Properties and Distribution

04204-08
Sheet Metal Duct
Fabrication Standards

04203-08
Fabrication Two —
Radial Line Development

04202-08
Plans and Specifications

04201-08 Trade Math Two

LEVEL TWO

SHEET METAL LEVEL ONE

CORE CURRICULUM:
Introductory Craft Skills

207CMAP.EPS

1.0.0 ◆ INTRODUCTION

Solder is a nonferrous alloy, which means it contains no iron. Because solder has a fairly low melting point, it can be heated enough to fasten two pieces of metal without also melting the metal. For this reason, soldering is classified as a nonfusion joining method. Soldering can be used to join iron, nickel, lead, tin, copper, zinc, aluminum, and galvanized sheet metal.

2.0.0 ◆ TYPES OF SOLDER

There are two main types of solder: hard and soft. Soft solder is an alloy of lead and tin. Hard solder is an alloy of silver, copper, and zinc. Solder, like all alloys, melts over a range of temperatures. The temperature at which solder begins to melt is called the **solidus**. The temperature at which solder is completely molten is called the **liquidus**.

Soft soldering is done at temperatures that do not exceed 840°F. It is not suitable for applications where great strength is required, and it is therefore used as a sealing process. Hard soldering, or brazing, is done at temperatures higher than 840°F. This process is sometimes called silver soldering because the solder alloy used for this process is at least five percent silver. Like soft soldering, hard soldering can be used as a sealer. However, hard soldering results in greater strength and resistance to corrosion, so it is used mainly in applications where those qualities are required, such as in making electrical and refrigeration piping connections. This module will focus on the soft soldering process and its application to sealing and fastening sheet metal.

WARNING!
You must be especially careful when using solders that contain cadmium. Cadmium fumes are very dangerous if inhaled. These fumes emit no odor and may not have an immediate effect. They can be found in silver solder, medical gas lines, large copper pipe, and plumbing. Signs that you have been exposed to cadmium fumes may not show up for hours after inhalation. Cadmium fumes can make you very ill and in some cases can be lethal. Be sure to read the MSDS when using cadmium-based solders, wear the appropriate respiratory protection, and work in a well-ventilated area.

DID YOU KNOW?
Tinning

Although wetting is also called tinning, the term tinning is not correct when applied to solders that do not contain tin, such as some hard solders. The term tinning comes from a time when sheet metal workers applied thin layers of tin to steel to produce tin cans. The process of tinning the steel required heat to bond the tin to the steel and to prevent corrosion.

ASTM International classifies tin-lead solders by their ratio of tin to lead (*Table 1*). These ratios range from 2A (2% tin and 98% lead) to 70A (70% tin and 30% lead). Solder with 50% tin and 50% lead is used for applications at room temperature. The 60A and 63A solders, also called fine solders, are used whenever temperature requirements are critical. The 70A solder is used in applications where a high tin content is necessary, such as soldering zinc.

Solders with a high percentage of tin flow more easily than those with a low percentage of tin and have a greater **wetting** ability. When you apply solder, you must first coat the surface of the base metal with the melted solder before filling in the gap between the surfaces to be joined. This coating process is called wetting, or **tinning**, the metal.

Solders are commercially available in various sizes and shapes. Two commonly used types are bar solder and wire solder (*Figure 1*). Manufacturers can also supply custom solders to suit a special application. The major groups of solders are listed in *Table 2*.

DID YOU KNOW?
Lead Solders and Drinking Water Systems

In 1986, the Safe Drinking Water Act Amendment banned lead solder, flux, and pipe from use in drinking water installations. Solders that contain more than 2% lead may still be used for drainage, heating, fire sprinklers, and for air conditioning and machine cooling. A number of manufacturers have developed lead-free solders, which may be applied with the same techniques used for tin-lead solders.

Table 1 ASTM International Alloy Grades for Tin-Lead Solders

ASTM Alloy Grade	Tin % Desired	Lead % Nominal	Completely Solid Solidus		Completely Liquid Liquidus	
			°C	°F	°C	°F
70A	70	30	183	361	192	378
70B	70	30				
63A	63	37	183	361	183	361
63B	63	37				
60A	60	40	183	361	190	374
60B	60	40				
50A	50	50	183	361	216	421
50B	50	50				
45A	45	55	183	361	227	441
45B	45	55				
40A	40	60	183	361	238	460
40B	40	60				
40C	40	58	185	365	231	448
35A	35	65	183	361	247	477
35B	35	65				
35C	35	63.2	185	365	243	470
30A	30	70	183	361	255	491
30B	30	70				
30C	30	68.4	185	364	250	482
25A	25	75	183	361	266	511
25B	25	75				
25C	25	73.7	184	364	263	504
20B	20	80	183	361	277	531
20C	20	79	184	363	270	517
15B	15	85	227	440	288	550
10B	10	90	268	514	299	570
5A	5	95	270	518	312	594
5B	5	95				
2A	2	98				

Table 2 Major Groups of Solders

Type	How Sold
Pig	Available in 50- and 100-pound pigs
Cake or Ingot	Available in rectangular or circular 3.5- and 10-pound shapes
Bars	Available in many lengths, cross sections, and weights
Paste	Available as a mixture of powdered solder and flux paste
Foil, Sheet, or Ribbon	Available in many thicknesses and weights
Segment or Drop	Available in triangular bars or as wire that can be cut into required lengths
Solid Wire	Available in spools 0.010 to 0.250 inch in diameter
Cored Wire	Available in spools 0.010 to 0.250 inch in diameter with flux in the center, or core, of the wire

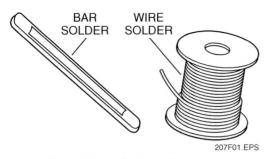

BAR SOLDER   WIRE SOLDER

207F01.EPS

Figure 1 ◆ Bar solder and wire solder.

3.0.0 ◆ FLUX

Flux is a fusible substance used in welding, brazing, and soldering. Flux chemically cleans the metal and helps the solder to make a good bond. Flux performs the following actions:

- Removes oxide films and other impurities that may be present at soldering temperatures
- Promotes the wetting or tinning action of the solder
- Prevents **oxidation** of the surfaces during heating
- Allows the solder to flow easily into the joint

Flux is divided into three main groups: corrosive, intermediate, and noncorrosive. This classification is based on the flux's residue, which is the material left behind when the flux cools. You must always choose the mildest flux that will meet the requirements of a specific application.

3.1.0 Corrosive Flux

Corrosive flux is more versatile than the other two types of flux and can be applied as a solution, dry salt, or paste. Because corrosive flux does not char, or burn, it works well with all heating methods. However, the residue of corrosive flux may cause severe corrosion at a joint. You must remove the residue of corrosive flux with water. If the water alone does not remove the residue, you may remove it with water and a damp rag. Corrosive fluxes are suited to applications that require a quick fluxing action. Examples of these applications include soldering sheet metal, copper, brass, and cast iron.

The main ingredient in most types of corrosive flux is zinc chloride, which is made by adding zinc to concentrated hydrochloric acid. Hydrochloric acid is also identified by its commercial name **muriatic acid**, but it may be called raw acid in some sheet metal shops. Zinc chloride has a melting temperature well above the melting temperature of most solders. When zinc chloride is used alone, unmelted salt particles may become

trapped in the joint. Therefore, zinc chloride must be mixed with other inorganic chlorides to lower the melting temperature of the flux.

> **NOTE**
>
> A weak solution of water and baking soda can also be used to remove the residue of corrosive flux from a joint.

3.2.0 Intermediate Flux

Intermediate flux is weaker than corrosive flux and consists mainly of organic acids such as lactic, citric, benzoic, and oxalic acids. This type of flux is used for quick spot-soldering operations. Its residue is easily removed with water.

3.3.0 Noncorrosive Flux

The chief ingredient of noncorrosive flux is rosin, a residue from the sap of pine trees; therefore, this type of flux is often called rosin flux. Rosin melts at 260°F and remains active in the molten state up to 600°F.

Because the residue of rosin flux is noncorrosive and nonconductive, this type of flux is widely used in the electrical and electronics industries. Noncorrosive fluxes are also effective for soldering connections and joints of copper, brass, bronze, nickel, and silver.

4.0.0 ◆ SOLDERING IRONS

A tool commonly used in soft soldering is the soldering iron. Because the commonly used soldering iron has a copper tip, it is often called a soldering copper. The typical soldering iron is a steel rod or shank with a handle on one end and a copper head on the other (*Figure 2*). The soldering iron performs two functions: it heats the metal to the melting point of the solder. and it melts the solder and carries it along the seam to be joined.

The soldering iron provides a constant flow of heat to the seam in a short amount of time. The minimal contact time protects the areas near the seam from becoming overheated. Soldering irons are divided into six broad groups:

- Light-duty irons
- Medium-duty irons
- Heavy-duty irons
- Temperature-controlled irons
- Transformer pencil irons
- Soldering guns

The light-duty iron is designed for light-duty soldering. It is available with a copper- or iron-plated tip so that you can match the tip to the soldering operation.

The medium-duty iron is designed for continuous production operations. This type of iron can withstand high-speed production use.

The heavy-duty iron is designed for continuous, fast production use. It is available in a number of sizes and wattages to ensure heat stability under heavy soldering loads (*Figure 3*).

The temperature-controlled iron has a sensor in the tip that reacts to small temperature changes. This enables the operator to match the heat of the iron to the requirements of the soldering project.

The transformer pencil iron is designed for use in light soldering repair work and production operations. It operates with less than 12 volts and has a rheostat, or voltage tip, to regulate heat output.

The soldering gun is used for light spot soldering of electrical connections. A soldering gun must be used with care. Because it does not allow you to control the heat output, you could potentially overheat the area being soldered.

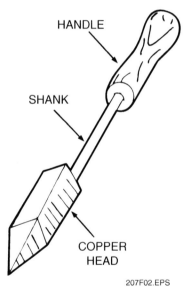

Figure 2 ❖ Soldering iron.

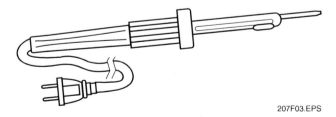

Figure 3 ❖ 25-watt heavy-duty soldering iron.

4.1.0 Tips

The tip of a soldering iron must:

- Conduct heat well
- Be easy to tin because the molten solder will adhere easily to the tip
- Have low oxidation properties in order to ensure good heat transfer and prevent the tip from freezing
- Be corrosion resistant
- Be resistant to erosion from the molten solder

There are four basic types of tips: copper, iron-plated with a coated shank, iron-plated with a stainless steel shank, and calorized. Copper tips conduct heat very well and have excellent tinning properties, but they quickly oxidize and erode from the molten copper. Iron-plated tips with a coated shank are made of copper with iron that has been electrodeposited over the entire tip. This type of tip has a long life (20 to 50 times that of copper), but it is not as efficient at conducting heat. Iron-plated tips with stainless steel shanks have the same characteristics as the iron-plated tip with a coated shank. In addition, they prevent the shank from freezing to the iron. Calorized tips are made by diffusing aluminum into a copper tip. Calorized tips resist high-temperature oxidation and also keep the shank of the soldering iron from freezing.

4.2.0 Heat Sources

The heat source for a soldering iron depends on the type of soldering iron being used. The basic soldering iron has no heat source of its own, so it must be heated in a gas-fired soldering furnace. The furnace may also be used to treat, temper, case harden, or forge metal.

Electrically heated soldering irons are also available. This type of iron is handy for continuous soldering and maintains a uniform heat. A 200-watt electric soldering iron is appropriate for most sheet metal applications. Periodically inspect the electrical cord to check for fraying or accidental scorching from contact with the iron.

Some soldering irons are designed for use with a propane torch. The flame from the torch heats the tip of the iron to melt the solder. Soldering torches consist of a fuel gas cylinder, a hose to deliver the gas, a regulator to control the flow of gas, and the torch itself. Gases used to fuel a soldering torch include oxyfuel, natural gas, and propane. Torch soldering is commonly used to seal pipes and structural joints.

Riveting Ideas

Dip Soldering

The dip soldering method is both cost-effective and labor saving. It is used to rustproof surfaces by coating them with solder or to solder an entire fixture. The solder is melted in a tank. The items to be soldered are cleaned in an acid bath, washed, and dried. They are then dipped into a flux bath and next into the molten solder. Finally, the items are allowed to drain so that excess solder runs off.

In electronics applications, dip soldering is used to coat silicon boards. Before dipping, the boards are coated with a nonstick substance in such a way as to allow the solder to stick to certain parts of the board. Presoldering is a type of dip soldering where metal sheets are fed through a vat of solder, and then another sheet is laid on top of the solder, bonding the two sheets.

WARNING! Parts to be dip soldered must be thoroughly dry. If a part has even a small amount of water on it, high-temperature steam will be produced as soon as it enters the solder bath. The steam can, in turn, cause the molten solder to erupt, potentially injuring workers or damaging equipment.

A soldering torch flame will heat large areas of metal quickly. However, the flame may also burn or carbonize the flux. Preheating the assembly before applying the flux and solder will prevent this problem. The high temperature of the flame may also damage sensitive components or areas of the workpiece that are near the area being soldered. To prevent this from happening, carefully control the flame.

The tips used with soldering torches come in different sizes. You can control the amount of heat by changing the tip size. A larger tip will deliver more heat, and a smaller tip will deliver less heat. Note that, although the amount of heat varies with different tip sizes, the temperature remains the same.

5.0.0 ◆ THE SOLDERING PROCESS

There are four basic steps in the soldering process. Each of these steps is discussed in the following sections:

• Fitting-up the joints
• Preparing the surface
• Applying flux and solder
• Cooling and cleaning the joints

5.1.0 Fitting-Up the Joints

When you fit-up the joints, you must pay attention to the clearance or space between the parts you are joining. The clearance must be large enough for the solder to be drawn into it, but not so large that the solder cannot fill the gap. A clearance of 0.005 inch is suitable for most work. On precoated metals, a clearance of 0.001 inch is required.

5.2.0 Preparing the Surface

One of the most important steps in the soldering process is proper surface preparation. If the surface of the metal is not properly cleaned, the solder will not adhere to it. As you have learned, flux cleans metal chemically. However, flux does not clean off grease, oil, dirt, rust, or scale that may be present on the surface of the metal. These substances prevent the wetting action of the flux and solder, so, before applying the flux, you must also clean the metal physically. The most common physical cleaning methods are:

• Degreasing
• Acid cleaning or pickling
• Abrasion and etching

The best way to degrease the metal is to suspend it in the vapors of an approved cleaning solvent. Because the metal is colder than the vapors, the vapors condense and drip off, leaving no residue.

Acid cleaning or pickling removes rust, scale, sulfides, and oxides from the base metal. Hydrochloric acid and sulfuric acid are commonly used for acid cleaning. After acid cleaning, the metal must be thoroughly washed in hot water and dried as quickly as possible.

Abrasive cleaning with materials such as steel wool or emery cloth is effective and inexpensive. However, this method has one disadvantage.

After being cleaned with steel wool or emery cloth, the metal may look clean, but tiny abrasive particles may have become stuck in its surface. These particles decrease the effectiveness of the solder. Therefore, you may have to etch the metal to remove them. Etching also removes a small amount of the base metal. This technique is used most often on nonplated copper surfaces. The metal may be etched with copper chloride, ferric chloride, or ammonium presulfide.

5.3.0 Applying Flux and Solder

Depending on the type of solder, flux and solder can be applied in one or two steps. If the solder does not contain flux, apply the flux first to chemically clean the metal. Then, apply the solder. Applying the flux and solder separately will allow you to more carefully control each step for the best result. The soldering flux must be applied with care to avoid dropping flux anywhere except on the joint (*Figure 4*).

If the solder does contain flux, as with a cored wire solder, the flux and solder flow onto the metal at the same time. The metal is chemically cleaned and soldered in one step. Regardless, you must first heat the joint to be soldered and the surrounding metal and then apply the flux and solder. Wet the surface of the metal, and then fill the gap between the surfaces to be joined with solder. Wetting the metal surface allows you to quickly and evenly apply the solder. It also prevents the formation of strong acid fluxes at the joint. A coating thickness of 0.0002 inch to 0.0005 inch of solder is generally recommended to ensure a well-soldered joint.

Melted solder flows to the hottest point on the metal. The greatest heat is produced in the body and base of the point of the soldering iron, not in the tip. Therefore, for most soldering operations, you must position the iron so that the head (not just the tip) of the soldering iron comes into contact with the solder (*Figure 5*). By positioning the

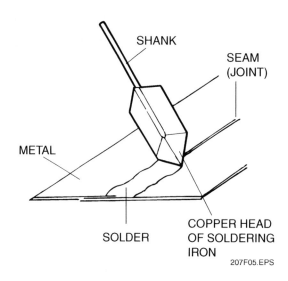

Figure 5 ◆ Proper positioning of the soldering iron head.

soldering iron correctly, you allow the maximum amount of heat to be transferred to the joint connections in the least amount of time. However, some soldering procedures, such as **pointing up** and **tacking**, require using only the tip of the soldering iron.

5.4.0 Cooling and Cleaning the Joints

Once you have soldered the joint, you must allow it to cool to room temperature. If the solder is cooled too slowly, it may become brittle. If the solder is cooled too quickly, it may warp and develop small fractures. Some of the cooling happens naturally as heat is conducted away to the parts being joined. You can assist the cooling process by spraying the joint with water or air.

Once the joint has cooled, you must remove the flux residue. Flux residue cleaning solvents are available from manufacturers of soldering tools and equipment. Be sure to read and follow the manufacturer's instructions for using these sol-

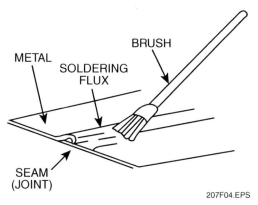

Figure 4 ◆ Applying flux to the joint with a brush.

Riveting Ideas

Working with Cored Flux Solder

Do not melt cored flux solder on the soldering iron and then carry it to the connection. Instead, touch the solder to the soldering iron tip to begin the heat transfer and melt it directly on the workpiece. When you are finished making the connection, wipe the tip of the soldering iron clean with a wet sponge or clean, damp cloth.

vents. Note that some flux residues can be removed with water or with a weak solution of water and baking soda. Generally you do not have to remove the residue of most noncorrosive (rosin-based) fluxes unless the joint is to be painted. You must always remove or neutralize the residue of corrosive fluxes.

6.0.0 ◆ JOINT DESIGN

Solders have less strength than the materials to which they are joined. Therefore, soldered joints must be designed so that they do not depend on the strength of the solder alone. It is important to know how the soldered joint will function so that you can select the correct joint design and solder to support the load stress that will be placed on a joint. When making a connection that will be subject to high load stress, you must do the following:

- Locate the joint in a low-stress area
- Support the joint mechanically
- Use a high-strength solder

The most widely used joint is the **lap joint** in which one surface slightly overlaps another surface. Examples of different joint designs are shown in *Figure 6*. Well-designed joints have the following qualities:

- The mating surfaces meet evenly with no gaps or flares
- The mating surfaces are supported throughout the joint area
- Tubular (round) pieces to be joined are properly aligned and fit snugly together

7.0.0 ◆ SAFETY

When the appropriate precautions are taken, soldering is a safe operation. However, you must be aware of the potential hazards involved so that

you can protect yourself and your co-workers. If you don't understand a safety procedure, be sure to ask about it. Here are some general safety guidelines you must follow when soldering:

NOTE

Safety instructions may vary from manufacturer to manufacturer and from product to product. Never assume that one manufacturer's instructions apply to another manufacturer's product.

- Always wear the appropriate personal protective equipment, which includes, but is not limited to, respiratory protection, safety glasses, and leather or **KEVLAR**® gloves.
- Always read and follow the manufacturer's material safety data sheet (MSDS) for detailed handling and safety information about solders, fluxes, chemicals, and soldering equipment.
- Always work in a clean, well-ventilated area.
- Do not touch a hot soldering iron. Do not set the iron down where it could damage a surface or cause a fire. Make sure the iron is completely cooled before storing it. Soldering iron stands are available to hold the iron while it cools.

CAUTION

Don't lay a hot iron down on a surface where it could cause damage or cause nearby materials to catch fire. It is best to place the iron in a holder to allow it to cool off.

- Do not touch recently soldered joints. They will be hot, and you could be burned.
- Wash your hands thoroughly with soap and water after handling solders, fluxes, cleaners, and chemicals.
- Never eat, drink, or smoke while soldering or in the presence of a soldering operation.
- Maintain soldering torches in good working condition by keeping them clean, replacing worn hoses, and ensuring that connections are tight and leakproof.

CAUTION

As an apprentice, you may not be qualified to mix acids to chemically clean metal or to purge tanks. You also may not be qualified to independently check the condition of soldering torches or hoses. Always check with your supervisor.

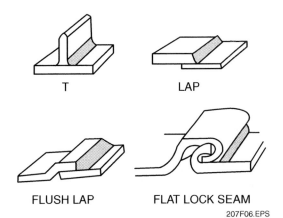

T   LAP

FLUSH LAP   FLAT LOCK SEAM

207F06.EPS

Figure 6 ◆ Examples of joint designs.

- Always keep soldering torch hoses away from flames and heated metal.
- Never use a tank that has contained flammable material until it has been thoroughly cleaned, purged, and vented.

WARNING!
In addition to causing rashes or burns to your skin and eyes, chemicals used in soldering can also damage equipment. Follow the manufacturer's guidelines for promptly cleaning up spills and properly disposing of cleaning materials.

8.0.0 ◆ PRACTICE TASKS

The following practice tasks will help you build several different soldering skills. You will practice cleaning, forging, and tinning a soldering iron, and have opportunities to solder different seam and metal types.

Some of these tasks involve situations that are potentially hazardous. You will need to protect your skin, eyes, and lungs from burns and fumes that have a real potential for harm or could even be fatal. So be certain to pay attention to your instructor's direction concerning the appropriate personal protective equipment for the task.

8.1.0 Task 1: Preparing a Soldering Iron

In this task, you will clean and forge (shape) a soldering iron, then tin the iron. Before proceeding with this task, inspect the point of the soldering iron. File the point to a smooth finish if you see any of the following:

- Holes or pits
- Scale or corrosion
- An oxidized surface
- Old tinning

DID YOU KNOW?
KEVLAR® Gloves

KEVLAR®, the material used in bulletproof vests, was invented by chemist Stephanie Kwolek. She invented it in 1964 while working at the DuPont Chemical Company. This lightweight material, which is five times stronger than steel, was first sold in 1971. In addition to bulletproof vests, it is used in canoes, tires, sports equipment, brake pads, firefighter clothing, helmets, and cut-resistant gloves.

Tools and Materials
- Appropriate personal protective equipment
- 1- or 1½-pound soldering iron
- Coarse-cut flat file and fine file
- **Sal ammoniac block**
- Wire or bar 50/50 tin-lead solder
- Soldering furnace
- Ball peen hammer
- Vise
- Anvil
- Soap, water, and a clean cloth

WARNING!
Sal ammoniac (ammonium chloride), used as a flux, releases toxic and irritating fumes when heated, including ammonia and hydrogen chloride. Always use sal ammoniac in a well-ventilated area and avoid directly inhaling its vapors.

If the soldering iron is so misshapen that filing the point will not form it properly, it must be forged to the proper shape. To clean and forge the soldering iron, follow these steps:

Step 1 Heat the soldering iron until it is cherry red.

Step 2 Place the soldering iron in a vise and file the part to be forged with the coarse-cut flat file to remove all scale (*Figure 7*).

Step 3 Place the shank end of the soldering iron against the anvil and force the misshapen point back into the body of the iron with a ball peen hammer (*Figure 8*).

Step 4 Reheat the soldering iron and place it on the anvil. Using a ball peen hammer, strike it with heavy blows to forge the point and body (*Figure 9*).

Step 5 Repeat the forging operation, if necessary.

Step 6 Reheat the body of the soldering iron and clamp it in the vise again. File off any remaining pits and old tinning.

DID YOU KNOW?
Sal Ammoniac

Sal ammoniac is a mineral containing ammonium that forms on volcanic rocks near fume-releasing vents. It is sold in blocks and is used to clean and tin soldering iron tips.

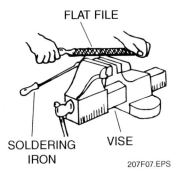

Figure 7 ◆ Filing the soldering iron.

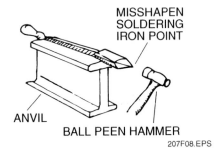

Figure 8 ◆ Forging a misshapen point.

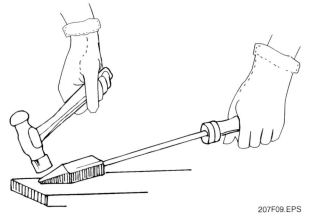

Figure 9 ◆ Forging the point and body of the soldering iron.

Tinning a soldering iron means covering the point with a film of solder. A correctly tinned soldering iron is a must for a good soldering job. Tinning keeps scale and corrosion from forming on the soldering iron and, therefore, allows more heat to flow. To tin a soldering iron, follow these steps:

Step 1   Heat the soldering iron. Do not overheat or underheat. Overheating will cause the solder to burn off; underheating will prevent the solder from adhering.

Step 2   Place the iron in the vise.

Step 3   File the iron with a fine file to remove the rough edges of the corners of the point.

Step 4   Reheat the iron enough to melt the solder.

Step 5   After applying a small amount of solder to the iron, rub the tip back and forth on the sal ammoniac block until the iron is tinned (*Figure 10*). Repeat the process, if necessary.

WARNING!

The heat from the soldering iron causes toxic fumes to rise from the sal ammoniac block. Do not inhale these fumes. You must wear the appropriate respiratory equipment and solder only in a well-ventilated area.

Step 6   Clean up your work area.

8.2.0 Task 2: Flat Lap Seam

In this task, you will solder a lap seam in the horizontal position (*Figure 11*). To hold the pieces together, you will tack the lap seam. Tacking is a process that temporarily holds pieces in position until you begin soldering. You must tack a lap seam as often as necessary to hold the pieces in position, usually at 1- or 1½-inch intervals. As an alternative, you may rivet or spot-weld the seam to hold it in position. Sometimes, lap seams and flat bars may be held in place by a needle-nosed vise and clamps.

Tools and Materials

- Appropriate personal protective equipment
- Two pieces of 2" × 6" galvanized iron, 24-gauge or lighter
- Properly tinned 1-pound soldering iron
- Wire or bar 50/50 tin-lead solder
- Soldering flux and acid swab or brush
- Soldering furnace

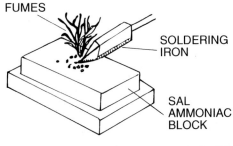

Figure 10 ◆ Rubbing the soldering iron on the sal ammoniac block.

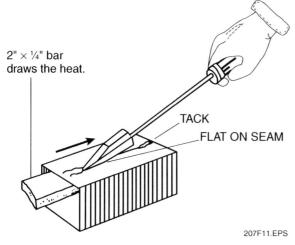

2" × ¼" bar draws the heat.

TACK

FLAT ON SEAM

207F11.EPS

Figure 11 ◆ Soldering a lap seam in the horizontal position.

- Sal ammoniac block
- **Transite** or other insulated backing sheet material
- Soap, water, and a clean cloth
- Flat metal bar, 2" × ¼"
- C-clamp vise grip

To complete this task, follow these steps:

Step 1 Place the item to be soldered in the proper position on the metal bar.

Step 2 Dip an acid swab or brush in the flux and apply it to the seam area.

Step 3 Heat the soldering iron in the furnace and re-tin it if necessary.

Step 4 Pick up some solder with the top of the soldering iron.

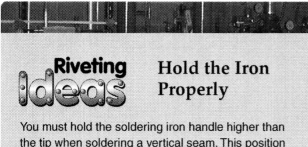

Riveting Ideas **Hold the Iron Properly**

You must hold the soldering iron handle higher than the tip when soldering a vertical seam. This position allows you to control the flow of solder. If you hold the tip higher than the handle, molten solder will run down the seam and pool at the bottom. Molten solder can also run down the soldering iron and over the handle. Holding the soldering iron improperly results in an incorrect seam and a big mess.

Step 5 Hold the soldering iron at one end of the seam until the heat penetrates the metal.

Step 6 Use a piece of wood as a weight to hold the seam together while tacking.

Step 7 Tack at 1- or 1½-inch intervals until the metal is securely positioned.

Step 8 Start at one end of the seam and hold the soldering iron with a tapered side flat along the seam until the solder flows freely into the seam.

Step 9 Draw the soldering iron slowly along the seam and add as much solder as necessary without raising the soldering iron from the metal. You may have to use two soldering irons for a long seam. It may be necessary to raise the soldering iron from the seam to reheat it, but it is best to limit the number of times you have to raise the iron before the seam is done. You must learn how to move the iron along the seam at the right speed: fast enough to keep from having to constantly reheat the iron but slow enough to ensure a strong, tight seam.

Step 10 Cool the seam and then clean the area using an appropriate cleaning method to remove all flux residue and prevent corrosion.

Step 11 Clean up your work area. Wash your hands with soap and water.

8.3.0 Task 3: Vertical Lap Seam

Vertical or upright seams are more difficult to solder than flat or horizontal seams because the liquid solder flows away from the top of the seam and downward. When working with vertical seams, it is best to **sweat solder** the seams before assembly. In this process, heat the metal and allow the solder to cover the area, then assemble the parts, clamp them together, and reheat the metal. The reheated metal causes the solder to flow and complete the joint (*Figure 12*).

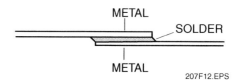

METAL

SOLDER

METAL

207F12.EPS

Figure 12 ◆ Position of the solder (shaded area) in a lap seam.

Tools and Materials

- Appropriate personal protective equipment
- Two pieces of 2" × 6" galvanized sheet metal, 24-gauge or lighter
- Properly tinned and sized soldering iron of appropriate weight
- Wire or bar 50/50 tin-lead solder
- Soldering flux and swab or brush
- Soldering furnace
- Sal ammoniac block
- Nails or clamps and wood (for backing)
- Soap, water, and a clean cloth

To complete this task, follow these steps:

Step 1 Place the item to be soldered on a flat surface and swab the surfaces with soldering flux.

Step 2 Heat the soldering iron in the furnace and re-tin the surface, if necessary.

Step 3 Add some solder to the soldering iron and lay the side flat along the seam of one of the surfaces to be joined until the solder begins to flow freely onto the surface.

Step 4 Draw the soldering iron along the seam until it is completely covered with solder.

Step 5 Repeat the procedure with the other piece of metal until it is also completely covered with solder.

Step 6 Place the tinned point of the heated soldering iron toward the bottom edge of the seam and keep the handle higher than the tip. This process is called pointing up the seam.

Step 7 Rub the tip of the soldering iron back and forth across the seam. This process is the sweating operation.

Step 8 When the sweating operation is completed, place the tip of the reheated soldering iron at the top of the seam. Keep the handle of the soldering iron above the tip (*Figure 13*).

Step 9 Add a small amount of solder and move the soldering iron back and forth across the seam, making small ridges. This process is called ridging (*Figure 14*).

Step 10 Cool the seam and then clean the area using an appropriate cleaning method to remove all flux residue and prevent corrosion.

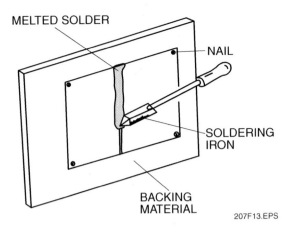

Figure 13 ◆ Vertical seam being soldered on backing material.

207F13.EPS

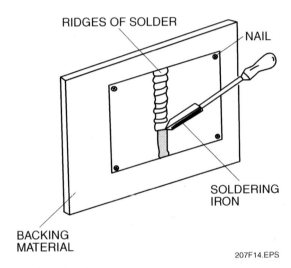

Figure 14 ◆ Ridging.

207F14.EPS

Step 11 Clean up your work area. Wash your hands with soap and water.

8.4.0 Task 4: Grooved Lock Seam

In this task, you will form, set, and sweat solder a grooved lock seam. Tacking is not necessary because the grooved lock holds the seam together. It is necessary to correctly apply flux to the mating surfaces and to clean the lock area physically and chemically.

Tools and Materials

- Appropriate personal protective equipment
- Two pieces of 2⅜" × 6" galvanized sheet metal, 24-gauge or lighter
- A number-three hand groover and anvil
- Ball peen or sheet metal hammer
- A properly tinned 1-pound soldering iron

- Wire or bar 50/50 tin-lead solder
- Soldering flux and brush or swab
- Soldering furnace
- Sal ammoniac block
- Piece of wood or transite (for backing)
- Hand seamer
- Soap, water, and a clean cloth
- Wire brush or an emery cloth

To complete this task, follow these steps:

Step 1 Place the pieces to be formed and soldered on a workbench and physically clean the mating surface. Use a wire brush or an emery cloth to remove oxidation. If the pieces are dry, use a clean cloth to remove oils.

Step 2 Bend and form the marked edges of the grooved seam.

Step 3 Swab the mating surfaces of the edges with soldering flux.

Step 4 Place the formed edges together and set the seam.

Step 5 Place a properly tinned and heated soldering iron on the seam with the flat edge of the soldering iron against the seam (*Figure 15*). Heat the metal until solder melts along the seam.

Step 6 Move the soldering iron slowly until you can see solder along the seam area. Reheat the soldering iron if necessary.

Step 7 Cool the seam and then clean the area using an appropriate cleaning method to remove all flux residue and prevent corrosion.

Step 8 Clean up your work area. Wash your hands with soap and water.

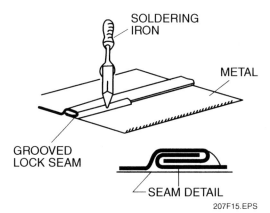

Figure 15 ◈ Soldering a grooved lock seam.

8.5.0 Task 5: Riveted Seam

In this task, you will sweat solder a riveted seam. Riveted seams are often soldered to make them watertight, and sweat soldering is the most effective method to use. It is best to solder the mating surfaces before placing the rivets for proper solder penetration and bonding. Pay special attention to solder on the rivet heads. Brush on flux for good solder adhesion.

Tools and Materials

- Appropriate personal protective equipment
- Two pieces of 2½" × 6" galvanized sheet metal, 24-gauge
- A rivet set and six 1¼-inch pop rivets
- Sheet metal hammer
- Drill with ⅛-inch drill bit
- Center punch
- Metal scribe
- Combination square
- Properly tinned medium-weight soldering iron
- Wire or bar 50/50 tin-lead solder
- Soldering flux and brush or swab
- Soldering furnace
- Sal ammoniac block
- Piece of wood or transite (for backing)
- Anvil
- Soap, water, and a clean cloth

To complete this task, follow these steps:

Step 1 Place the workpieces on a workbench, lay out the rivet holes, and center punch the marked locations.

Step 2 Drill through both pieces of metal and remove the metal burrs left from the drill.

Step 3 Physically clean the mating edges of the seam.

Step 4 Apply soldering flux, wipe clean, and place the rivets in the proper positions.

Step 5 Heat a soldering iron and clean it on a sal ammoniac block, if necessary.

Step 6 Place the soldering iron on the seam and allow the metal to heat until it melts the solder (*Figure 16*).

Step 7 Move the flat edge of the soldering iron slowly along the seam until the solder is drawn into the seam all along the length. Inspect to ensure that solder has penetrated the seam.

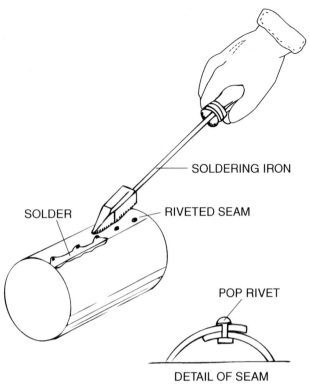

Figure 16 ◈ Soldering a riveted seam.

Step 8 Cool the seam and then clean the area using an appropriate cleaning method to remove all flux residue and prevent corrosion.

Step 9 Clean up your work area. Wash your hands with soap and water.

8.6.0 Task 6: Soldering a Bottom Seam

In this task, you will make solder beads and solder a bottom seam on a round container. (Your instructor may have you work on a square or rectangular container.) Solder beads, or shots, are sometimes used to solder seams on containers. You make them in advance and allow them to cool. You then place the beads along the seam and apply heat so that the solder will flow into the seam. Using beads is a neater and more efficient method of soldering in tight or curved seams.

Tools and Materials

• Appropriate personal protective equipment
• Prefabricated metal container, preferably round
• Wire or bar 50/50 tin-lead solder
• Properly tinned medium-weight soldering iron

• Soldering furnace
• Soldering flux and brush or swab
• Sal ammoniac block
• Soap, water, and a clean cloth

To complete this task, follow these steps:

Step 1 Properly heat, clean, and tin a soldering iron.

Step 2 Hold the solder against the hot soldering iron and allow the beads to drop onto a clean surface and cool (*Figure 17*).

Step 3 Apply flux to the seam and wipe it into the seam opening.

Step 4 Reheat, clean, and place the soldering iron against the seam (*Figure 18*).

Step 5 Drop one of the cold beads of solder into the bottom of the container and position it against the seam.

Step 6 Hold the flat side of the soldering iron in one position until the solder bead begins to melt and flow freely into the seam area.

Step 7 Draw the soldering iron slowly along the seam, adding more solder beads as needed.

Step 8 Reheat the soldering iron if necessary.

Step 9 Continue this procedure until the seam is filled and capillary action has drawn enough solder into the seam.

Step 10 Cool the seam and then clean the area using an appropriate cleaning method to remove all flux residue and prevent corrosion.

Step 11 Clean up your work area. Wash your hands with soap and water.

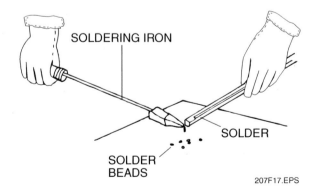

Figure 17 ◈ Making solder beads.

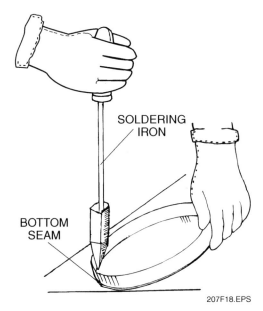

Figure 18 ◆ Soldering a bottom seam.

8.7.0 Task 7: Soldering Aluminum

In this task, you will prepare and properly solder a selected aluminum sheet. Aluminum alloys are generally more difficult to solder than many other metals. The difficulty is that aluminum alloys are covered with a layer of oxide, which must be removed. Many aluminum alloys can, however, be successfully soldered if proper techniques are followed.

Tools and Materials

- Appropriate personal protective equipment
- Two pieces of 2" × 6" aluminum sheet, 0.025-inch thick
- Wire or strip aluminum solder (tin-zinc or tin-cadmium)
- Aluminum soldering flux and brush or swab
- Properly tinned medium-weight soldering iron
- Soldering furnace
- File, metal scraper, or stainless steel wire brush
- Piece of wood or transite (for backing) and clamps
- Soap, water, and a clean cloth

To complete this task, follow these steps:

Step 1   Physically clean the mating surfaces completely and remove the oxide with a file, metal scraper, or stainless steel wire brush.

Step 2   Use a corrosive soldering flux to chemically clean the mating surfaces.

 **Riveting Ideas** **Capillary Action**

Solder is drawn into joints by a process called capillary action. To observe capillary action, conduct this simple experiment. Pour a small amount of water into a dish. Notice that after flowing out a bit, the water forms an irregular blob. This happens because the water molecules stick to one another and flow together. Put the corner of a paper towel into the water blob. The water immediately flows onto the towel. This is capillary action. The same thing happens when solder is heated. The molecules of solder, like those of water, stick to one another and flow together into the gap that you are filling.

Step 3   Tin the surface of the area to be joined with the aluminum solder.

Step 4   Apply flux to the solder and again to the work area.

Step 5   Apply a properly heated soldering iron to the work surfaces and work the aluminum solder well into the surfaces until both surfaces are tinned.

Step 6   Place the two tinned surfaces together and clamp in place.

Step 7   Sweat the solder by working the flat edge of the soldering iron over the surfaces.

Step 8   Cool the seam and clean the area using an appropriate cleaning method to remove all flux residue and prevent corrosion.

Step 9   Clean up your work area. Wash your hands with soap and water.

8.8.0 Task 8: Soldering Stainless Steel

In this task, you will prepare stainless steel surfaces for soldering and solder a selected stainless steel sheet. Like aluminum, stainless steel is a difficult metal to solder. The chromium in stainless steel forms an oxide that is very hard to remove.

 CAUTION

It is important to ensure that the stainless steel is not overheated. Excess heat will discolor the metal.

Tools and Materials

- Appropriate personal protective equipment
- Two pieces of 2" × 6" stainless steel sheets, 24-gauge
- Wire or bar 50/50 tin-lead solder
- Stainless steel flux
- Hydrochloric acid
- Fiber brush or swab
- Properly tinned and sized soldering iron of appropriate weight
- Soldering furnace
- Emery cloth or sandpaper
- Soap, water, and a clean cloth

To complete this task, follow these steps:

Step 1   Physically clean the surfaces to be joined.

Step 2   Roughen the surface with emery cloth or sandpaper. After the surfaces have been cleaned and roughened, do not touch them with your hands to prevent oils from your fingers from contaminating the surfaces.

Step 3   Cover the surfaces thoroughly with hydrochloric acid and wipe clean with a damp cloth.

Step 4   Coat the surfaces with a stainless steel flux.

DID YOU KNOW?

Metal Food Containers

Metal food containers must be soldered with tin and antimony solder (95 % tin, 5 % antimony). Solder for food containers cannot contain lead, cadmium, or zinc. These metals, which are associated with serious health risks, could leach into the food.

Step 5   Use a properly cleaned, tinned, and heated soldering iron to tin the prepared surfaces.

Step 6   Lay the flat edge of the soldering iron on the top surface of the area to be joined and allow the metal to melt the solder. Finish sweat soldering the seam.

Step 7   Cool the seam and then clean the area using an appropriate cleaning method to remove all flux residue and prevent corrosion.

Step 8   Clean up your work area. Wash your hands with soap and water.

1. The temperature at which solder begins to melt is called the _____.
   a. liquidus
   b. solidus
   c. fusion point
   d. heat fusion index

2. The process of coating metal with solder is called _____.
   a. floating or wetting
   b. flooding or wetting
   c. wetting or tinning
   d. wetting or laminating

3. The chief ingredient of noncorrosive flux is _____.
   a. rosin
   b. dry salt
   c. muriatic acid
   d. hydrochloric acid

4. _____ is one of the most important steps in the soldering process.
   a. Cooling the metal
   b. Applying the flux
   c. Surface preparation
   d. Joint fit-up

5. The most widely used joint in soldering is the _____.
   a. T-joint
   b. flat lock seam
   c. L-joint
   d. lap joint

6. For soldering purposes, tanks that have contained flammable material must _____.
   a. be cleaned, purged, and vented
   b. never be used for a soldering operation
   c. be cleaned with a solution of baking soda and water
   d. be tagged with a "Do Not Use" sticker

7. If the point of a soldering iron has holes or pits, you must _____.
   a. replace the point
   b. discard the iron
   c. re-tin the point
   d. file the point

8. To reshape a soldering iron, you will _____ it.
   a. chemically clean
   b. physically clean
   c. forge
   d. re-tin

9. Generally it is best to tack lap seams at _____.
   a. every 0.75-inch
   b. 1- to 1.5-inch intervals
   c. 2- to 3-inch intervals
   d. every 6 inches

10. A layer of oxide that is difficult to remove makes _____ difficult to solder.
    a. iron and aluminum
    b. copper and tin
    c. aluminum alloys and stainless steel
    d. galvanized steel and stainless steel

Summary

This module introduced you to the tools, equipment, and techniques used in soldering, a method of joining metal pieces with a tin-lead alloy. You learned about different types of solders and fluxes and how to properly tin and use a soldering iron.

You also had an opportunity to put what you learned to practical use in a number of soldering tasks. As you practiced these tasks, you saw first-hand how solder behaves when it is heated. You learned why paying attention to safety guidelines is important as you worked with the hot soldering iron and saw the fumes rising from the sal ammoniac block. By paying attention to proper techniques and safety, you will soon be soldering joints safely, efficiently, and effectively.

Notes

Flux: A fusible substance used in welding and soldering that helps fuse the metals and that prevents surface oxidation.

KEVLAR®: A silky, man-made fiber that is stronger than steel and is used in a wide variety of products including tires, sporting goods, and protective gloves.

Lap joint: A joint formed by overlapping the edges of metal sheets.

Liquidus: The temperature at which a metal or an alloy becomes completely molten or liquid. Compare with *solidus.*

Muriatic acid: The commercial name for a diluted solution of hydrochloric acid.

Oxidation: The reaction of metal to a chemical compound, such as oxygen, that can weaken a metal weld.

Pointing up: The process of using the tip of a soldering iron to distribute molten solder across a seam. Most often used with vertical seams.

Sal ammoniac block: A material used to clean and tin soldering iron tips.

Solidus: The temperature at which a metal or an alloy first begins to melt. Compare with *liquidus.*

Sweat solder: The process of coating metal pieces with solder, pressing them together, and then reheating the metal to melt the solder and form a bond.

Tacking: A soldering or welding method used to hold metal pieces in place temporarily until a solid seam can be made.

Tinning: The process of applying a thin coat of solder to metal parts to be joined. Also called wetting.

Transite: A heat-resistant board made from Portland cement and nonasbestos fibers.

Wetting: The process of applying a thin coat of solder to metal parts to be joined. Also called tinning.

Additional Resources

This module is intended to present thorough resources for task training. The following reference work is suggested for further study. This is optional material for continued education rather than for task training.

Principles of Soldering and Brazing, 1992. Giles Humpston et al. Materials Park, OH: ASM International.

References

ASTM International, 100 Barr Harbor Drive, West Conshohocken, Pennsylvania, USA

NCCER makes every effort to keep these textbooks up-to-date and free of technical errors. We appreciate your help in this process. If you have an idea for improving this textbook, or if you find an error, a typographical mistake, or an inaccuracy in NCCER's Contren® textbooks, please write us, using this form or a photocopy. Be sure to include the exact module number, page number, a detailed description, and the correction, if applicable. Your input will be brought to the attention of the Technical Review Committee. Thank you for your assistance.

Instructors – If you found that additional materials were necessary in order to teach this module effectively, please let us know so that we may include them in the Equipment/Materials list in the Annotated Instructor's Guide.

Write: Product Development and Revision
National Center for Construction Education and Research
3600 NW 43rd St, Bldg G, Gainesville, FL 32606

Fax: 352-334-0932

E-mail: curriculum@nccer.org

Craft _____ Module Name _____

Copyright Date _____ Module Number _____ Page Number(s) _____

Description _____

_____

_____

_____

(Optional) Correction _____

_____

_____

(Optional) Your Name and Address _____

_____

_____

04208-08

Basic Piping Practices

04208-08
Basic Piping Practices

Topics to be presented in this module include:

Overview

Pipe is made from a variety of metal alloys such as carbon steel, stainless steel, and aluminum, as well as from several types of plastic. No matter what material the pipe is made of, the basic piping tasks include measuring, cutting, threading, joining, and supporting pipe in a piping system. In this module, you will learn the basic skills for working with pipe, as well as the characteristics of different pipe materials. Knowing these characteristics will help you prepare and install each type of pipe correctly.

Objectives

When you have completed this module, you will be able to do the following:

1. Measure, cut, and join selected types of pipe.
2. Describe the materials from which pipe is made.
3. List some of the applications for selected pipe materials.
4. Describe the common types of pipe fittings, hangers, and supports.

Trade Terms

ABS (acrylonitrile-
  butadiene-styrene)
Bearing plate
Black iron pipe
CPVC (chlorinated
  polyvinyl chloride)
Downhand welding
Flame cutting
Galvanized pipe
Gouging
Laminate construction
Machining
MIG (metal inert gas)
Plain end (PE)

Polyethylene (PE)
Polypropylene (PP)
PVC (polyvinyl chloride)
Radiant heating system
Shielded metal-arc
  welding
Slag
Solvent weld
Thermoplastic
Thermoset
Thread engagement
Threaded and coupled
  (T&C)
TIG (tungsten inert gas)

Required Trainee Materials

1. Appropriate personal protective equipment

Prerequisites

Before you begin this module, it is recommended that you successfully complete *Core Curriculum; Sheet Metal Level One; Sheet Metal Level Two*, Modules 04201-08 through 04207-08.

This course map shows all of the modules in the second level of the Sheet Metal curriculum. The suggested training order begins at the bottom and proceeds up. Skill levels increase as you advance on the course map. The local Training Program Sponsor may adjust the training order.

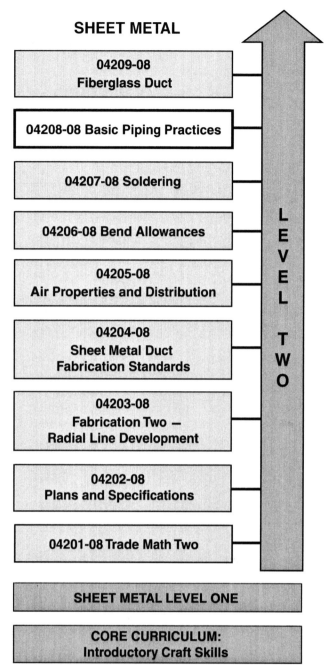

SHEET METAL

04209-08
Fiberglass Duct

04208-08 Basic Piping Practices

04207-08 Soldering

04206-08 Bend Allowances

04205-08
Air Properties and Distribution

04204-08
Sheet Metal Duct
Fabrication Standards

04203-08
Fabrication Two —
Radial Line Development

04202-08
Plans and Specifications

04201-08 Trade Math Two

LEVEL TWO

SHEET METAL LEVEL ONE

CORE CURRICULUM:
Introductory Craft Skills

208CMAP.EPS

1.0.0 ◆ INTRODUCTION

Knowing the characteristics of different types of piping will help you to choose the right technique, regardless of what type of pipe is called for by the job. For example, you will know not to cut threads into pipe that is designed to be **solvent welded**. Certain types of plastic piping should not be threaded. Learning about pipe materials will help you select the right pipe for the application. Some pipe, for example, is designed to resist corrosion or high pressures and temperatures. Other kinds of pipe will not react with chemicals, so they are suitable for applications in which chemicals are distributed through the piping system.

In working with pipe, you must always review the engineering drawings; be sure you have the latest revision, and read the manufacturer's instructions and specifications. The instructions will describe the recommended joining and supporting methods. The specifications will describe available pipe sizes, the application for which the pipe is designed, and the temperature and pressure ratings for the piping material. You must also consult the local building code, which may place restrictions on everything from the type of pipe material allowed to acceptable methods of joining.

2.0.0 ◆ MEASURING PIPE

Pipes are usually sized according to their inside diameter (ID). For example, a plastic pipe with a ⅞-inch outside diameter is called a ½-inch pipe because it has a ½-inch inside diameter.

In measuring pipe for installation, you must measure the length of a pipe running from one fitting to the next, as well as the **thread engagement**. This is the part of a threaded pipe that is screwed into a fitting. Pipes must fully extend into fixture and fitting sockets, or the joint could leak. Socket depths vary from one pipe size and material to another, so account for the depth of each fitting's socket in the total length of pipe needed between fittings.

The methods for measuring pipe include end-to-end, end-to-center, end-to-face, center-to-center, and face-to-face (*Figure 1*). It is important to allow for fitting dimensions and for the length of the thread engagement or the depth of the pipe end that will be inserted into a socket joint.

- *End-to-end* measures the full length of the pipe, including the threads on both ends.
- *End-to-center* measure is used for a length of pipe with a fitting screwed on one end only. The measurement is equal to the length of the pipe plus the length of the thread engagement minus the end-to-center dimension of the fitting.

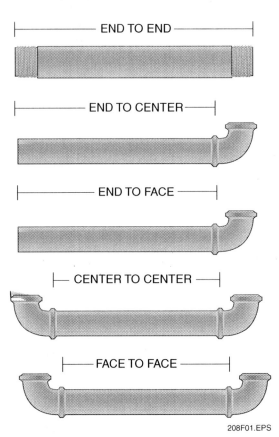

Figure 1 ◆ Pipe-measuring methods.

- *End-to-face* measure is also used for a length of pipe with a fitting screwed on one end only. The measurement is equal to the length of pipe plus the length of the thread engagement.
- *Center-to-center* measure is used to measure pipe that has a fitting screwed to both ends. The measurement is equal to the length of the pipe plus two times the length of the thread engagement minus the sum of the end-to-center dimensions of the fittings.
- *Face-to-face* measure is also used to measure pipe that has a fitting screwed to both ends. The measurement is the length of pipe plus two times the length of the thread engagement.

3.0.0 ◆ CUTTING AND REAMING PIPE

You can cut metal pipe with a hacksaw and miter box (*Figure 2*) or a tube cutter (*Figure 3*). The tube cutter allows you to cut pipe quickly and efficiently. However, the cutting process leaves burrs (rough or sharp edges) on the inside of cut pipe. These burrs can restrict flow through the pipe, so you must use a reamer to remove them (*Figure 4*). Some tube cutters come with a reamer built into the tool.

To cut and ream pipe, follow these steps:

Step 1 Measure and mark the pipe to be cut.

Step 2 Turn the cutter around the pipe, tightening the cutting wheel one-quarter revolution with each turn.

CAUTION
Avoid overtightening the cutting wheel on a tube cutter. Overtightening could damage or break the wheel.

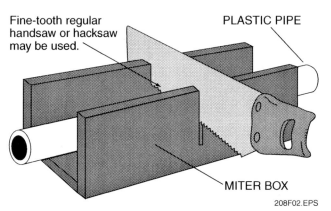

Fine-tooth regular handsaw or hacksaw may be used.

PLASTIC PIPE

MITER BOX

208F02.EPS

Figure 2 ◆ Cutting with a saw and miter box.

208F03.EPS

Figure 3 ◆ Tube cutter.

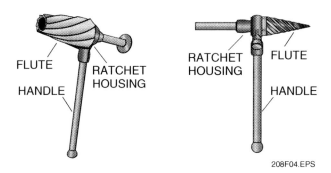

FLUTE

HANDLE

RATCHET HOUSING

RATCHET HOUSING

FLUTE

HANDLE

208F04.EPS

Figure 4 ◆ Pipe reamers.

WARNING!
Wear appropriate personal protective equipment when cutting and reaming pipe. The chips of pipe material are sharp and can cut you. Always use a rag to wipe off chips. Never use your bare hand.

Step 3 Ream the burrs from the inside of the pipe.

To cut plastic pipe, use a hacksaw. Place the pipe into a miter box to hold it in place while you cut. This cutting method also produces burrs on the cut end of the pipe. When you push the pipe into the fitting, burrs can scrape away the solvent cement used to join the pipe, which can weaken the bond. Before you push plastic pipe into the fitting, use a knife or a file to remove all burrs from the pipe.

CAUTION
When cutting plastic pipe, avoid making diagonal cuts. Diagonal cuts reduce the bonding area at the deepest part of the fitting socket and can result in a weak joint.

4.0.0 ◆ THREADING PIPE

Pipes are available in two styles, **plain end**, which is abbreviated PE, and **threaded and coupled**, which is abbreviated T&C. You may see these abbreviations used on specifications, drawings, inventories, and order forms. To produce a threaded assembly, you must cut threads on unthreaded metal or plastic pipe. You can cut pipe threads by hand, with a die and stock, or with an electric pipe-threading machine (*Figure 5*). To cut threads using a hand die and stock, follow these steps:

Step 1 Select the correct size die for the pipe being threaded.

CAUTION
Not all sizes of plastic pipe can be threaded. Check the manufacturer's instructions before threading plastic pipe.

Step 2 Inspect the die to make sure that the cutters are free of nicks and wear.

Step 3 Lock the pipe securely in the vise.

Step 4 Slide the die over the end of the pipe, guide end first.

Step 5 Push the die against the pipe with the heel of one hand. Take three or four short, slow, clockwise turns. Keep the die pressed firmly against the pipe. When enough thread is cut to keep the die firmly against the pipe, apply some thread cutting oil. Friction can cause the pipe to overheat; the oil prevents overheating and lubricates the die. Oil the threading die after every two or three downward strokes.

Step 6 Back off a quarter turn after each full turn forward to clear out the metal chips. Continue until the pipe projects one or two threads from the die end of the stock. Too few threads are as bad as too many threads.

Step 7 To remove the die, rotate it counter-clockwise.

Step 8 Wipe off excess oil and any chips with a rag.

WARNING!

When wiping away chips or excess oil, use a rag or brush, not your bare hand or an air hose. The chips are sharp and can cut you.

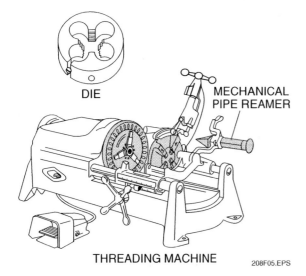

DIE

MECHANICAL PIPE REAMER

THREADING MACHINE

208F05.EPS

Figure 5 ◆ Pipe die and pipe-threading machine.

Threading machines differ from manufacturer to manufacturer. You must read the manufacturer's operating procedures before you operate the machine. Pay special attention to the safety and maintenance instructions. A poorly maintained machine can become a safety hazard. To cut threads using a threading machine, follow these steps:

Step 1 Select and install the correct size die and inspect it for nicks.

Step 2 Mount the pipe into the machine chuck. Be sure to support longer pipe lengths.

Step 3 Check the pipe and die alignment.

Step 4 Cut threads until two threads appear at the other end of the die, and then stop the threading action. (Apply cutting oil during the threading operation.)

Step 5 Back off the die until it is clear of the pipe.

Step 6 Remove the pipe from the machine chuck. Be careful not to mar the threads.

Step 7 Wipe the pipe clean of oil and chips with a rag.

5.0.0 ◆ PIPE JOINING METHODS

The method you will use to join pipe depends on the material the pipe is made from and how the pipe is to be used. You must follow the manufacturer's instructions and specifications for joining pipe. This section discusses several commonly used pipe joining methods. As always, before you do any tasks in construction work, you must be aware of the potential dangers involved and know how to protect yourself and others from injury or death.

5.1.0 Pipe Joining Safety

Pipe joining involves the use of potentially hazardous materials and substances. In welding, for example, you will use intense heat sources, and the molten metals may give off harmful fumes. Solvent welding involves the use of toxic and flammable materials, so it is very important to pay attention to safety. Here are some general safety guidelines you must follow when joining pipe by welding:

• When you are working with welding equipment, materials, and solvents, read the manufacturer's MSDS. The MSDS will give details on known hazards, exposure limits, safe handling practices, and what to do in case of an accident.

- Read and follow the manufacturer's operating instructions for mechanical equipment and tools.
- Always wear the appropriate personal protective equipment. For welding operations, this includes proper clothing in addition to face and respiratory protection. For example, cuffless trousers, a shirt with a button-down collar, and pockets with button-down flaps all ensure that no welding sparks can fall into folds or pockets where they could smolder and burn you (*Figure 6*).
- If others are in your work area during welding operations, you must set up welding shields and ensure that everyone nearby is wearing goggles to protect their eyes from the sudden bright flash that occurs when a welding operation begins.
- When the welding shield on your welding headgear is down, it will restrict your field of vision. You must have a co-worker watch you and the surrounding work area in case of fire or other emergency. The co-worker should also rope off the area to keep other workers away.

- Stay alert. Do not do any pipe-joining operations if you are taking medication that could make you feel drowsy. Never use any substance that could impair your concentration.
- Be aware of your work area. Do not weld in the presence of sparks or flame.

The following are general safety guidelines that you must adhere to when joining pipe with solvents:

- Solvent cements and primers are flammable. The vapors from these products can ignite or explode. Never use solvents or primers in the presence of open flames, sparks, or sources of ignition.
- Store solvent cements and primers, with their lids tightly fastened, in a cool, dry area. While using these products, keep the container closed as much as possible. Avoid breathing the vapors. Always work in a well-ventilated area. If the MSDS requires it, you must wear an approved respirator.
- Do not smoke, eat, or drink while using solvent cements and primers.

EYE PROTECTION

HOOD

VENTILATOR

POCKET WITH
BUTTON-DOWN FLAP

LONG-SLEEVED SHIRT

GAUNTLET-TYPE LEATHER
WELDING GLOVES

PANTS (NO CUFF)

HIGH-TOP LEATHER BOOTS

208F06.EPS

Figure 6 ◆ Personal protective equipment for welding.

- When working with solvents and primers, avoid contact with skin, eyes, and clothing. Wash any clothing that has been splashed with solvents or primers before wearing it again.
- Stay alert. Do not do any pipe-joining operations if you are taking medication that could make you feel drowsy. Never use any substance that could impair your concentration.
- Be aware of your work area.

5.2.0 Welded Joints

Welding is usually the best method for joining large-diameter pipe. When properly made, welded joints are leakproof and stronger than the material the pipe is made from. However, because you must not weld in the presence of flammable materials, you can only weld in those areas that present no danger of fire. The welding process also causes the formation of icicles on the inside of the pipe joint. Icicles are irregular bits of weld metal that drip through the joint and harden inside the pipe. Icicles decrease the pipe's diameter and can obstruct flow inside the pipeline. In most cases, though, this is not a serious condition.

There are several types of welds used in the welding process. In working with pipe, you will use either the butt weld or the socket weld. Butt welding is used to join pipe of equal diameters. The equal diameters allow you to butt one end of pipe snugly against another pipe end. Socket welding is used to join pipes of unequal diameter. Insert a straight length of pipe into a pipe fitting with an opening that is slightly larger than the outside diameter of the straight pipe, then weld the circumference.

The edges of the base metal must be prepared, or shaped, for welding. The type of edge preparation depends on the thickness of the metal. Edge preparation techniques include **machining**, **gouging**, **flame cutting**, shearing, and bending. If the base metal is thin, the edges can simply be squared without additional machining or cutting. When the base metal is thicker (about ³⁄₁₆-inch), the edges are beveled by machining or flame cutting.

Pipes can be welded in any position. When possible, rotate pipes so that all welding can be done in the **downhand welding**, or horizontal position. More often, though, welds must be made with the pipe in a fixed position, so the weld itself may change position as it moves around the pipe. The weld may move from vertical to horizontal, to flat, to overhead. The American Welding Society provides recommended positions, called test positions, for all welding positions.

There are several types of welding methods. Some of the more commonly used methods are discussed in the following sections.

5.2.1 Shielded Metal-Arc Welding

Shielded metal-arc welding (SMAW) is also called stick welding. The stick is a consumable metal electrode connected to an electric current. The metal to be welded is also connected to the current. The welder touches the tip of the electrode to the metal and then draws it away to form an electric arc. The heat generated by the arc melts the surface of the base metal. It also completely melts the electrode, which fills the weld with molten metal. The electrode and the metal are shielded with flux to prevent oxidation, which is caused by oxygen and nitrogen in the surrounding atmosphere. Oxidation can decrease the effectiveness of the final weld. During the welding process, **slag** forms. The slag further protects the weld against oxidation, but it must be removed after each pass of the electrode to ensure a good weld. The shielded metal-arc welding process is fairly simple and requires only a power supply, power cables, and an electrode holder. It is commonly used in construction, shipbuilding, and pipeline work, especially in remote locations.

 WARNING!
During arc welding, even a reflected arc can harm your eyes. Be sure to wear eye protection at all times. You can seriously injure your eyes or even go blind if you fail to protect your eyes. Also, if you wear contact lenses, you should not wear them during the welding process. The ultraviolet rays produced by the process may dry out the moisture beneath the lenses, causing them to stick to your eyes.

5.2.2 TIG Welding

TIG (tungsten inert gas) welding is also known as gas tungsten-arc welding. This method uses a nonconsumable tungsten electrode. A chemically inert gas, such as argon or helium, is used to shield the metal from oxidation. The heat from the arc formed between the electrode and the metal melts the edges of the metal. If necessary, metal for the weld is added by placing a wire filler rod in the arc or at the point of the weld. This welding method can be used on almost all metals, including aluminum, magnesium, and titanium. TIG is a precise method. It allows the operator to control the heat so that thin sections of metal can be joined without burn-through. TIG produces a weld of high purity with the best appearance of any welding process. Usually, no slag is produced, and no cleanup is required. However, TIG takes more time than the other welding processes.

The hot filler rod is exposed to the air and can become contaminated. It also tends to cost more than other welding methods.

5.2.3 MIG Welding

MIG (metal inert gas) welding is also known as gas metal-arc welding. In this process, the electrode is shielded from the air by surrounding it with an inert gas such as argon or carbon dioxide or by coating it with flux. In addition, deoxidizers in the electrode prevent oxidation in the molten weld, making multiple weld layers possible at the joint. MIG welding is faster than shielded metal-arc welding and TIG. The temperatures involved in MIG welding are relatively low and are therefore suitable for thin sheet and metal sections less than ¼-inch wide. The MIG process can easily be automated, and it is operated robotically in welding operations in manufacturing plants.

5.2.4 Submerged-Arc Welding

Submerged-arc welding is similar to gas metal-arc welding, but no gas is used to shield the weld. Instead, the arc and tip of the filler wire are submerged beneath a layer of granular, fusible flux. The flux forms a thick protective layer that completely covers the molten area and prevents spatter and sparks. It also acts as a thermal insulator, thus allowing deeper heat penetration. The process is limited to welding in a horizontal position and is used for high-speed sheet or plate steel welding with automatic or semiautomatic machines. In the submerged-arc welding process, the flux can be recovered, treated, and reused. Submerged-arc welding is a high-productivity process and is used in the construction of large ships and oil rigs. It is also used for fabricated plate, pressure valve tool and die, pipe fabrication, tube mill, and rolling machinery applications.

5.2.5 Resistance Welding and Thermite Welding

In the resistance-welding method, electrodes are clamped on each side of metal parts that are to be welded. The parts are put under a great amount of pressure, and a heavy current is quickly and briefly run through the electrodes. Resistance occurs at the spot where the two metals meet. This resistance generates the heat to melt the metal and create the weld. Resistance welding is suitable for the repetitive welds made by machines.

In thermite welding, heat is generated by a chemical reaction. When a mixture of aluminum powder and iron oxide (known as thermite) is

ignited, the aluminum and oxygen unite, generating the heat that melts the metal. Thermite welding is used mainly in welding breaks or seams in heavy iron and steel sections. It is also used to weld the rail for railroad tracks.

5.3.0 Threaded Joints

Threaded pipe allows for quick installation, and the joints can be made leak resistant with sealants. A wide range of fittings is available for threaded-pipe systems, making this method of joining pipe easy, versatile, and practical. Most of the materials from which pipe is made, including plastic, can be threaded. Threads are located on the inside or the outside of the pipe. Threads on the inside of a pipe are often called female threads. Threads on the outside are often called male threads.

There are some disadvantages, however, to threaded pipe. Because the threads are cut into the pipe surface, threading reduces the outside diameter of the pipe, which, in turn, decreases the strength of the pipe at the joint. In addition, of all pipe joints, threaded joints are the most prone to leaking.

You can use either pipe joint compound or Teflon® tape to join threaded metal pipe. Note that some local codes may restrict the use of Teflon® tape on certain applications. Always check the local code for restrictions on any joining materials.

Apply the joint compound to the male threads before assembling a pipe connection. Do not apply the compound to the threads of the fitting. When using Teflon® tape, apply the tape in a clockwise direction, the same direction the fitting turns.

Start the fitting onto the threaded pipe by hand. Make sure that the threads on the pipe and the fitting mesh smoothly to avoid cross-threading. Cross-threading will cause the threads to lock up. Turn the fitting in a clockwise direction. Finish tightening the fitting with a pipe wrench (*Figure 7*).

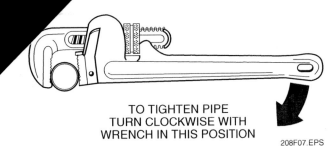

TO TIGHTEN PIPE
TURN CLOCKWISE WITH
WRENCH IN THIS POSITION

208F07.EPS

Figure 7 ◆ Use a pipe wrench to tighten the fitting.

5.4.0 Flanged Joints

Flanges are used to join pipe on installations that require easy access for repairs or service. Installation is easy. You place a gasket between two flanges and bolt the flanges together. To repair or service the pipe, all you have to do is unbolt the flanges and remove the gasket. Because the flange does not come into contact with the material that flows through the pipeline, it does not have to be made of the same material as the pipe.

5.5.0 Soldered Joints

Soldering, sometimes called sweating, is used to join copper pipe. In soldering, capillary action draws molten solder into the joint, making a strong leak-resistant bond (*Figure 8*). To solder joints, first brush a light, even coating of flux or soldering paste onto the surfaces to be joined. Then heat the metal with a soldering iron and apply the solder to the joint. The flux helps fuse the metal joint and prevents oxidation. As molten solder flows into the joint, the flux burns away. Remove the heat when solder drips, or sweats, from the pipe. Lightly brush the joint with a damp rag, taking care not to burn your fingers.

5.6.0 Solvent-Cemented Joints

Most plastic and fiberglass reinforced pipe and fittings are joined with solvent cements. Not all types of plastics can be cemented, and some types of plastic require specially formulated cements. Before using solvent cement, you must read the manufacturer's instructions for the following information:

- Safe handling practices.
- The types of plastic pipe on which you can use the solvent cement.
- The recommended temperature for application. Most solvent cements are designed for application within a temperature range of 40°F to 90°F, but some permit application in temperatures as low as 10°F.

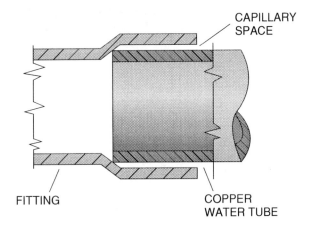

CAPILLARY SPACE

FITTING

COPPER WATER TUBE

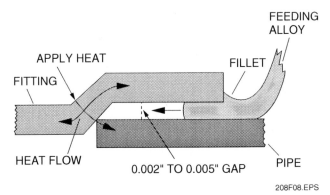

APPLY HEAT

FITTING

HEAT FLOW

FEEDING ALLOY

FILLET

0.002" TO 0.005" GAP

PIPE

208F08.EPS

Figure 8 ◆ Capillary action.

- The size limits. Some solvent cements are designed to join pipe up to six inches in diameter; a stronger cement may be required on pipes of larger diameters.
- Cleaning requirements. You must remove surface dirt, grease, and grime that will interfere with the joint. Follow the manufacturer's instructions for the proper cleaning solution.
- Priming requirements. The primer has two purposes. It cleans the portion of the pipe and fitting to be connected, and it conditions the plastic so that the solvent cement will adhere. Primer is often a different color from the solvent cement. You must not use a primer designed for one plastic on another type of plastic. Note that some types of plastic do not require the use of a primer.

No matter what type of plastic you are joining, you should always dry-fit the pipe before applying the solvent cement. Because of the close tolerances between pipe and fittings, a dry pipe may not slide to the full depth of the socket. You should be able to push it in about one-third of the way. Remember that the solvent cement will lubricate and soften the pipe to permit a full-depth fit. Pipes are dry-fitted to ensure that they will not wobble in the fitting socket. If the pipe feels loose, try another fitting. Solvent cements set-up (dry) fairly quickly, so you might not have a chance to correct a potentially costly mistake if you skip the dry-fitting step. Follow these steps to join plastic pipe (*Figure 9*):

Step 1 Apply primer, if required.

Step 2 Apply a thick coat of cement to the outside of the pipe (*Figure 9A*).

Step 3 Apply a thin coat of cement to the inside of the fitting (*Figure 9B*).

Step 4 Insert and twist the pipe one-quarter turn into socket. Hold for at least 30 seconds (*Figure 9C*).

Step 5 Look at the joint. A bead of solvent cement should form around the completed joint.

Step 6 Allow the cement to set-up. Set-up time may vary depending on the type of the cement and the temperature. Read the manufacturer's directions for drying time.

Step 7 Wipe excess cement from the joint (*Figure 9C, inset*).

Step 8 Allow the joints to completely dry for at least one hour before the pipes are tested or put into service.

5.7.0 Fusion-Welded Joints

Fusion welding may only be done on certain types of **thermoplastic** pipe. In this process, a special welding tool is heated either electrically or by a fuel gas. A filler plastic of the same composition as the pipe is added and fused to the pipe and the fitting.

5.8.0 Grooved Joints

The concept of grooved piping dates to World War I when armies in the field needed a quick and efficient way to join pipe. Since then, the pipe-grooving method has undergone continuous development, and today it is considered a primary means of joining pipe.

A grooved pipe joint consists of grooved pipe ends machine-cut to exact manufacturer's standards, a synthetic rubber gasket, a lubricant, and a coupling consisting of two housings (*Figure 10*). A grooved-pipe system includes pipe fittings and valves with grooved ends, as well as specially designed gaskets.

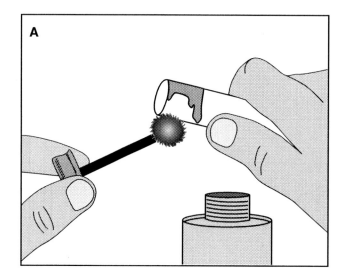

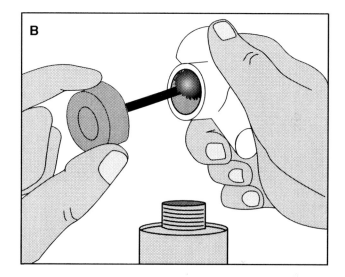

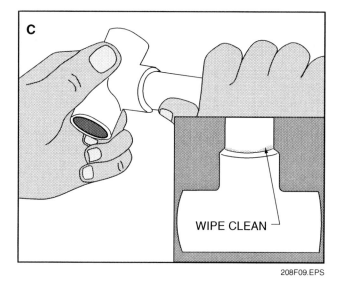

WIPE CLEAN

208F09.EPS

Figure 9 ◆ Joining pipe and fittings using solvent cement.

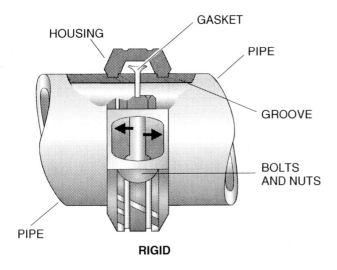

RIGID

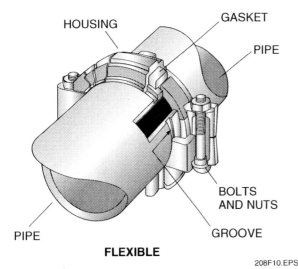

FLEXIBLE

208F10.EPS

Figure 10 ◆ Rigid and flexible grooved pipe couplings.

To join grooved pipe, follow these steps:

Step 1 Review the manufacturer's instructions.

Step 2 Apply a thin coat of lubricant to the gasket lips and the outside of the gasket.

Step 3 Check the pipe ends. They must be free from indentations, projections, or roll marks to ensure a leakproof seal.

Step 4 Install the gasket over the pipe end. Be sure the gasket lip does not hang over the pipe end.

Step 5 Align and bring the two pipe ends together. Slide the gasket into position and center it between the grooves on each pipe. Make sure that no part of the gasket extends into the groove on either pipe.

Step 6 Assemble the housing segments loosely, leaving one nut and bolt off to allow the housing to swing over the joint.

Step 7 Install the housing, swinging it over the gasket and into position into the grooves on both pipes.

Step 8 Insert the remaining bolt and nut. Be sure that the bolt track head engages the recess in the housing.

Step 9 Tighten the nuts alternately and equally to maintain metal-to-metal contact at the angle bolt pads.

6.0.0 ◆ PIPE MATERIALS

Throughout history, people have made pipes from materials that were readily available. The ancient Babylonians, for example, made pipes of clay baked with straw. Other civilizations used terra cotta, copper, or lead to fabricate their piping systems. In the United States, pipes made of wood were used until the early 1800s, when the rapid growth of cities made it necessary to switch to a more durable material—cast iron. No matter what type of piping was used in the past, people had to deal with the same issues that pipe installers deal with today: cutting and fitting the pipe sections together, changing the direction of the pipe, supporting it at the proper pitch, and making the pipes as leakproof as possible.

The type of pipe used for piping jobs is chosen with several factors in mind: suitability for the application, ease of maintenance, durability, and cost. It would be wasteful, for example, to use an expensive, specialized pipe material when a more economical pipe will meet all the requirements of the application. Also, some types of pipe materials are not suitable for use in systems that must carry certain chemicals. The job specifications will detail what type of pipe is to be used. You should also follow the engineering drawings and manufacturer's instructions for specific information about how the pipe is to be used, installed, and supported.

6.1.0 Carbon Steel Pipe

Carbon steel is an alloy of iron and carbon. The term carbon steel is often used to avoid confusion with stainless steel. Carbon steel pipe is used for a wide variety of applications. The most common include the following:

- Hot and cold water distribution
- Steam and hot water heating systems
- Air-piping systems
- Gas-piping systems, using flexible pipe (*Figure 11*)
- Drainage, waste, and vent (DWV) systems

The two common types of carbon steel pipe are **black iron pipe** and **galvanized pipe**. Black iron pipe gets its color from the carbon in the steel. Black iron pipe is used where corrosion will not affect its uncoated surfaces. It is most often used in gas or air-pressure piping.

Galvanized steel pipe is dipped in molten zinc. This process coats the pipe—protecting the surfaces from abrasive, corrosive materials—and gives the pipe a dull, grayish color.

Carbon steel pipe is manufactured in three weights or strengths: standard (Schedule 40), extra strong (Schedule 80), and double extra strong (Schedule 120). Some manufacturers use the terms extra heavy instead of extra strong and double extra heavy instead of double extra strong. The cross sections in *Figure 12* show the inside diameters (IDs) and outside diameters (ODs) for 1-inch nominal-size steel pipe in three weights.

Because all weights have the same outside diameter, the same threading dies will fit all three weights of pipe. The wall thickness and the inside diameter differ with each weight. The thicker the wall, the smaller the inside diameter, but the more pressure the pipe will withstand. Generally, stan-

dard weight is adequate for most plumbing situations.

The width of carbon steel pipe is stated as a nominal size, which is based on the inside diameter. Nominal size is measured in inches. Carbon steel pipe is manufactured in sizes ranging from ⅛-inch up to 12 inches (ID). Carbon steel pipes wider than 12 inches are measured by their outside diameter.

Carbon steel pipe is manufactured in lengths of 21 feet, threaded or unthreaded. Larger sizes (6 inches and larger) are made in lengths between 18 and 22 feet. For ease of handling, this pipe is shipped in bundles. However, the number of lengths per bundle is determined by the size of the pipe. A tag is attached to each bundle, labeling the number of feet within the bundle. Consult a manufacturer's catalog to determine the number of feet contained in specific-sized bundles.

6.2.0 Copper Pipe

Copper pipe is also called copper tubing. Copper pipe is classified into color-coded types and uses (*Table 1*).

Copper tubing may be soft or hard. Soft copper tubing is commercially available in sizes from ⅛-inch OD to 1⅝-inch OD. It is usually sold in 25-, 50-, or 100-foot coil lengths. Soft copper tubing is joined by soldering or with flared or other mechanical fittings. It is easily bent or shaped, but it must be held in place by clamps or other hangers because it cannot support its own weight.

Hard copper tubing is widely used in commercial refrigeration and air conditioning systems (*Figure 13*). It is also used for inlet and outlet piping in water heaters (*Figure 14*) and for natural gas. It comes in straight lengths of 20 feet and in sizes from ⅜-inch OD to over 6 inches OD. Hard copper tubing is designed for use with formed fittings to make the necessary bends or changes in directions. It is more self-supporting and therefore needs fewer supports than soft copper tubing.

Copper pipe is also used in **radiant heating systems**. A radiant heating system is a network of hot water pipes embedded in the floor or ceiling (*Figure 15*). Radiant heating systems warm a room through radiation, which is the discharge of heat through space from one body to another. Tubing for a radiant heating system can be laid in a concrete slab or hung underneath a wood floor. Because they are located in or below the floor or ceiling, radiant heating systems do not get in the way when the room is painted or cleaned. However, they are more complex to design and install than other heating systems.

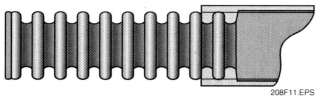

208F11.EPS

Figure 11 ◆ Flexible steel tubing for gas.

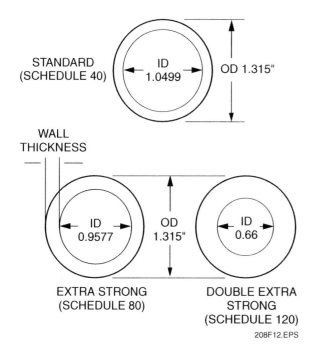

STANDARD (SCHEDULE 40)

ID 1.0499 OD 1.315"

WALL THICKNESS

ID 0.9577 OD 1.315"

ID 0.66

EXTRA STRONG (SCHEDULE 80)

DOUBLE EXTRA STRONG (SCHEDULE 120)

208F12.EPS

Figure 12 ◆ Inside and outside diameters.

Table 1 Types of Copper Pipe

Tube Type	Color Code	Standard	Application [1]	Commercially Available Lengths [2]		
				Nominal or Standard Sizes	Drawn	Annealed
Type K	Green	ASTM B 88 [3]	Domestic Water Service and Distribution Fire Protection Solar Fuel/Fuel Oil HVAC Snow Melting Compressed Air Natural Gas, Liquified Petroleum (LP) Gas Vacuum	**Straight Lengths** ¼-inch to 8-inch 10-inch 12-inch **Coils** ¼-inch to 1-inch 1¼-inch and 1½-inch 2-inch	20 ft. 18 ft. 12 ft. — — — —	20 ft. 18 ft. 12 ft. 60 ft. 100 ft. 60 ft. 40 ft. 45 ft.
Type L	Blue	ASTM B 88	Domestic Water Service and Distribution Fire Protection Solar Fuel/Fuel Oil Natural Gas, Liquified Petroleum (LP) Gas HVAC Snow Melting Compressed Air Vacuum	**Straight Lengths** ¼-inch to 10-inch 12-inch **Coils** ¼-inch to 1-inch 1¼-inch and 1½-inch 2-inch	20 ft. 18 ft. — — — —	20 ft. 18 ft. 60 ft. 100 ft. 60 ft. 40 ft. 45 ft.
Type M	Red	ASTM B 88	Domestic Water Service and Distribution Fire Protection Solar Fuel/Fuel Oil HVAC Snow Melting Vacuum	**Straight Lengths** ¼-inch to 12-inch	20 ft.	N/A
DWV	Yellow	ASTM B 306	Drain, Waste, Vent HVAC Solar	**Straight Lengths** 1¼-inch to 8-inch	20 ft.	N/A
ACR	Blue	ASTM B 280	Air Conditioning Refrigeration Natural Gas, Liquified Petroleum (LP) Gas Compressed Air	**Straight Lengths** ⅜-inch to 4⅛-inch **Coils** ⅛-inch to 1⅝-inch	20 ft. —	Special Order 50 ft.
OXY, MED, OXY/MED, OXY/ACR, ACR/MED	(K) Green (L) Blue	ASTM B 819	Medical Gas Compressed Medical Air Vacuum	**Straight Lengths** ¼-inch to 8-inch	20 ft.	N/A

[1] There are many other copper and copper alloy tubes and pipes available for specialized applications. For information on these products, contact the Copper Development Association, Inc.

[2] Individual manufacturers may have commercially available lengths in addition to those shown in this table.

[3] Tube made to other ASTM International standards is also intended for plumbing applications, although ASTM B 88 is by far the most widely used. ASTM Standard Classification B 698 lists six plumbing tube standards including B 88.

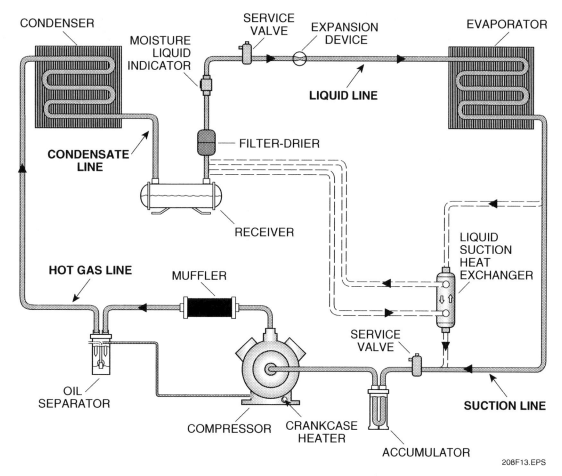

Figure 13 ◆ Refrigeration system pipeline layout.

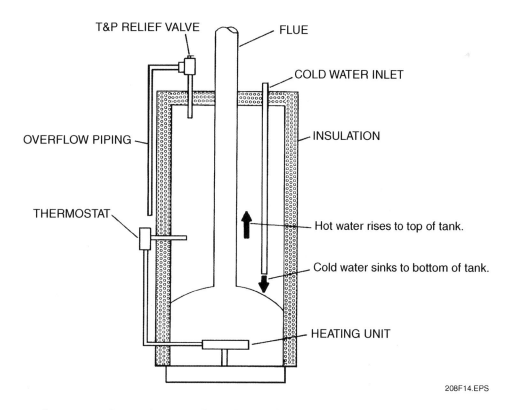

Figure 14 ◆ Cross section of a gas water heater showing inlet and outlet lines.

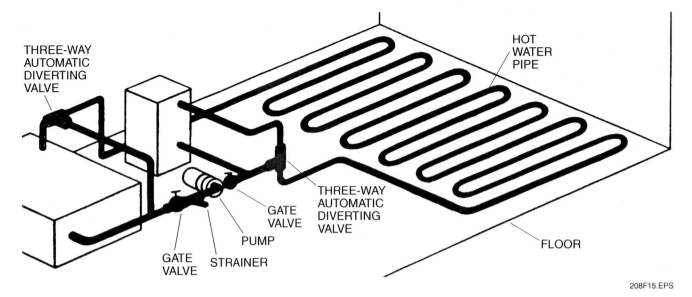

Figure 15 ◆ A radiant heating system installation in a floor.

6.3.0 Copper-Nickel Pipe

The addition of copper to nickel creates an alloy that is stronger than nickel and able to resist many types of corrosion. Because of its high resistance to rapidly flowing seawater, copper-nickel pipe is used in piping systems for shipboard services, coastal installations, desalination plants, and off-shore oil production. Copper-nickel pipes that contain a small percentage of iron are used to fabricate pipes for heat interchangers and condensers.

Copper-nickel pipe is joined by welding. The most consistent quality and higher productivity is obtained when the pipe can be rotated for downhand welding. Downhand welding is done with the weld and base metal in an almost horizontal position. It is sometimes called flat position welding.

6.4.0 Stainless Steel Pipe

Stainless steel is an iron alloy made up chiefly of chromium and small amounts of nickel. In addition to its ability to resist staining, stainless steel offers high resistance to corrosion. It is used in applications where high purity is required, such as in hospitals, research labs, and nuclear power plants. Schedule 40 steel pipe is the most commonly used piping material in natural gas systems. It is able to retain its strength at both high and low temperatures. Stainless steel pipe can be joined by welding, threading, or flanging. However, welding is the most common method used to join stainless steel pipe.

 DID YOU KNOW?

Solderless Copper Piping Connections

The ProPress™ system for connecting copper piping uses proprietary fitting connections and a special tool to allow a connection of copper tubing without the use of solder. Fittings are available for types K, L, and M hard copper tuning from ½" to 2", and for soft copper tubing from ½" to 1½".

To preserve its corrosion-resistant qualities, you must not allow stainless steel to come into contact with material other than stainless steel during its installation. Handle the steel with special slings made for use only with stainless steel, and support it with hangers that will not scratch the surface of the metal. Often stainless steel **bearing plates** are installed between the pipe support and the pipe to protect the pipe's surface.

6.5.0 Chromium-Molybdenum Pipe

The addition of molybdenum to chromium steels with low carbon content increases the ability of the steel to harden as well as its ability to resist corrosion. This makes chromium-molybdenum (usually shortened to chrome-moly) pipe suitable for high-pressure, high-temperature applications such as steam lines in power plants. Chrome-moly pipe is joined by welding.

6.6.0 Aluminum Pipe

The term *aluminum* has come to stand for a large group of alloys, each having a particular application. Aluminum is corrosion resistant, so it is often used in industrial HVAC circulating systems. In addition, because it retains its strength at very low temperatures, aluminum is well suited for low-temperature applications.

Aluminum pipe is most commonly joined by TIG welding, but it can also be threaded or flanged. To retain the corrosion-resistant qualities of the metal, welding is best.

As with stainless steel pipe, you must take care when supporting aluminum pipe. You should use only aluminum or padded hangers. It is possible to use galvanized hangers, but they will be good only as long as the galvanized coating lasts.

6.7.0 Other Pipe Alloys

The use of specialized alloys is increasing to meet the demands of new developments in technology. While there are far too many alloys to include in this module, the following three have become common.

6.7.1 Hastelloy®

Hastelloy® is a trade name of the Union Carbide Corporation. This alloy is well-suited for pipes used in the chemical processing industry to carry highly corrosive materials, such as boiling sulfuric acid. Pipe made of Hastelloy® is available in sizes up to 30 inches. It can be joined by welding, threading, or flanging.

6.7.2 Monel®

Monel® is a trade name of the International Nickel Corporation. This alloy has good resistance to salt, lye, and sulfuric acid. It is used for pipes that handle salt water. Pipe made of Monel® is available in sizes up to 12 inches. It can be joined by welding, threading, or flanging. It is most commonly welded with the TIG process.

6.7.3 Inconel®

Inconel® is a trade name of the International Nickel Corporation. Because this alloy has good resistance to acids in foods, such as fruit juices, it is used for pipes in the food processing industry. It also has applications in the pharmaceutical industry because it does not contaminate the contents of the pipeline. Inconel® is also used for pipes that carry dry chlorine gas, which is highly corrosive. Pipe made of Inconel® can be joined by welding or threading.

6.8.0 Plastic Pipe

Many different types of plastic are used to make plastic pipe. Like pipes made of metal, plastic pipes must be chosen to suit the application. It is important to follow the manufacturer's recommendations as well as local code when selecting and installing plastic pipe. Although each type of plastic has its own characteristics, in general the main benefits of plastic pipe are as follows:

- Resists corrosion, chemicals, and biological matter
- Offers low thermal conductivity and friction loss
- Is lightweight and flexible

Plastic pipe also has drawbacks, including its tendency to expand and contract more than other pipe materials. It can give off harmful fumes when burned.

Plastic piping can be joined by several methods, including solvent cementing, heat fusion, threading, and flanging. Like copper pipe, plastic pipe is color coded for ease of identification (*Table 2*).

6.8.1 PVC Pipe

PVC (polyvinyl chloride) pipe is used widely in plumbing applications. PVC offers high-impact tensile strength and resistance to most salt solutions, acids, and bases. It can be installed above or below ground. It is usually white. It is often used in high-efficiency gas furnaces to supply outdoor air for combustion and to exhaust the combustion products to the outdoors (*Figure 16*).

Table 2 Plastic Pipe Uses and Color Codes

Use	Color
Gas	Yellow or black with yellow stripes (formerly bright orange or tan)
Water	Black, light blue, white, clear, or gray
DWV	Black, white, tan, or gray
Sewers	Green, white, black, or gray
Hot and cold water	Tan, red, white, blue, silver, or clear
Fire sprinklers	Orange
Reclaimed water	Purple or brown

* Do not depend on color as the only means of identification of a pipe or tube for an application, or a pipe or tube currently in service. Roll marking and other methods of system identification must also be used.

PVC pipe is suitable for service up to 140°F. However, at a constant internal or external temperature of 110°F, PVC has a tendency to sag. Horizontal installation at this temperature or higher requires continuous support along the length of the run.

Even though PVC's rate of thermal conductivity (ability to conduct heat) is lower than that of most of the other plastic piping materials, it rapidly expands or contracts in response to alternating hot and cold temperatures. Therefore, expansion joints or offsets may be required in the system.

PVC pipe and fittings can be joined using solvent cement or by threading or flanging. PVC is available in sizes from ¼-inch to 16 inches in diameter and is classified as either Schedule 40 or Schedule 80. Only Schedule 80 PVC should be threaded; conventional pipe dies can be used for this purpose. However, the dies must be sharp and should only be used to thread plastic pipe. Threaded connections to Schedule 40 PVC pipe can be made with threaded pipe adapters. PVC flanges are available for both Schedule 40 and Schedule 80 pipe. They may be joined with threaded or solvent-cemented joints.

6.8.2 CPVC Pipe

CPVC (chlorinated polyvinyl chloride) pipe is a variation of PVC pipe. It is a corrosion-resistant pressure piping designed to withstand higher temperatures than PVC. It was introduced in 1960, and since then has gained wide acceptance.

CPVC is used primarily for piping hot water and chemicals. It is sometimes used in radiant heating systems. It is rated at 215°F. CPVC is self-insulating. This means that its thermal conductivity is a tiny fraction of that of copper tubing. Thus, heat remains inside the pipe and hot water is delivered more quickly. This reduces the danger of burns from touching hot water lines. The problem of condensation on cold water pipes is reduced. There is also less likelihood of sludge or scale buildup within the water distribution system.

CPVC is available in the same sizes as copper tubing. Sizes range from ½-inch to 2 inches in diameter. The pipe, which is tan, is often manufactured to a standard dimension ratio, which ensures that the pipe and fittings have the same pressure rating regardless of size.

CPVC is also available in the same sizes as industrial iron pipe. Sizes range from ½-inch to 10

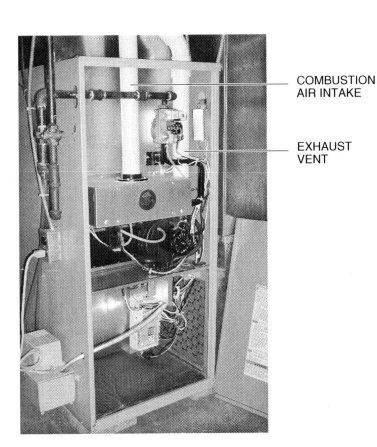

COMBUSTION AIR INTAKE

EXHAUST VENT

208F16.EPS

Figure 16 ◆ A high-efficiency gas furnace, showing PVC intake and exhaust lines.

inches in diameter. This type of CPVC pipe is gray and is used primarily with corrosive chemicals.

CPVC has the same chemical resistance as PVC, but it is not recommended for use with certain solvents, acids, and alcohols. CPVC pipe and fittings can be joined using solvent cement, threading, and flanges.

6.8.3 ABS Pipe

ABS (acrylonitrile-butadiene-styrene) pipe and fittings are used primarily for DWV systems, sewer mains, and some water distribution systems. Two grades of ABS are manufactured: DWV and service. The grade is identified by the label on the pipe. Most building codes require DWV grade ABS pipe within a structure. ABS pipe is black.

ABS pipe is nontoxic and resistant to household liquids. It retains its chemical properties over a temperature range of –40°F to 200°F, but it should not be used in applications where the temperature rises to over 180°F.

ABS pipe has a low rate of thermal transfer. In a DWV system, this means that hot water going to the sewer line retains most of its heat, allowing greasy waste materials to flow through the pipe and not stick to the sides.

ABS pipe may be joined with a designated solvent cement that does not require a primer. It may also be joined with mechanical connections. Threading is not recommended, but adaptors are available to make the transition to threaded materials.

6.8.4 Polyethylene Pipe

Polyethylene (PE) pipe is often used for underground applications such as lawn sprinklers, wells, and portable ice-skating rinks. It is also used in natural gas, irrigation, and potable water systems, and in radiant heating systems. It is more resistant to corrosives and chemicals than many other types of plastic and it will not deteriorate when exposed to the ultraviolet rays of the sun. It can be installed outdoors without a protective coating (*Figure 17*).

Because polyethylene is highly resistant to chemicals, this kind of pipe cannot be joined by solvent welding. It can be joined using heat fusion or with mechanical joints and clamps. Heat fusion is similar to a weld on steel; the joined parts essentially become one piece.

Polyethylene pipe is available in low, medium, and high density. Medium- and high-density pipe can be used in systems that operate under pressure. Polyethylene pipe is available in coils for sizes up to two inches in diameter. Larger sizes are made in rigid lengths.

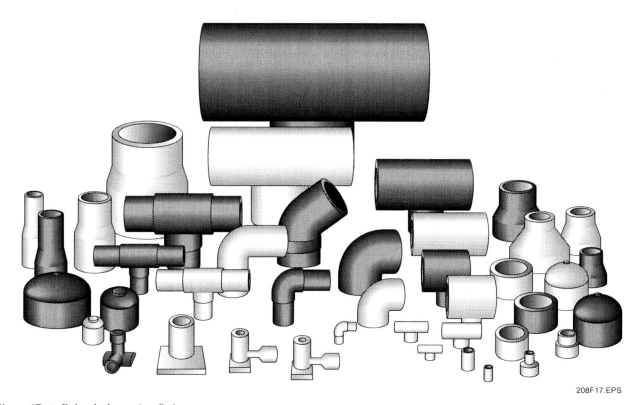

208F17.EPS

Figure 17 ◆ Polyethylene pipe fittings.

6.8.5 Polypropylene Pipe

Polypropylene (PP) pipe is the lightest of the plastic pipe materials discussed in this module. Polypropylene pipe is often used in irrigation, water treatment, and water distribution systems because of its resistance to soil conditions.

Even though the surface hardness of polypropylene is greater than that of other plastic piping materials, the surface tends to mar more easily than the surface of metal piping. Polypropylene also is less resistant to punctures than metal piping is. Therefore, even though polypropylene pipe requires no protection against soil corrosion, it should be protected from direct contact with hard materials during trenching operations.

You can join polypropylene pipe by socket fusion or fillet welding. The socket-fusion process uses a special 110-volt tool that provides the correct temperature for the fusion process to take place. The tool heats the joint for a specified amount of time at a temperature of 540°F. For ½-inch pipe, the time is about 10 to 15 seconds; for 4-inch pipe, the time is about 25 to 35 seconds. During this time, a bead of melted material appears around the complete circumference of the pipe.

The welding process uses a heated gas, not a burning gas, as is common with soldering or brazing. Generally, compressed air is used, but it is also possible to use nitrogen equipment. The welding rod, which must be of the same material as the pipe being welded, is held at a 75-degree angle to the joint while pressure is applied to the rod. The operator moves the torch in an arcing motion. For a full-fillet weld, three passes are required:

- Pipe to fitting
- Pipe to bead
- Fitting to bead

Because the weakest part of any plastic welded bead is the beginning, the end of the bead should overlap the beginning by at least ⅜- to ½-inch. Welding of polypropylene pipe is not recommended for pressure-rated systems.

6.8.6 Fiberglass Pipe

Several types of pipe are made with **thermoset** material. Two common types are fiberglass reinforced polyester resins and fiberglass epoxy resins. Both types are manufactured in a similar way. Fiberglass pipe is built layer on layer in a type of construction known as **laminate construction**. Fiberglass serves as the bonding agent for the resin. Properties of such pipes vary according to the material and the method of construction. Fiberglass pipe cannot be reshaped or bent after it is cured.

6.8.7 Plastic Pipe Standards and Labeling

Labels on plastic pipe indicate the quality of the product and the type of material. The four-digit code on plastic pipe identifies the pipe material, type, grade, and strength of the pipe to the nearest 100 pounds. An example of labeling is shown in *Figure 18*.

To assure users that plastic pipe and fittings are of high quality and suitable for specified systems, the National Sanitation Foundation (NSF) tests them. The American Society for Testing and Materials International (ASTM International) sets size and material standards for plastic pipe, and the NSF uses these standards in its testing program. Before a product can bear the NSF seal (*Figure 19*), the following criteria must be met:

- Each product line must pass the ASTM International test for size and material performance.
- The manufacturer must allow the NSF to make random inspections of the production steps, inspection techniques, and warehouse facilities.
- The manufacturer must show the NSF production records documenting the quality of the materials used.

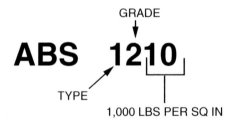

208F18.EPS

Figure 18 ◈ Plastic pipe label.

208F19.EPS

Figure 19 ◈ National Sanitation Foundation seal.

- If the NSF finds that a product is defective, the manufacturer must correct the problem and prove that the defective product has been removed from the market. Failure to do so will result in loss of the NSF seal.

7.0.0 ◆ PIPE FITTINGS

Pipe fittings are standardized parts used to join two or more sections of pipe together. They may be used to extend sections of pipe, to change the direction of a pipe, or to close off a section of pipe. Fittings are available in all of the same materials as pipe. They may be smooth edged, or they may come equipped with internal or external threads. As with pipe, you can also cut threads into fittings to suit a particular application.

7.1.0 Tees

Tees are used to make a branch connection that is 90 degrees to the main pipe. Tees have three openings or outlets: two runs (straight through) and one branch (side opening). If all three outlets are the same size, the fitting is called a regular tee (*Figure 20*). If the outlet sizes vary, the fitting is called a reducing tee (*Figure 21A*).

Tees are identified by their outlet sizes. The run dimensions are shown with the larger run stated first. Then the branch outlet dimension stated. For example, a tee with a run outlet inches, a run outlet of 1 inch, and a branch outlet of ¾-inch is a 2 × 1 × ¾ tee (*Figure 21B*).

7.2.0 Elbows

Elbows are often called ells for short. This fitting is used to change the direction of the pipe. Common ells include the 90-degree elbow, 45-degree elbow, and the street elbow (*Figure 22*). Elbows are also available to make 11¼-, 22½-, and 60-degree bends.

7.3.0 Unions

Unions make it possible to quickly disassemble a threaded-piping system. After disconnecting the union, you can unscrew the length of pipe on either end of the union. There are various types of pipe unions. The two most common are the ground joint and the flange.

The ground-joint union (*Figure 23*) connects two pipes by screwing the thread and shoulder pieces onto the pipe. Then, both the shoulder and thread parts are drawn together by the collar. This union creates a gastight or watertight joint.

The flange union (*Figure 24*) connects two separate pipes. Screw flanges onto the ends of the pipes to be joined, then connect the flanges using nuts and bolts. A gasket between the flanges makes the connection gastight and watertight.

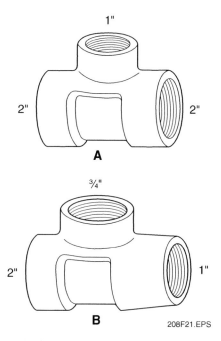

208F21.EPS

Figure 21 ◆ Reducing tees.

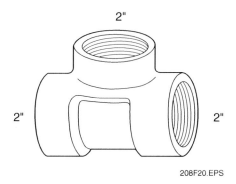

208F20.EPS

Figure 20 ◆ Regular tee.

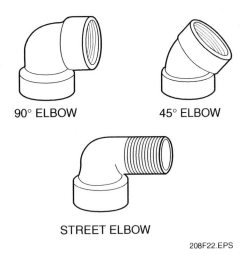

90° ELBOW   45° ELBOW

STREET ELBOW

208F22.EPS

Figure 22 ◈ Elbows.

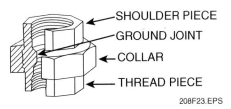

SHOULDER PIECE
GROUND JOINT
COLLAR
THREAD PIECE

208F23.EPS

Figure 23 ◈ Ground-joint union.

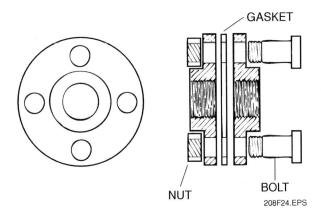

GASKET

NUT   BOLT

208F24.EPS

Figure 24 ◈ Flange union.

7.4.0 Couplings

Couplings (*Figure 25*) are short fittings with female threads in both openings. They are used to connect two lengths of pipe when making straight runs. Couplings cannot be used in place of unions because they cannot be disassembled.

7.5.0 Nipples

Nipples are pieces of pipe 12 inches or less in length, threaded on both ends and used to make extensions from a fitting or to join two fittings. Nipples are manufactured in many sizes. Examples of common types are shown in *Figure 26*.

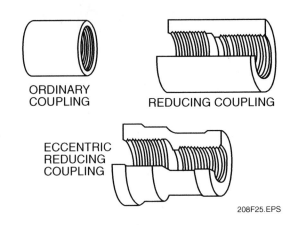

ORDINARY COUPLING

REDUCING COUPLING

ECCENTRIC REDUCING COUPLING

208F25.EPS

Figure 25 ◈ Types of couplings.

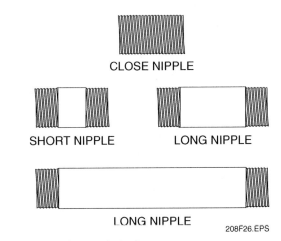

CLOSE NIPPLE

SHORT NIPPLE   LONG NIPPLE

LONG NIPPLE

208F26.EPS

Figure 26 ◈ Types of nipples.

7.6.0 Crosses

Crosses, also called pipe crosses, are four-way distribution devices (*Figure 27*). This fitting has four openings at right angles to each other. Each of the four openings is threaded for ease in making connections.

7.7.0 Plugs and Caps

Plugs are male-threaded fittings used to plug or close openings in other fittings. Plugs are available with a variety of heads (*Figure 28*).

Caps are fittings with a female thread (*Figure 29*). Like plugs, caps are used to close openings in other fittings. The cap fits on the male end of a pipe or nipple.

7.8.0 Bushings

Bushings are fittings with a male thread on the outside and a female thread on the inside (*Figure 30*). They are used to connect the male end of a pipe to a fitting of a larger size. A bushing has a hexagon nut at the female end.

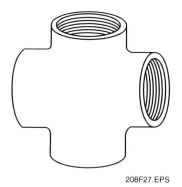

Figure 27 ◈ Pipe cross.

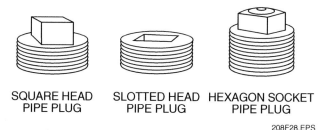

SQUARE HEAD PIPE PLUG

SLOTTED HEAD PIPE PLUG

HEXAGON SOCKET PIPE PLUG

208F28.EPS

Figure 28 ◈ Types of plugs.

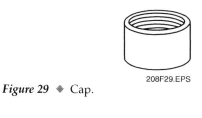

208F29.EPS

Figure 29 ◈ Cap.

208F30.EPS

Figure 30 ◈ Bushing.

8.0.0 ◈ PIPE HANGERS AND SUPPORTS

Heating, ventilating, and air conditioning (HVAC) systems, plumbing systems, and electrical conduits and fixtures require hangers and supports. Technically, the words hanger and support describe two different methods of holding pipe in place. A hanger suspends a single line of pipe from structural steel or concrete. A support carries the weight of one or several pipes and is made from structural steel or a combination of steel and concrete. In actual practice, however, the term *support* covers both methods.

There are several types of pipe supports and pipe support systems. Several styles of hangers used for supporting piping from the ceiling horizontally are shown in *Figure 31*. Piping can be sup-

ported on wood frame construction with several styles of pipe attachments. These include pipe hooks, U-hooks, hold-down clips, tin straps, half clamps, and pipe clamps (*Figure 32*). Several styles of pipe attachments are available for supporting pipes from the walls or the sides of beams and columns (*Figures 33* and *34*).

Support vertical pipe at each floor level using riser clamps (*Figure 35*). Mount the riser clamp so that the bracket supports the pipe weight directly on the floor. Sometimes it will be necessary to install hangers that are attached between floors to the walls or vertical structural members. These hangers will maintain vertical alignment and also support part of the pipe's vertical load. Use either an extension ring hanger and wall plate or a one-hole strap.

Other pipe attachments include universal pipe clamps (*Figure 36*) for use with standard 1⅝-inch or ½-inch channels (*Figure 37*). Insert the notched steel clamps by twisting them into position along the slotted side of the channel. The pipes can be aligned as close to one another as the couplings allow (*Figure 38*).

CLEVIS HANGER

ADJUSTABLE SWIVEL RING HANGER

STIRRUP

Anvil International
208F31.EPS

Figure 31 ◈ Hangers used to support pipe from the ceiling horizontally.

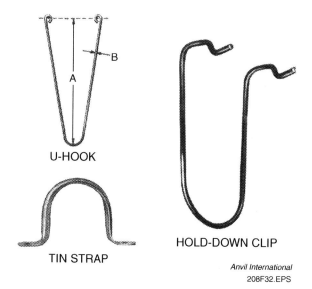

U-HOOK

TIN STRAP

HOLD-DOWN CLIP

Anvil International
208F32.EPS

Figure 32 ◈ Pipe supports for wood frame construction.

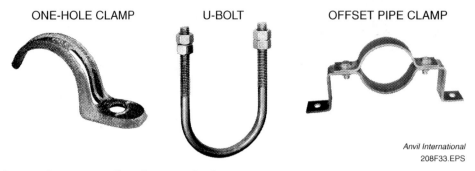

ONE-HOLE CLAMP   U-BOLT   OFFSET PIPE CLAMP

Anvil International
208F33.EPS

Figure 33 ◆ Attachments for use on walls or beams and columns.

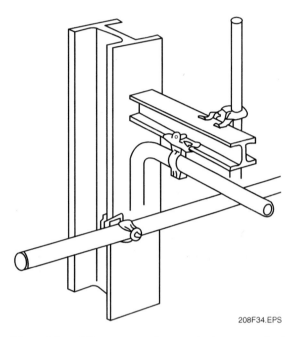

208F34.EPS

Figure 34 ◆ Clip-type attachments for beams and columns.

Install spring hangers at hanger points where vertical thermal movement occurs (*Figure 39*). There are two types of spring hangers: the constant load and the variable load. Constant load hangers have a coil spring and a lever that rest inside a housing. Movements of the piping system, provided they are not excessive, do not change the force this spring exerts on the pipe. Variable spring hangers and supports operate on the same principle, with the exception that the load exerted by the spring is not constant. This means that an external load may be put on the pipe by the support system itself.

In addition, many manufacturers make specialty hangers and clamps designed for use with their systems. One such specialty system is called Unistrut®. This system is used for noninsulated pipe in plumbing applications as well as in air conditioning and refrigeration systems. The system is made up of a series of slotted channels made from low-carbon strip steel, spring nuts, and bolts. The nuts and bolts have coarse screw threads. Unistrut® eliminates welding and drilling and is easily adjustable and reusable.

No matter what type of pipe support system is used, it must satisfy the following requirements:

- It must carry the total weight of the system with an adequate margin of safety. Total weight includes the weight of the liquid within the system and the weight of any insulation. Total weight also includes external factors, such as wind loads or ice buildup, that the piping system may be exposed to if it is located outdoors.

- It must handle stress. If the material the pipe is made from is subject to more stress than it was designed for, structural damage may result.

- It must allow for draining. Proper pitch must be maintained in the installation.

Anvil International
208F35.EPS

Figure 35 ◆ Riser clips.

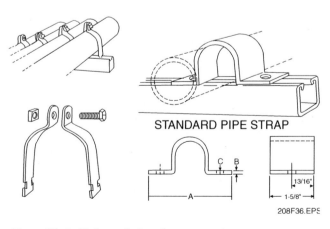

STANDARD PIPE STRAP

208F36.EPS

Figure 36 ◆ Universal pipe clamps.

- It must allow for thermal expansion and contraction. Depending on the piping material used, operating temperatures can cause piping materials to expand or contract a little or a lot.
- It must be able to withstand vibration caused by pumps or compressors within the system.
- It must have properly spaced supports. Poorly spaced supports can cause stresses to build up within the system that, in extreme cases, can cause structural failure of the pipe. Proper spacing takes into account such factors as piping material, operating temperatures, rates of thermal expansion or contraction, and weight.

When selecting pipe supports, you must be aware that not all supports can be used with all pipes. For example, if carbon steel comes into contact with stainless steel, the stainless steel will corrode. Therefore, bearing plates made of stainless steel are usually placed between the pipe and the hanger. The same holds true for aluminum pipe. It is possible to support aluminum pipe with hangers made of galvanized steel, but these hangers protect the pipe only as long as the galvanized coating lasts.

Insulated systems also require special consideration. The insulation must be protected from cuts and abrasions that will make the insulation ineffective. You must use supports that will not cut into the insulation (*Figure 40*).

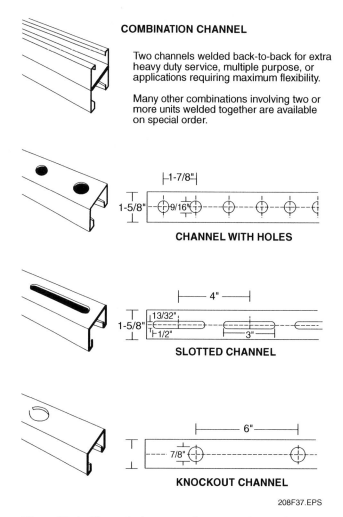

COMBINATION CHANNEL

Two channels welded back-to-back for extra heavy duty service, multiple purpose, or applications requiring maximum flexibility.

Many other combinations involving two or more units welded together are available on special order.

CHANNEL WITH HOLES

SLOTTED CHANNEL

KNOCKOUT CHANNEL

208F37.EPS

Figure 37 ◆ Channels for use with universal pipe clamps.

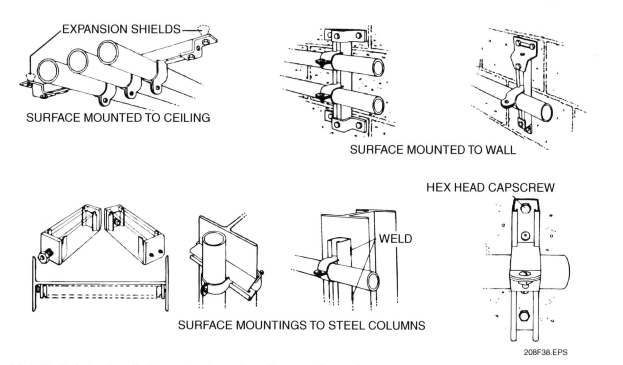

208F38.EPS

Figure 38 ◆ Typical pipe installations using channels and universal pipe clamps.

Plastic-piping systems require support at more frequent intervals than systems using metal pipe, especially in applications with elevated temperatures. The hangers should have broad, smooth surfaces rather than narrow, sharp surfaces. In areas where a plastic-piping system must be used at elevated temperatures, support can be supplied by using continuous lengths of angle iron or channel under the pipe.

Always consult your local code before installing hangers and supports. The local code will specify the proper spacing, materials that can be used, and the correct installation practices in your area. Refer, also, to the manufacturer's instructions and the job specifications. As you gain experience, you will become more familiar with proper installation standards. You must always check these sources before proceeding with an installation.

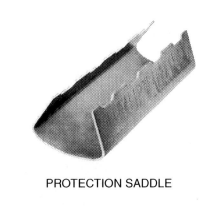

PROTECTION SADDLE

INSULATION SHIELD

Anvil International
208F40.EPS

VARIABLE
SPRING HANGER

VIBRATION
CONTROL HANGER

LIGHT DUTY
SPRING HANGER

SPRING
CUSHION HANGER

Anvil International
208F39.EPS

Figure 39 ◆ Spring hangers.

Figure 40 ◆ Protection saddle and shield for insulated pipe.

1. Pipes are usually sized according to their _____.
   a. outside diameter
   b. inside diameter
   c. radius
   d. circumference

2. The _____ method measures the full length of the pipe, including the threads on both ends.
   a. face-to-face
   b. end-to-center
   c. end-to-end
   d. end-to-face

3. _____ measure is the length of pipe plus two times the length of the thread engagement.
   a. End-to-face
   b. End-to-end
   c. Face-to-face
   d. End-to-center

4. The pipe-cutting process leaves _____ on the inside of cut pipe.
   a. threads
   b. burrs
   c. grooves
   d. pits

5. To cut threads in pipe by hand, you use a(n) _____.
   a. groover
   b. electric thread cutter
   c. pipe die and stock
   d. tube cutter

6. Flux helps to fuse metal and protect against _____.
   a. burn-through
   b. flame up
   c. calcification
   d. oxidation

7. To retain the corrosion-resistant qualities of the metal, it is best to join aluminum pipe by _____.
   a. welding
   b. flanging
   c. threading
   d. solvent welding

8. PVC pipe is classified as _____.
   a. Schedule 40 or Schedule 80
   b. Schedule 40 or Schedule 120
   c. Schedule 80 or Schedule 120
   d. Schedule 120 only

9. _____ is plastic pipe is designed to with-stand higher temperatures.
   a. PE
   b. PP
   c. CPVC
   d. ABS

10. Support for piping on wood frame con-struction can include _____.
    a. universal pipe clamps and U-bolts
    b. stirrups and clevis hangers
    c. riser clamps
    d. U-hooks, hold down clips, and half clamps

Summary

Pipe and pipe fittings are made from a variety of metal alloys and plastics. These materials are designed to meet the needs of specific applications. No matter what material the pipe is made of, it must be carefully measured and cut, reamed or cleaned, securely joined, and then properly supported in a piping system.

This module introduced you to the common piping materials, showed you how to measure pipe, and reviewed the most common methods used to join pipe. It also provided an overview of commonly used pipe fittings and outlined the basic methods for supporting pipe. Even though many tasks involved in working with pipe are fairly simple to learn, you must do them carefully, with attention to both safety and craftsmanship. Some of the pipe-joining methods involve the use of potentially hazardous equipment and substances, so you must learn how to protect yourself and others.

Now that you are familiar with the basic piping practices, the best way to gain skill is to observe the work habits of skilled workers and to practice these tasks during your apprenticeship.

Notes

Trade Terms Introduced in This Module

ABS (acrylonitrile-butadiene-styrene): A plastic used to make pipes for drainage systems, storm sewers, and underground electrical conduit.

Bearing plate: A piece of steel placed between a support and a pipe to protect the surface of the pipe from corrosion.

Black iron pipe: A noncoated pipe made of carbon steel.

CPVC (chlorinated polyvinyl chloride): A plastic used in hot and cold water piping systems and in drainage systems.

Downhand welding: A welding process that is done with both the weld axis and the base metal in a nearly horizontal or flat position.

Flame cutting: The process of cutting metal using an oxyacetylene torch.

Galvanized pipe: A steel or wrought iron pipe that has been coated with a thin layer of zinc to protect it from corrosion.

Gouging: The process of cutting a groove in a metal surface.

Laminate construction: A building process in which a product is made by bonding together two or more layers of material—for example, wood or plastic.

Machining: The process of grinding metal by hand or by machine to shape the edge or grind it to a desired size.

MIG (metal inert gas): A type of welding that uses an inert gas such as argon or carbon dioxide to shield the weld.

Plain end (PE): Pipe manufactured without threads.

Polyethylene (PE): A plastic used to make flexible plastic pipe.

Polypropylene (PP): A tough plastic with excellent resistance to heat and chemicals.

PVC (polyvinyl chloride): A thermoplastic resin that resists chemicals and corrosion and is used for pipe fittings and piping in cold water systems and in sewage and waste lines.

Radiant heating system: A space heating system that consists of a network of hot water pipes installed in the floor or ceiling.

Shielded metal-arc welding: A type of welding that uses the heat produced by an arc between a covered metal electrode and the items to be welded.

Slag: A grayish aggregate resulting from heat applied to metal—for example, from a blast furnace or from a welding process.

Solvent weld: A pipe-joining method in which a liquid solvent adhesive is brushed onto the pipe to form a strong bond between pipe sections or between a pipe and a fitting.

Thermoplastic: A plastic that, when heated, becomes soft and pliable without losing any of its other properties and becomes hard and rigid when cooled.

Thermoset: A material such as a synthetic resin that hardens when heated and does not soften when reheated.

Thread engagement: On a length of threaded pipe, the area of threaded pipe that is screwed into the fitting.

Threaded and coupled (T&C): Pipe manufactured with either threads or coupling grooves.

TIG (tungsten inert gas): A welding method that uses a nonconsumable tungsten electrode to create an arc and weld metal together.

Additional Resources

This module is intended to present thorough resources for task training. The following reference works are suggested for further study. These are optional materials for continued education rather than for task training.

Pocket Guide to Flanges, Fittings, and Piping Data, 1999. R.R. Lee. Houston, TX: Gulf Professional Publishing Company.

Practical Shielded Metal-Arc Welding, 1997. Mike Gellerman. Upper Saddle River, NJ: Prentice Hall.

Figure Credits

Copper Development Association, 208F15, Table 1

Gerald Shannon, 208F16

Anvil International, 208F31-208F33, 208F35, 208F39, 208F40

ERICO, 208F34

Power-Strut, 208F36, 208F37

Plastic Pipe and Fittings Association, Table 2

NCCER makes every effort to keep these textbooks up-to-date and free of technical errors. We appreciate your help in this process. If you have an idea for improving this textbook, or if you find an error, a typographical mistake, or an inaccuracy in NCCER's Contren® textbooks, please write us, using this form or a photocopy. Be sure to include the exact module number, page number, a detailed description, and the correction, if applicable. Your input will be brought to the attention of the Technical Review Committee. Thank you for your assistance.

Instructors – If you found that additional materials were necessary in order to teach this module effectively, please let us know so that we may include them in the Equipment/Materials list in the Annotated Instructor's Guide.

Write: Product Development and Revision
National Center for Construction Education and Research
3600 NW 43rd St, Bldg G, Gainesville, FL 32606

Fax: 352-334-0932

E-mail: curriculum@nccer.org

Craft _____ Module Name _____

Copyright Date _____ Module Number _____ Page Number(s) _____

Description _____

_____

_____

_____

(Optional) Correction _____

_____

_____

(Optional) Your Name and Address _____

_____

_____

04209-08

Fiberglass Duct

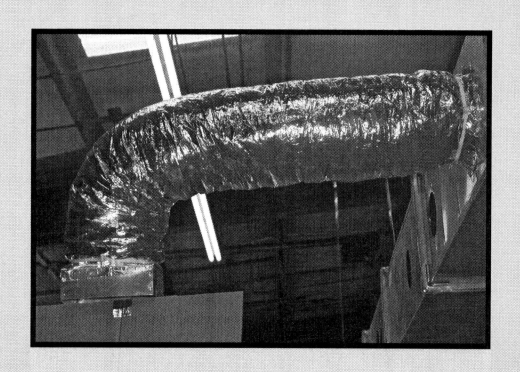

04209-08
Fiberglass Duct

Topics to be presented in this module include:

Overview

Fiberglass duct is often used in the place of sheet metal duct because it has some advantages over sheet metal. By its very nature, fiberglass duct is already insulated. Fiberglass duct is quiet, absorbing sound generated by the air handling equipment and wind noises from the duct itself.

Because fiberglass duct is made of a different material, it also has different fabrication methods, tools, and standards. There are also different personal safety considerations when working with fiberglass than there are when working with sheet metal. In this module, you will learn about working with fiberglass duct, including some of its advantages over sheet metal, and how to work safely when fabricating and installing fiberglass ductwork.

Objectives

When you have completed this module, you will be able to do the following:

1. Identify types of fiberglass duct.
2. Demonstrate fiberglass duct layout and fabrication methods.
3. Demonstrate the various closure methods for sealing fiberglass duct.
4. Fabricate selected duct modules and fittings using the appropriate tools.
5. Demonstrate hanging and support methods for fiberglass duct.
6. Repair major and minor damage to fiberglass duct.

Trade Terms

Closure system	R-value
EI rating	Standard tooling
End cap	Staple flap
Fatigue test	Stretchout
FSK (foil-scrim-kraft)	Tap-ins
Plenum	Torsion
Reverse tooling	Turning vanes
Runout	Vertical riser

Required Trainee Materials

1. Appropriate personal protective equipment
2. Rule
3. Pencil and paper

Prerequisites

Before you begin this module, it is recommended that you successfully complete the following: *Core Curriculum; Sheet Metal Level One; Sheet Metal Level Two*, Modules 04201-08 through 04208-08.

This course map shows all of the modules in the second level of the Sheet Metal curriculum. The suggested training order begins at the bottom and proceeds up. Skill levels increase as you advance on the course map. The local Training Program Sponsor may adjust the training order.

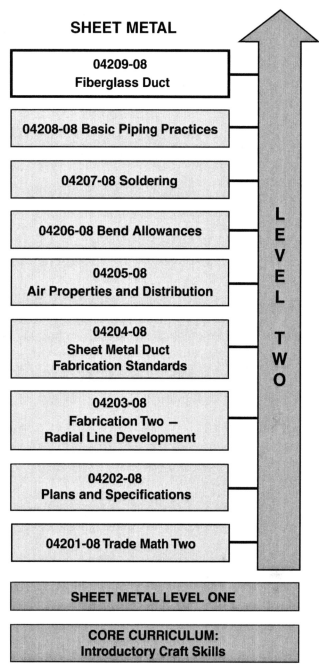

SHEET METAL

04209-08 Fiberglass Duct

04208-08 Basic Piping Practices

04207-08 Soldering

04206-08 Bend Allowances

04205-08 Air Properties and Distribution

04204-08 Sheet Metal Duct Fabrication Standards

04203-08 Fabrication Two — Radial Line Development

04202-08 Plans and Specifications

04201-08 Trade Math Two

LEVEL TWO

SHEET METAL LEVEL ONE

CORE CURRICULUM: Introductory Craft Skills

209CMAP.EPS

1.0.0 ◆ INTRODUCTION

Fiberglass duct material, like sheet metal, is used to make the main trunk lines, with branch ducts or **runouts**, to the diffuser or grille in an air distribution system. Because fiberglass duct is already insulated, it is not necessary to wrap or line it with insulation as you must do with sheet metal duct. In addition, it is lightweight and easy to transport and handle. The following are some additional benefits of fiberglass duct:

- Provides consistent heating and cooling performance through controlled insulation thickness
- Absorbs noise caused by air handling equipment and turbulence
- Reduces expansion, contraction, and vibration noise
- Has a lower air leakage rate than sheet metal duct

In this module, you will learn how fiberglass duct is fabricated into straight duct sections and fittings using both hand and machine fabrication tools. You will practice techniques for fabricating straight duct sections and fittings using rigid duct board, round duct, and flexible duct. You also will learn how to properly hang and support fiberglass duct and how to identify and repair different types of damage.

The standards in this module apply to air distribution systems designed for use in residential situations, including constant air volume systems operating at 0.5-inches water gauge. For systems operating above 0.5-inches water gauge, up to 2-inches water gauge, refer to *Fibrous Glass Duct Construction Standard* for additional reinforcing requirements and details. You will find a complete reference to this publication in the References and Acknowledgments section of this module.

The material and figures in this module have been adapted from *Fibrous Glass Duct Construction Standard* (fourth edition) and *Fibrous Glass Residential Duct Construction Standard* (second edition) with the permission of the North American Insulation Manufacturers Association (NAIMA). Be sure to refer to the most recent editions of these standards when installing fiberglass duct.

CAUTION

Take extra precautions when working with fiberglass duct. Always wear the appropriate personal protective equipment. Gloves, dust masks, and goggles will protect your hands, lungs, and eyes.

2.0.0 ◆ TYPES OF FIBERGLASS DUCT

Fiberglass duct (*Figure 1*), sometimes called fibrous glass duct, is made from flame-resistant glass fibers that are bonded with a heat-setting resin. The types of fiberglass duct include rigid duct board and flexible round duct.

Rigid duct board comes in three thicknesses: 1 inch, 1½ inches, and 2 inches. Fabricators use these rigid boards to make rectangular and 10-sided duct sections and air distribution boxes.

Rigid round duct is available in a variety of diameters. The exterior is covered with a tough, abuse-resistant **FSK (foil-scrim-kraft)** vapor barrier. FSK is a flame-retardant barrier made from a

---

DID YOU KNOW?

Fiberglass

Fiberglass has been commercially manufactured and marketed for more than 50 years. It is used to make energy-conserving products that help reduce pollution and conserve the environment. Fiberglass can absorb sound, control heat flow, remove impurities from liquids and gases, reinforce other materials, and, with a vapor barrier, help control condensation. These qualities make fiberglass an important material in the heating, ventilating, and air conditioning (HVAC) industry.

 **The Health and Safety Aspects of Fiberglass**

In October 2001, the International Agency for Research on Cancer (IARC) removed fiberglass from its list of materials that may cause cancer. The U.S. National Academy of Sciences reached a similar decision in 2000. This means that the fiberglass materials used in thermal and acoustical installation are believed to pose no risk of cancer in people who are exposed to them on the job. The IARC report specifically stated:

"Epidemiologic studies... provide no evidence of increased risk of lung cancer or mesothelioma (cancer of the lining of the body cavities) from occupational exposures during manufacture of these materials, and inadequate evidence overall of any cancer risk."

Figure 1 ◆ Fiberglass duct.

209F01.EPS

sandwich of aluminum foil, fiberglass yarn (called scrim), and kraft paper. The smooth interior surface offers minimal resistance to airflow.

Insulated flexible round duct has a reinforced inner air barrier core and an outer jacket that is usually made from polyvinyl chloride (PVC) plastic. Between the inner and outer jackets is a layer of insulation. Flexible round duct is available in standard 25-foot lengths that can be cut to the required length. This type of duct gets its flexibility from a Mylar®-coated inner steel coil. It is used to connect the main duct to registers and diffusers.

2.1.0 Requirements and Standards for Rigid Duct Board

Underwriters Laboratories (UL) tests the materials that are manufactured for use in air distribution systems. NAIMA recommends standards for their fabrication and installation. These organizations work to ensure that the materials used are safe and durable. Some of the areas for which fiberglass duct is tested and the organizations that do the testing are discussed in this section.

2.1.1 Strength, Deflection, and Fatigue

Fiberglass duct board must be strong enough to withstand building stress. It must also be able to span certain distances without deflecting (deforming), and it must stand up under the pressure of varying air cycles.

The strength and stiffness of rigid duct board are identified by its **EI rating**. The EI rating is the result of a calculation based on two tests: Young's modulus of elasticity (E) and the moment of inertia (I). Engineers and testing organizations use this calculation to determine the level of stress the duct board can withstand. There are three EI rat-

DID YOU KNOW?

EI Rating

Young's modulus of elasticity—the E in the EI rating—is a test. During this test, the duct board is placed under stress and strain to test its rigidity. The result is a ratio of stress to strain. The I part of the rating is based on a mathematical calculation. The results of the test and the calculation are combined to assign the appropriate numerical rating to rigid duct board.

ings for fiberglass duct: 475-EI, 800-EI, and 1400-EI. These ratings indicate the increasing stiffness of the duct.

Organizations such as NAIMA test duct systems under normal service conditions. These **fatigue tests** gauge the ability of rigid duct to withstand normal operating pressures. The tests are important in establishing the strength and performance of rigid duct board over time.

2.1.2 Moisture Control

It is important to avoid exposing duct board to liquid water, ice, or water vapor. If the insulation gets wet, it will become ineffective. Therefore, when working with fiberglass duct, you must follow these guidelines:

- Use a drip pan and sheet metal sleeve to protect the immediate area near evaporative coolers or humidifiers.
- Tightly close duct systems used for cooling that run through nonconditioned space to prevent water vapor from accumulating in the duct system during the heating season.
- Do not install wet fiberglass duct.
- Replace any sections of fiberglass duct that get wet during installation.

CAUTION

Fiberglass duct is often installed in tight spaces. If you are working in a small or crowded space, pay special attention to your surroundings. Wear the appropriate personal protective equipment and protect yourself against falls and cuts. Before touching uninsulated duct, check its temperature to avoid burns. Hold your hand about one inch away from the duct to make sure it is not too hot to touch.

Fiberglass Duct and Building Codes

Fiberglass duct systems must meet the requirements of several model codes. Builders must also refer to local codes that may differ from model codes to reflect local conditions. The following are the organizations and model codes that address requirements for fiberglass duct:

- Building Officials and Code Administrators International (BOCA)
  - National Building Code
- Council of American Building Officials (CABO)
  - One and Two Family Dwelling Code
- International Code Council (ICC)
  - International Mechanical Code
- International Conference of Building Officials (ICBO)
  - Uniform Mechanical Code
- Southern Building Code Congress International (SBCCI)
  - Standard Building Code
  - Standard Mechanical Code

- Find and correct the source of any liquid water that appears in installed duct sections during normal service. Contact the manufacturer for help or additional information, if needed.

2.1.3 Fabrication and Installation Requirements

The National Fire Protection Association (NFPA) and UL have set the following fabrication and installation requirements for fiberglass duct:

- Fiberglass duct shall be constructed of Class 1 duct materials as tested in accordance with the UL standard for factory-made air ducts and air connectors (*UL 181*).
- Such duct shall be installed in accordance with the conditions of its listing. This means that you must follow the installation recommendations of the testing organization.
- Such duct may not be used in air duct systems that operate with an air temperature higher than 250°F (121°C) entering the ducts.
- Such duct shall not be used as a **vertical riser** in air duct systems serving buildings that are more than two stories high.

UL tests rigid fiberglass duct for its ability to withstand a wide variety of conditions. The UL testing program covers the following areas:

- Surface burning characteristics
- Flame penetration
- Burning (other than surface burning)
- Corrosion
- Mold growth and humidity
- Temperature
- Puncture
- Static load
- Impact
- Leakage
- Erosion
- Pressure
- Collapse

2.1.4 Thermal and Acoustical Performance

The American Society of Heating, Refrigerating, and Air Conditioning Engineers (ASHRAE) has set standards for thermal performance of fiberglass duct. These standards, which have been made part of many local building codes, recommend **R-values** for duct board based on climate zones. The recommended R-values are also based on where in a building the duct board is installed. For example, the recommended R-value for a roof or building exterior is higher than that for an attic or garage.

Fiberglass insulation absorbs fan and air turbulence noise and reduces the popping noises caused by expansion and contraction. ASTM International has established typical acoustical performance standards for fiberglass duct. Because acoustical performance is a fairly complex subject, it is best to contact the fiberglass manufacturer for detailed acoustical performance information.

2.1.5 Air Leakage

The industry provides standards for air leakage, which is the expected amount of leakage along seams and joints. For example, a system fabricated to the Leak Class 6 standard will have an expected rate of leakage of no more than 6 cubic feet per minute (cfm) for every 100 square feet of fiberglass duct that is put under 1 inch water gauge pressure. The same amount of unsealed sheet metal duct under the same pressure is expected to leak up to 48 cfm.

Limiting the amount of air leakage has the following benefits:

- Better noise reduction
- More economical performance
- More efficient operation

2.1.6 Other Standards

In addition to the standards just discussed, standards exist for connections between duct and runouts and between fiberglass duct and sheet metal, for sealing seams and joints, for hangers and supports, and for repairs. Each of these topics will be discussed in more detail later in this module.

2.2.0 Requirements and Standards for Flexible Duct

Flexible duct is also tested in accordance with *UL 181*. In addition to all the tests conducted for rigid duct, flexible duct is tested for tension, **torsion**, and bending. The Air Diffusion Council (ADC) tests other properties of flexible duct. These properties include thermal performance, friction loss, acoustical performance, static pressure performance, temperature performance, and air leakage.

2.3.0 Restrictions on the Use of Fiberglass Duct

Fiberglass duct systems have many benefits. However, fiberglass duct is not suitable for some applications:

- In kitchen or fume exhaust ducts
- For solids or corrosive gases
- For installations in concrete or buried below grade
- For outdoor installations
- Near high-temperature electric heating coils without radiation protection
- For vertical risers in air duct systems serving more than two stories

- With equipment of any type that does not include automatic maximum temperature controls
- With equipment fueled by coal or wood
- As penetrations in construction where fire dampers are required, except when the fire damper is installed in a sheet metal sleeve extending through the wall

3.0.0 ◆ EXTENDED PLENUM SUPPLY SYSTEM

The extended **plenum** supply system is standard for residential applications. It is the ideal system for basement installations. It is sturdy and efficient and provides excellent acoustical and thermal performance. You do not need to change the size after each runout or branch.

Study *Figure 2*, which is a sample extended plenum supply system. The plenum is installed on the air handler. Runouts extend from the plenum to the registers or diffusers. The runouts may be made from rigid duct board, 10-sided duct, flexible round duct, or insulated metal duct. Note that an actual installation would probably not combine all the types of duct shown in the figure and that a complete illustration would include the required hangers and supports. These elements have been left out of the illustration for the sake of simplicity. The following items are identified in the figure:

- Rigid duct board (plenum extension)
- Connection of the plenum to the central air equipment
- Plenum
- **Tap-ins** for rectangular, round, and 10-sided duct runouts (three locations)
- Registers or diffusers
- Connections to registers or diffusers using flexible duct runouts

4.0.0 ◆ CUTTING AND FORMING DUCT BOARD

Duct modules are basically boxes formed from flat stock. However, depending on the needs of the job, there are several different ways to cut and form these boxes. Study *Figure 3* and note how the edges of each panel are cut to form the following duct modules:

- One-piece
- Two-piece L-shape
- Two-piece C-shape
- Two-piece U-shape

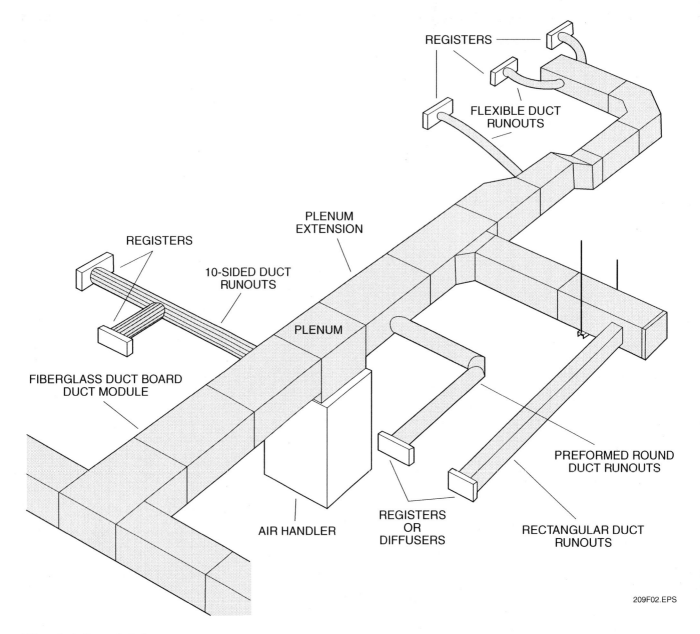

Figure 2 ◆ Extended plenum system components.

REGISTERS

FLEXIBLE DUCT
RUNOUTS

PLENUM
EXTENSION

REGISTERS

10-SIDED DUCT
RUNOUTS

PLENUM

FIBERGLASS DUCT BOARD
DUCT MODULE

PREFORMED ROUND
DUCT RUNOUTS

AIR HANDLER

REGISTERS
OR
DIFFUSERS

RECTANGULAR DUCT
RUNOUTS

209F02.EPS

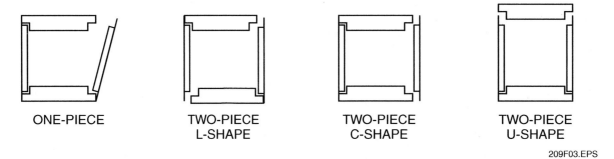

ONE-PIECE

TWO-PIECE
L-SHAPE

TWO-PIECE
C-SHAPE

TWO-PIECE
U-SHAPE

209F03.EPS

Figure 3 ◆ Methods for making a duct system.

Working Safely with Fiberglass

According to NAIMA, which has done a great deal of research, fiberglass is safe to manufacture, install, and use when recommended work practices are followed. Recommended work practices and safety guidelines vary from one manufacturer to another. Therefore, you must always read the material safety data sheet (MSDS) for the latest information and recommended actions regarding the fiberglass product being used. In general, however, NAIMA recommends the following:

- If there is a question as to the fiber count in the air, use a respirator. Because it is important to use a respirator properly, you must read and follow the manufacturer's recommendations for proper fit and use. The manufacturer may also recommend a program for training and testing.

- Wear a cap, gloves, long-sleeved shirt, and long pants. To help keep glass fibers from being trapped between your clothing and your skin, it is best if your work clothes fit loosely. Fiberglass can irritate your skin. This is not an allergic reaction but a temporary mechanical response of the skin when the glass fibers rub against or become embedded in it.

- Wear safety glasses with side shields, safety goggles, or a face shield to protect your eyes.

- Don't rub or scratch your skin. Remove fiberglass particles from your skin by washing the area gently and thoroughly with warm water and mild soap. Using a skin lotion may also help.

- Wash work clothing separately from other household laundry to prevent the transfer of glass fibers to other clothing. Rinse the washing machine thoroughly before using it again. If your work clothes have lots of fibers on them, it is best to soak and rinse them separately before placing them in the washing machine.

- Keep your work area clean. Avoid handling scrap fiberglass by keeping disposal bins as close to the working area as possible. Don't let scrap material accumulate on the floor or other surfaces.

- Prevent airborne dust. Do not dry sweep or use a compressed air line to clean your work area. Use a filtered vacuum or wet sweeping technique.

Generally, workers engaged in such operations as sawing, machining, and blowing fiberglass products may be exposed to more airborne glass fibers than installers. Therefore, NAIMA recommends using dust collection systems whenever exposure to glass fibers may exceed recommended levels. Currently, NAIMA recommends that airborne exposures be kept below one fiber per cubic centimeter (f/cc). A pamphlet and video describing recommended work practices are available from NAIMA. (See the References and Acknowledgments section of this module.)

The method you choose to make the duct depends on the duct size, total **stretchout** dimensions, and the best use of materials. As you gain experience, you will learn to choose methods that provide the best balance between your labor and the amount of material waste, called cut-off, that is created.

You can fabricate fiberglass duct board using one of two types of corner joints: the modified shiplap or the V-groove (*Figure 4*). The modified shiplap is considered the industry standard for machine-fabricated 1-inch duct board. The V-groove is an alternate method generally used for hand grooving duct board.

Depending on how you use your tools, you can produce a closure flap on either the right side or the left side of the duct board. A closure flap is a piece of the duct's covering jacket from which you remove the insulation. Pull the flap over a seam or joint and seal it with staples, tape, or both. The layout techniques shown in this module will produce duct sections with shiplapped panels on the width

and square-edged panels on the height of the duct section. You can reverse this setup when needed.

Both hand tools and machine tools can be used to fabricate duct sections and fittings from rigid duct board. Flexible duct components are fabricated entirely with hand tools. Duct fabrication tools are designed to make specific types of cuts in the duct board and to ensure tight joints and connections.

CAUTION

Protect your hands against cuts by wearing gloves. Always wear safety goggles to protect your eyes.

4.1.0 Hand Grooving Tools

Hand grooving tools are used to fabricate straight sections of fiberglass duct board. These tools come in two sizes for use on 1-inch or 1½-inch duct

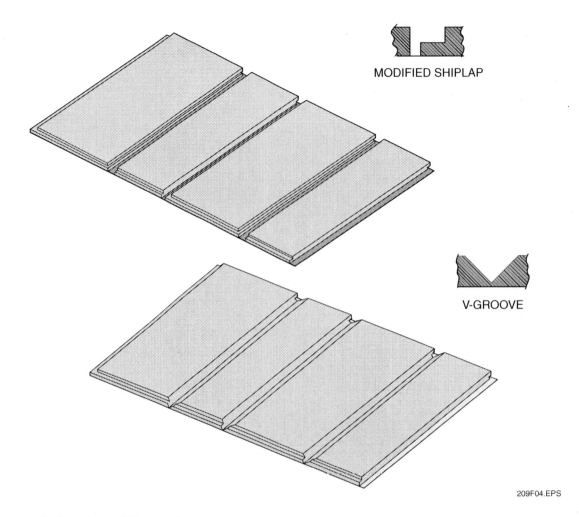

MODIFIED SHIPLAP

V-GROOVE

209F04.EPS

Figure 4 ◆ Modified shiplap and V-groove joints.

board (*Figure 5*). Usually these tools come from the factory with the blades preset. You must check the blade settings before you start work because they may not be in the proper position. If the blades are not properly set, you will not be able to make accurate cuts. Hand grooving tools include the following:

- Modified shiplap tool
- Female shiplap tool
- Male shiplap tool
- V-groove tool
- Shiplap and end cut-off tool

Use hand grooving tools to fabricate one-piece duct (*Figure 6*), two-piece L-shape duct (*Figure 7*), and two-piece C-shape duct (*Figure 8*). Note that the dimensions shown in the figures are nominal. Actual dimensions may vary from one manufacturer to another.

4.2.0 Machine Grooving Tools

Machine grooving tools are used for production work when many pieces must be fabricated quickly and accurately (*Figure 9*). These machines can cut all the corner folds to the correct stretchout dimensions and finish the long seam edges with a **staple flap** all in one pass. The staple flap is sometimes called a closure flap. The machine fabrications shown in this module are based on using machine grooving tools made by the Glassmaster Company. Dimensions may vary with tools made by other manufacturers; therefore, it is best to follow the manufacturer's instructions.

Numbered tabs on machine grooving tools match up to the location and width of the cutting area. To set up the machine, measure the interior dimensions of the duct between the numbered tabs on the tools. The tabs represent the required add-on allowances.

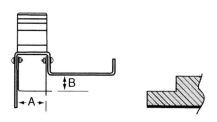

BOARD THICKNESS	A	B
1" (25mm)	$^7/_8$" (22mm)	$^1/_2$" (13mm)
1$^1/_2$" (38mm)	1$^3/_8$" (35mm)	$^3/_4$" (19mm)
2" (51mm)	1$^7/_8$" (48mm)	1" (25mm)

FEMALE SHIPLAP TOOL–Cuts female slip joint. Also cuts seating edge of duct board at longitudinal closure corner.

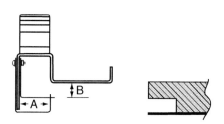

BOARD THICKNESS	A	B
1" (25mm)	$^7/_8$" (22mm)	$^1/_2$" (13mm)
1$^1/_2$" (38mm)	1$^3/_8$" (35mm)	$^3/_4$" (19mm)
2" (51mm)	1$^7/_8$" (48mm)	1" (25mm)

MALE SHIPLAP TOOL–Cuts male slip joint, which mates with female slip joint to connect two duct sections.

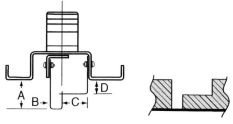

BOARD THICKNESS	A	B	C	D
1" (25mm)	$^{15}/_{16}$" (24mm)	$^3/_8$" (24mm)	$^7/_8$" (22mm)	$^1/_2$" (13mm)
1$^1/_2$" (38mm)	1$^3/_8$" (35mm)	$^9/_{16}$" (14mm)	1$^3/_8$" (35mm)	$^{11}/_{16}$" (17mm)
2" (51mm)	1$^7/_8$" (48mm)	$^7/_8$" (22mm)	1$^7/_8$" (48mm)	$^{15}/_{16}$" (24mm)

MODIFIED SHIPLAP TOOL–Removes insulation for shiplap corner folds. Reversing the tool allows both left-hand and right-hand shiplaps to be cut.

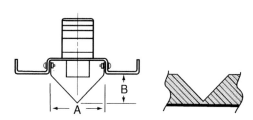

BOARD THICKNESS	A	B
1" (25mm)	1$^3/_4$" (44mm)	$^7/_8$" (22mm)
1$^1/_2$" (38mm)	2$^3/_4$" (67mm)	1$^3/_8$" (35mm)
2" (51mm)	3$^3/_4$" (95mm)	1$^7/_8$" (48mm)

V-GROOVE TOOL–Cuts 90° V-grooves for corner folds when modified shiplap grooving method is not used.

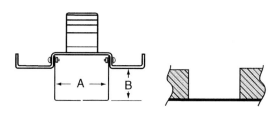

BOARD THICKNESS	A	B
1" (25mm)	1$^3/_4$" (44mm)	1" (25mm)
1$^1/_2$" (38mm)	2$^5/_8$" (67mm)	1$^1/_2$" (38mm)

STAPLE FLAP TOOL–Used with straight knife to make staple flap and end cut.

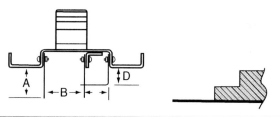

BOARD THICKNESS	A	B	C	D
1" (25mm)	$^7/_8$" (22mm)	1$^3/_8$" (35mm)	$^7/_8$" (22mm)	$^1/_2$" (13mm)
1$^1/_2$" (38mm)	1$^3/_4$" (35mm)	1$^{11}/_{16}$" (43mm)	1$^3/_8$" (35mm)	$^3/_4$" (19mm)
2" (51mm)	1$^7/_8$" (48mm)	2$^1/_8$" (25mm)	1$^3/_4$" (44mm)	1" (25mm)

SHIPLAP AND END CUT-OFF TOOL–Cuts shiplap on end of board for longitudinal corner closure, plus staple flap. May also be used in some fitting fabrication techniques.

209F05.EPS

Figure 5 ◆ Hand grooving tools for 1-inch and 1½-inch duct board.

The two methods for setting up the cutting tools in the grooving machine are **standard tooling** and **reverse tooling** (*Figure 10*). The tools are identified using both numbers and letters. Generally, the numbers are used for standard tooling and the letters are used for reverse tooling. However, the tool identification system may vary from one manufacturer to another.

In the standard tooling setup, the closure flap is produced on the right side of the duct board. As the board passes through the grooving machine, each groove is cut and the closure flap fabricated in one pass.

Reverse tooling is also called preferred tooling. In this setup, the closure flap is produced on the left side of the duct board as it passes through the machine. In the reverse tooling method, the blade that is closest to the side frame of the machine cuts the closure flap. This area is where the tool bars are supported. Therefore, it is possible to get a cleaner staple flap, and less hand labor is needed to remove the insulation.

Use machine grooving tools to fabricate one-piece duct (*Figure 11*), two-piece L-shape duct (*Figure 12*), and two-piece U- and C-shape duct (*Figure 13*). Note that the dimensions shown in the figures are nominal. Actual dimensions may vary from one manufacturer to another.

4.3.0 Other Tools

Manufacturers provide a number of tools specially designed for use in fabricating fiberglass duct. Read and follow the manufacturer's han-

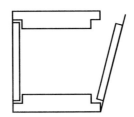

HAND FABRICATION—SHIPLAP METHOD

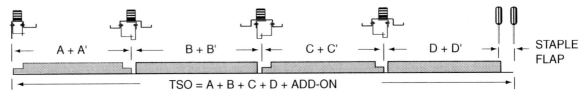

HAND FABRICATION—V-GROOVE METHOD

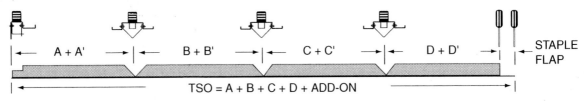

	DUCT BOARD THICKNESS	A'	PANEL ADD-ON DIMENSIONS B'	C'	D'	STAPLING FLAP	STRETCH-OUT ADD-ON
A, B, C, D	1" (25mm)	1¾" (44mm)	1¾" (44mm)	1¾" (44mm)	1⅜" (35mm)	1⅜" (35mm)	8" (203mm)
ARE INSIDE	1½" (38mm)	2¾" (70mm)	2¾" (70mm)	2¾" (70mm)	2⅛" (54mm)	2⅛" (54mm)	12½" (318mm)
DIMENSIONS	2" (51mm)	3¾" (95mm)	3¾" (95mm)	3¾" (95mm)	2⅞" (73mm)	2½" (64mm)	16½" (422mm)

209F06.EPS

Figure 6 ◆ Hand fabrication of one-piece duct.

dling, maintenance, and safety instructions when using the following tools:

- *Insulation knife* – Makes straight cuts in fiberglass duct board, cleans fibers from staple flaps, and cuts closure tapes.
- *Peeler knife* – Cleans fibers from staple flaps.
- *Round hole cutter* – Cuts round holes for tap-ins.
- *Duct layout squares and transition tools* – Measuring and cutting guides for hand layout and fabrication.
- *Staple guns* – Mechanically closes long seams, transverse joints, and staple flaps. May be hand- or power-operated.
- *Heat sealing irons* – Used with heat-activated tape. Some have long handles for use in areas with little clearance.

5.0.0 ◆ DUCT BOARD FABRICATION METHODS

To fabricate fiberglass duct, make a series of cuts or grooves that will allow you to form a flat piece of duct board into a duct section. The process is similar to making a box out of a flat piece of cardboard. However, because the duct is lined with insulation, the cuts must be made with special tools to create precisely measured grooves. You must also remove the insulation along the groove lines so that you can fold the duct board into the required shape. When making measurements to lay out duct board, add the allowances for your folds and the staple flap. There are two methods for laying out fiberglass duct. They are the centerline method or the guide edge method.

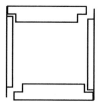

HAND FABRICATION—SHIPLAP METHOD

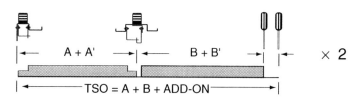

HAND FABRICATION—V-GROOVE METHOD

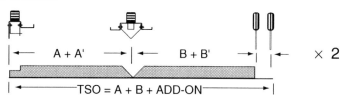

| | DUCT BOARD THICKNESS | PANEL ADD-ON DIMENSIONS | | | | STAPLING FLAP | TSO ADD-ON |
		A'	B'	C'	D'		
A, B, C, D	1" (25mm)	1¾" (44mm)	1¾" (44mm)	1¾" (44mm)	1¾" (44mm)	1½" (38mm)	4⅝" (117mm)
ARE INSIDE	1½" (38mm)	2¾" (70mm)	2⅛" (54mm)	2¾" (70mm)	2⅛" (54mm)	2⅛" (54mm)	7" (178mm)
DIMENSIONS	2" (51mm)	3¾" (95mm)	2⅞" (73mm)	3¾" (95mm)	2⅞" (73mm)	2½" (64mm)	9⅛" (232mm)

209F07.EPS

Figure 7 ◆ Hand fabrication of two-piece L-shaped duct.

This section shows examples of how to fabricate fiberglass duct with both hand and machine tools. The examples that use hand fabrication tools cover both the centerline method and the guide edge method. The instructions given in this module are general. Tools used to fabricate fiberglass duct board vary from manufacturer to manufacturer, so you should always follow the manufacturer's instructions. Stretchout dimensions can also vary. You must determine the actual stretchout dimensions that apply to the tools you are using. Again, follow the tool manufacturer's instructions.

5.1.0 Hand Fabrication, Centerline Method

In the centerline method you will form grooves and closure flaps along centerlines that you first draw on the duct board. The inside dimensions of the duct plus add-on allowances determine the location of each centerline. Add the total of these allowances to the inside duct dimensions to determine the total stretchout. The stretchout is the total board length you need to make a straight duct section with the correct inside dimensions. Here is the formula for calculating the stretchout:

$$SO = 2W + 2H + A$$

Where:

SO = Length of the stretchout

W = Width

H = Height

A = Allowances

The following steps are based on fabricating a 12-inch (W) by 10-inch (H) duct using 1-inch duct board. The add-on allowance (A) is calculated by adding $1\frac{3}{4}" + 1\frac{3}{4}" + 1\frac{3}{4}" + 1\frac{3}{8}" + 1\frac{3}{8}"$. *Table 1* indi-

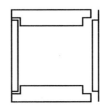

HAND FABRICATION—SHIPLAP METHOD

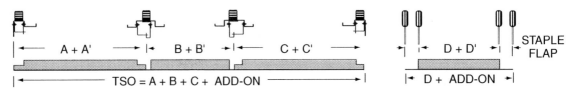

HAND FABRICATION—V-GROOVE METHOD

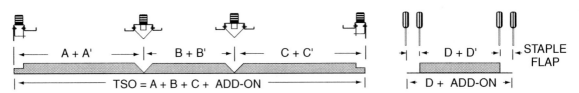

	DUCT BOARD THICKNESS	PANEL ADD-ON A', B', C'	U SECTION TSO ADD-ON	D'	STAPLING FLAP	PANEL D TSO ADD-ON
A, B, C, D	1" (25mm)	1¾" (44mm)	5¼" (133mm)	1" (25mm)	1⅛" (38mm)	4" (102mm)
ARE INSIDE	1½" (38mm)	2¾" (70mm)	8½" (210mm)	1½" (38mm)	2¼" (54mm)	5¾" (146mm)
DIMENSIONS	2" (51mm)	3½" (95mm)	11¼" (286mm)	2" (51mm)	2½" (64mm)	7" (178mm)

209F08.EPS

Figure 8 ◆ Hand fabrication of two-piece C-shaped duct.

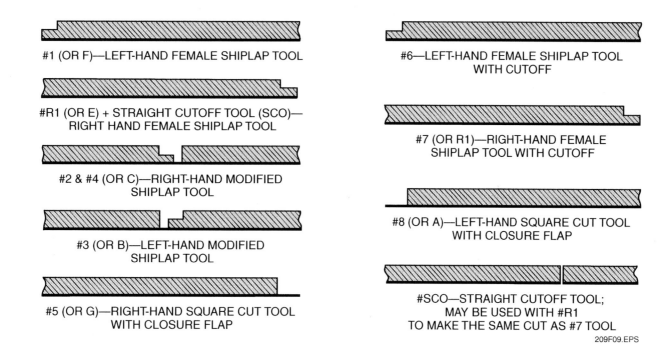

#1 (OR F)—LEFT-HAND FEMALE SHIPLAP TOOL

#R1 (OR E) + STRAIGHT CUTOFF TOOL (SCO)—
RIGHT HAND FEMALE SHIPLAP TOOL

#2 & #4 (OR C)—RIGHT-HAND MODIFIED
SHIPLAP TOOL

#3 (OR B)—LEFT-HAND MODIFIED
SHIPLAP TOOL

#5 (OR G)—RIGHT-HAND SQUARE CUT TOOL
WITH CLOSURE FLAP

#6—LEFT-HAND FEMALE SHIPLAP TOOL
WITH CUTOFF

#7 (OR R1)—RIGHT-HAND FEMALE
SHIPLAP TOOL WITH CUTOFF

#8 (OR A)—LEFT-HAND SQUARE CUT TOOL
WITH CLOSURE FLAP

#SCO—STRAIGHT CUTOFF TOOL;
MAY BE USED WITH #R1
TO MAKE THE SAME CUT AS #7 TOOL

209F09.EPS

Figure 9 ◆ Typical machine grooving tools.

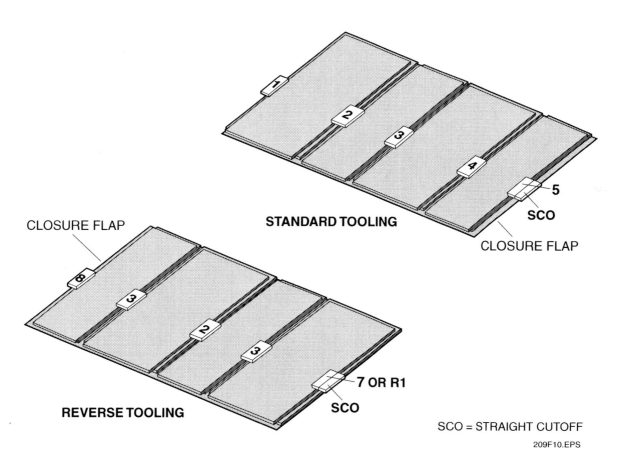

209F10.EPS

Figure 10 ◆ Standard and reverse tooling.

cates stretchout dimensions for both 1-inch and 1½-inch fiberglass duct. Plug these numbers into the formula to calculate the stretchout.

SO = 2W + 2H + A

SO = (2)(12) + (2)(10) + 8

SO = 24 + 20 + 8

SO = 52 inches

Once you have done this calculation and have double-checked your figures, you are ready to fabricate the duct board. To hand fabricate fiberglass duct board using the centerline method, follow these steps (*Figure 14*):

Step 1 With the factory-formed female shiplap facing you, measure 12 inches from the left edge of the board (first inside duct dimension) plus 1¾ inches for the add-on allowance. Draw the centerline for the first groove.

Step 2 Draw the second inside duct dimension. From the line just drawn, measure 10 inches plus 1¾ inches for the add-on allowance. Draw the centerline for the second corner groove.

Step 3 Draw the third inside duct dimension. From the line just drawn, measure 12

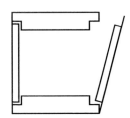

MACHINE FABRICATION—STANDARD

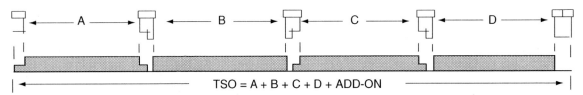

TSO = A + B + C + D + ADD-ON

MACHINE FABRICATION—PREFERRED

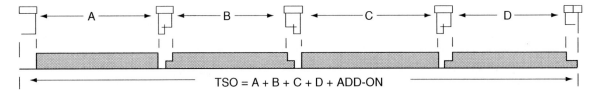

TSO = A + B + C + D + ADD-ON

Inside duct dimensions A, B, C, D are measured between tool tabs.

DUCT BOARD THICKNESS	STRETCH-OUT ADD-ON
1" (25mm)	8" (203mm)
1½" (38mm)	12" (305mm)
2" (51mm)	16" (406mm)

209F11.EPS

Figure 11 ◆ Machine fabrication of one-piece duct.

inches plus 1¾ inches for the add-on allowance. Draw the centerline for the third corner groove.

Step 4 Draw the fourth inside duct dimension. From the line just drawn, measure 10 inches plus 1¾ inches for the add-on allowance. Draw the centerline for the fourth corner groove.

Step 5 From the line just drawn, measure 1⅜ inches and draw a line to locate the edge of the closure flap. The duct board is now ready to be grooved.

Step 6 Use the female shiplap tool to form the shiplap edge along the left edge of the

first panel. Remove scrap from the groove. Lifting the board slightly makes it easier to remove the scrap.

Step 7 Use the modified shiplap tool to cut along the first corner centerline. The shiplap should be on the right side of the first panel. Remove the scrap.

Step 8 Rotate the modified shiplap tool end-for-end and cut along the second centerline. The shiplap should be on the left side of the third panel. (You can also use the next sequentially numbered tool instead of rotating the modified shiplap tool.) Remove the scrap.

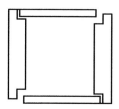

MACHINE FABRICATION—STANDARD

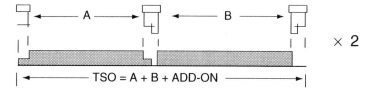

MACHINE FABRICATION—PREFERRED

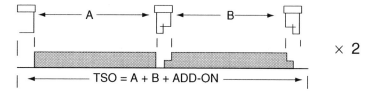

Inside duct dimensions A, B, C, D are measured between tool tabs.

DUCT BOARD THICKNESS	L-SECTION TSO ADD-ON
1" (25mm)	4½" (114mm)
1½" (38mm)	6¾" (171mm)
2" (51mm)	8⅞" (225mm)

209F12.EPS

Figure 12 ◆ Machine fabrication of two-piece L-shaped duct.

Step 9 Rotate the modified shiplap tool back to the position it was in at Step 7. Cut along the third centerline. The shiplap should be on the right side of the third panel. Remove the scrap.

Step 10 Using an insulation knife, cut along the fourth centerline. Cut through the insulation only. Do not cut through or score the facing.

Step 11 Using an insulation knife, cut through both the insulation and the facing on the last line. This line represents the outer edge of the closure flap. Peel the insulation from the flap. The duct board is now ready for assembly.

5.2.0 Hand Fabrication, Guide Edge Method

In the guide edge method, you cut grooves using a squaring tool. The inside dimensions of the duct determine how far you must move the tool after each cut is made. The usual method is to work from the right side of the guide edge, as shown in this module. Tool manufacturers also provide instructions for working from the left side of the guide edge. Follow the manufacturer's instructions for the specific tools you are using.

In the guide edge method, you do not have to draw layout lines and calculate the stretchout before cutting the grooves. The steps in the following example are for a one-piece fabrication (*Figure 15*):

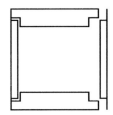

MACHINE FABRICATION—STANDARD

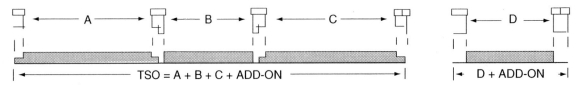

MACHINE FABRICATION—PREFERRED

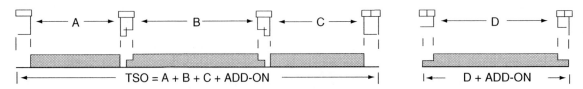

Inside duct dimensions A, B, C, D are measured bewteen tool tabs.

	STANDARD		PREFERRED	
DUCT BOARD THICKNESS	**U-SECTION ADD-ON**	**FILLER PANEL ADD-ON**	**U-SECTION ADD-ON**	**FILLER PANEL ADD-ON**
1" (25mm)	4¾" (121mm)	4¼" (108mm)	7¾" (197mm)	1½" (38mm)
1½" (38mm)	7⅜" (194mm)	5¾" (146mm)	11" (279mm)	2½" (64mm)
2" (51mm)	10½" (267mm)	7¼" (184mm)	14¼" (362mm)	3½" (89mm)

209F13.EPS

Figure 13 ◆ Machine fabrication of two-piece U- and C-shaped duct.

Table 1 One-Piece Stretchout Dimensions in Inches

One-Piece Stretchout Dimension in Inches, 1" Board

Duct Height	6	8	10	12	14	16	18	20	22	24	26	28	30	32	34	36
6	32	36	40	44	48	52	56	60	64	68	72	76	80	84	88	92
8	36	40	44	48	52	56	60	64	68	72	76	80	84	88	92	96
10	40	44	48	52	56	60	64	68	72	76	80	84	88	92	96	100
12	44	48	52	56	60	64	68	72	76	80	84	88	92	96	100	104
14	48	52	56	60	64	68	72	76	80	84	88	92	96	100	104	108
16	52	56	60	64	68	72	76	80	84	88	92	96	100	104	108	112
18	56	60	64	68	72	76	80	84	88	92	96	100	104	108	112	116
20	60	64	68	72	76	80	84	88	92	96	100	104	108	112	116	120
22	64	68	72	76	80	84	88	92	96	100	104	108	112	116	120	
24	68	72	76	80	84	88	92	96	100	104	108	112	116	120		
26	72	76	80	84	88	92	96	100	104	108	112	116	120			
28	76	80	84	88	92	96	100	104	108	112	116	120				
30	80	84	88	92	96	100	104	108	112	116	120					

One-Piece Stretchout Dimension in Inches, 1½" Board

Duct Height	6	8	10	12	14	16	18	20	22	24	26	28	30	32	34	36
6	36	40	44	48	52	56	60	64	68	72	76	80	84	88	92	96
8	40	44	48	52	56	60	64	68	72	76	80	84	88	92	96	100
10	44	48	52	56	60	64	68	72	76	80	84	88	92	96	100	104
12	48	52	56	60	64	68	72	76	80	84	88	92	96	100	104	108
14	52	56	60	64	68	72	76	80	84	88	92	96	100	104	108	112
16	56	60	64	68	72	76	80	84	88	92	96	100	104	108	112	116
18	60	64	68	72	76	80	84	88	92	96	100	104	108	112	116	120
20	64	68	72	76	80	84	88	92	96	100	104	108	112	116	120	
22	68	72	76	80	84	88	92	96	100	104	108	112	116	120		
24	72	76	80	84	88	92	96	100	104	108	112	116	120			
26	76	80	84	88	92	96	100	104	108	112	116	120				
28	80	84	88	92	96	100	104	108	112	116	120					
30	84	88	92	96	100	104	108	112	116	120						

STEP 1

STEP 2

STEP 3

STEP 4

STEP 5

STEP 6

STEP 7

STEP 8

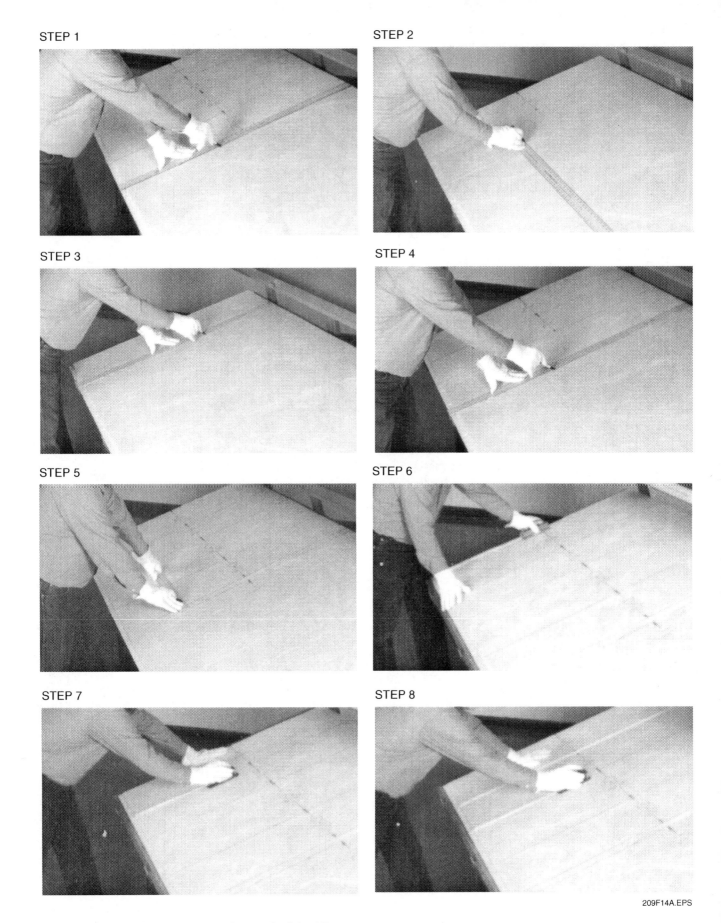

209F14A.EPS

Figure 14 ◆ Hand fabrication, centerline method (1 of 2).

STEP 9

STEP 10

STEP 11

209F14B.EPS

Figure 14 ◆ Hand fabrication, centerline method (2 of 2).

Step 1 For a right-hand shiplap and staple flap, set up the squaring tool so that the guide edge is on the right with the female shiplap facing you.

Step 2 Using the female shiplap tool, cut along the left edge of the duct board to form a shiplap edge without a closure flap. Remove the scrap.

Step 3 Place the squaring tool on the duct board so the edge of the rule is parallel with the factory-formed female shiplap. Line up

the inside duct width dimension on the rule with the right edge of the cut made in Step 2.

Step 4 Place the corner-grooving tool (either the modified shiplap or the V-groove) along the guide edge of the squaring tool and cut the first corner groove. Remove the scrap.

Step 5 Reposition the squaring tool to the right. This allows you to line up the desired inside duct dimension on the rule with the right edge of the first corner cut.

Step 6 Rotate the corner-grooving tool end for end and place it along the guide edge of the squaring tool to cut the second corner groove. (You can also use the next sequentially numbered tool.) Remove the scrap.

Step 7 Reposition the squaring tool to the right as you did in Step 3. Cut the third corner groove as you cut the corner groove in Step 4.

Step 8 Reposition the squaring tool to the right for the final cut. Use an insulation knife to cut through the insulation. Do not score or cut through the facing on the duct board.

Step 9 Using an insulation knife, cut through the insulation and facing along the right-hand cut. Peel the insulation from the closure flap. The duct board is now ready for assembly.

5.3.0 Machine Fabrication

Machine grooving tools vary from one manufacturer to another. Therefore, it is important to read and follow the manufacturer's instructions for setting up, using, and maintaining machine grooving tools. It is also important to carefully read and follow the manufacturer's safety recommendations. Be sure to wear the appropriate personal protective equipment when using these tools.

Once you have properly set up the machine, it takes far fewer steps and much less time to fabricate fiberglass duct. To fabricate a one-piece duct, follow these steps (*Figure 16*):

Step 1 Following the manufacturer's instructions, install the grooving tools in the machine. Depending on the desired outcome, you can use either standard tooling or reverse tooling.

STEP 1

STEP 2

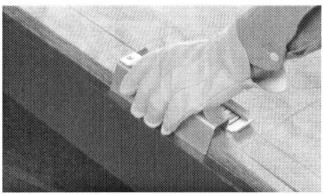

STEP 3

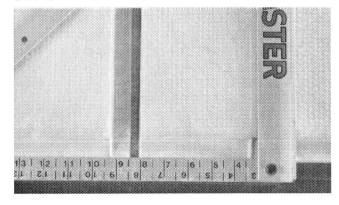

STEP 4

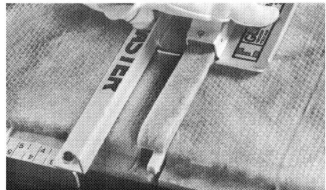

STEP 5

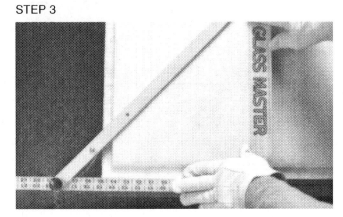

STEP 6

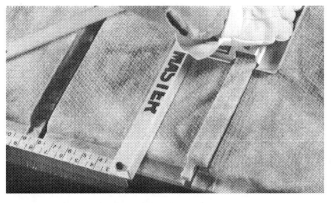

STEP 7

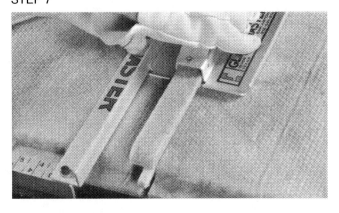

STEP 8

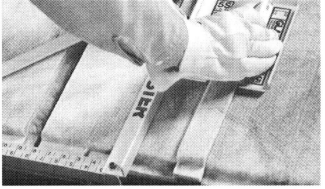

209F15A.EPS

Figure 15 ◆ Hand fabrication, guide edge method (1 of 2).

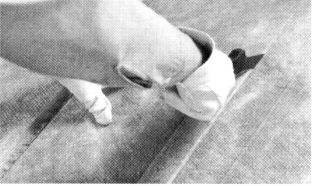

209F15B.EPS

Figure 15 ◆ Hand fabrication, guide edge method (2 of 2).

Step 2 Feed the duct board into the machine, female shiplap edge first. Hold the left edge firmly against the guide, parallel to the rollers. Turn on the machine and guide the board into it. Friction rollers will pick up the board and pull it through the cutting tools.

Step 3 Remove scrap from the grooves, check the dimensions. The duct board is now ready for assembly.

CAUTION

Let the machine do the work. Guide the duct board and allow the friction rollers to pick up the board. Do not push the board. Keep your fingers clear of the friction rollers.

6.0.0 ◆ CLOSURE SYSTEMS FOR DUCT BOARD

Once you have grooved and assembled the duct board, you must seal the joints using the appropriate **closure system**. You must follow the applicable local code and the job specifications when selecting the appropriate closure system. Only closure systems that comply with *UL 181A* are suitable for use with rigid fiberglass duct systems. These closure systems include the following:

- Pressure-sensitive aluminum foil tapes listed in *UL 181A, Part I (P)*.
- Heat-activated aluminum foil or scrim tapes listed in *UL 181A, Part II (H)*. This is the preferred method.

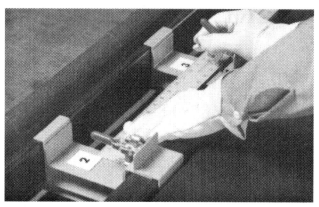

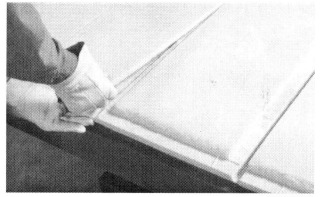

209F16.EPS

Figure 16 ◆ Machine fabrication.

- Mastic and glass fabric tape closure systems listed in *UL 181A, Part III (M)*.

6.1.0 Joints with Staple Flaps

Duct joints that have been closed with staple flaps must also be sealed with tape. Center the tape over the edge of the staple flap. The tape should completely cover the staples with a 1-inch minimum overlap (*Figure 17*).

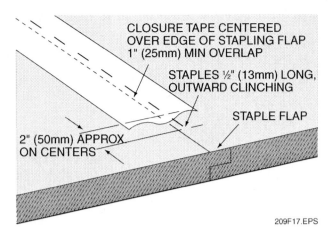

Figure 17 ◆ Tape closure with staple flap.

6.2.0 Joints without Staple Flaps

When there are no staple flaps, place cross tabs of either pressure-sensitive tape or heat-activated tape on the joints to add strength. Use 2-inch wide tape and center the tape so that one inch of tape is on either side of the joint. Tape tabs may be placed either over or under pressure-sensitive or heat-activated tape. Tape tabs should be placed under mastic and glass fabric tape. Tab length may vary, but in many cases it should be eight inches long (*Figure 18*). Refer to your local code.

6.3.0 Working with Closure Systems

Preparing the surface is an important step in working with closure systems. The surfaces to be bonded must be clean and dry. If there is any dust, dirt, oil, grease, or moisture present, the adhesives will not stick. In most cases, you can clean the area to be bonded with an oil-free, lint-free cloth or paper towel. For best results, follow the manufacturer's cleaning recommendations.

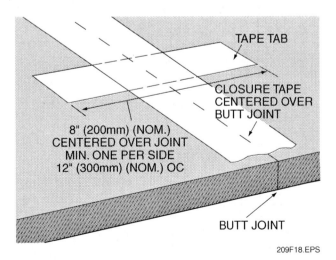

Figure 18 ◆ Tape closure without staple flap or staples.

Most tapes and mastics have storage requirements and a limited shelf life. It is important to make sure that these materials have been stored according to the manufacturer's instructions. Do not use any product that is past its shelf life. The adhesive material may have dried out, reducing its bonding ability.

6.4.0 Applying Pressure-Sensitive Tape

To properly apply pressure-sensitive tape, place it so that there is a 1-inch minimum overlap over the seam or joint. Hold the tape taut as you press it into place, taking care to avoid wrinkles and folds (*Figure 19*). Rub the tape firmly with a plastic sealing tool until the facing reinforcement shows through (*Figure 20*). When the temperature is below 55°F (13°C), apply additional heat to ensure that the tape will stick.

You can also use pressure-sensitive tape to seal fiberglass fittings to sheet metal. Be sure to read the manufacturer's metal cleaning instructions before using pressure-sensitive tape in this way.

If the temperature at the job site is below 50°F (10°C), you must apply heat to ensure a good bond. Use a heating iron set at approximately 400°F (200°C) and preheat the area to be taped. Quickly position the tape on the preheated area and press in place. Rapidly pass the iron two or three times over the taped area (*Figure 21*). Then rub the tape firmly with a plastic sealing tool until the facing reinforcement shows clearly through the tape.

CAUTION

When rubbing pressure-sensitive tape, avoid placing too much pressure on the sealing tool. Too much pressure could cause punctures wherever the tape covers staples.

6.5.0 Applying Heat-Activated Tape

When working with heat-activated tape, position the tape so that there is at least a 1-inch overlap. Use a heating iron with a temperature between 550°F and 600°F (288°C and 326°C). Move the iron slowly along the tape seam with enough pressure to activate the adhesive (*Figure 22*). Heat indicator dots on the tape will darken when the proper temperature is reached (*Figure 23*). Apply a squeegee action with the side of the iron to ensure that the tape has a good bond.

Make a second pass with the iron to complete the bond by applying pressure to the front edge of the iron. To avoid puncturing tape that covers staples, don't put too much pressure on the iron. Allow all joints and seams to cool below 150°F (66°C) before putting any stress on them.

WARNING!

Both the iron and the heated surfaces get very hot. To avoid serious burns, pay close attention to what you are doing. To protect your hands, wear KEVLAR® gloves. When positioning the tape, set the iron down on a heat-resistant surface. Work so that the hot iron is always in front of the electric wire. Work carefully to keep your feet from getting tangled in the wire.

6.6.0 Applying Mastic and Glass Fabric Tape

Mastic and glass fabric tape is a two-part closure system consisting of liquid mastic that comes in a can and a glass fabric tape that looks like a screen. Although there are some similarities among mastics, the application rate, safety precautions, shelf life limits, and minimum setup time will vary from one manufacturer to another. Therefore, you must always read the manufacturer's instructions and recommendations. Depending on the application, you can use either pressure-sensitive or heat-activated tape in combination with this closure system.

When working with this closure system, apply a thin coat of mastic about three inches wide over the center of the seam. Firmly press the glass fabric tape into the mastic and smooth it in place (*Figure 24*). Apply a second coat of mastic so that it fills the screen pattern in the glass fabric tape (*Figure 25*).

After applying the mastic, clean the area using soap and water. Drying times can vary greatly, so always refer to the manufacturer's instructions before starting the system.

 **Effective Cleaning**

When cleaning a surface to be sealed, move a clean cloth across the area in one smooth sweep. If necessary, fold the cloth and make another smooth sweep. Apply even pressure on the cloth because too little pressure is ineffective and too much pressure could damage the duct board. Do not rub the cloth back and forth or scrub at the surface. Doing so can redeposit some of the dirt or grease you just removed.

7.0.0 ◆ FABRICATING AND JOINING A DUCT MODULE

To fabricate a duct module of standard straight duct, you will cut shiplaps and corner grooves using either hand tools or machine tools. In either case, you can use the procedures outlined earlier in this module for using these tools. Study the figures in this section and follow these steps (*Figure 26*):

Step 1 Cut shiplaps and corner grooves with hand or machine tools. Refer to *Figures 6 through 13*. Remove scrap from the grooves.

Step 2 Fold the panels to form the duct. Make sure that the ends are flush and properly seated in the shoulder of the shiplap edges.

Step 3 Hold the duct at about a 30-degree angle. Staple the long flap with ½-inch outward clinching staples approximately 2 inches on center.

 **A New Standard for Duct Tape**

The Air Diffusion Council and Underwriters Laboratories worked with a tape manufacturer to develop and test a new standard for duct tape: *UL-181B Standard, Closure Systems for Use with Flexible Air Ducts and Air Connectors*. According to UL, this standard evaluates duct tape for "tensile strength; peel adhesion strength; shear adhesion strength under a variety of weights, temperatures, and humidity levels; long-term high-temperature effects; and surface burning characteristics." The ADC now recommends that all pressure-sensitive tape used on HVAC duct meet the new *UL-181B* standard.

Information for this feature was reviewed on the following Web site, May 2008: http://www.energy.ca.gov/title24/ducttape/documents/05-22-01_ADC_LETTER.PDF.

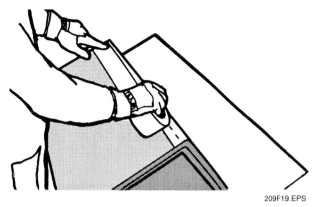

Figure 19 ◆ Applying pressure-sensitive tape.

Figure 20 ◆ Rub tape until facing reinforcement shows through.

Figure 21 ◆ Applying heat to the heat-seal tape.

Figure 22 ◆ Applying heat with an iron to heat-activated tape.

Figure 23 ◆ Heat indicator dots darken to show the proper temperature.

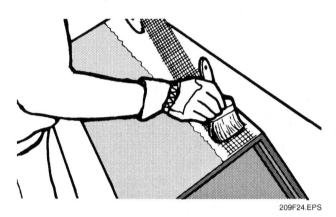

Figure 24 ◆ Apply a thin coat of mastic, then add mesh tape.

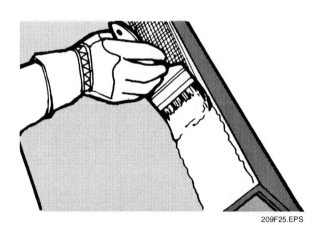

Figure 25 ◆ Apply a second coat of mastic over the mesh tape.

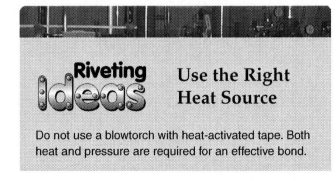

Riveting Ideas

Use the Right Heat Source

Do not use a blowtorch with heat-activated tape. Both heat and pressure are required for an effective bond.

Step 4 Close the long seam using one of the closure methods discussed earlier in this module.

Constructing a duct module is only one part of the task you will do on the job. You must also know how to join the duct modules together. This is a fairly simple task; however, you must do it carefully, paying attention to the joints to ensure that they fit tightly and evenly. Carefully seal the joints with approved tape. When standing on end, the large duct sections tend to wobble. Prop the sections against a wall to steady them while you join them together. To join duct modules together, follow these steps (*Figure 27*):

Step 1 Using an insulation knife, slit the facing tabs at each corner. Make sure you don't cut below the male shiplap shoulder or into the inside duct surfaces. The best way to make these slits is to rest the end of the knife blade on the shoulder of the shiplap and pull the blade outward to cut the facing.

Step 2 Push the two sections together. Make sure the male and female shiplaps fit tightly together. This step is easiest if you place one duct section on the floor and set the other section on top of it.

Step 3 Using a staple gun, staple the flaps on all four sides. Use ½-inch outward clinching staples about 2 inches on center.

Step 4 Make sure the duct board surface is clean and dry. Then tape the joint with pressure-sensitive tape. Using a plastic rubbing tool, rub the tape firmly until you can see the facing screen clearly through the tape.

8.0.0 ◆ TEN-SIDED FIBERGLASS DUCT

Ten-sided fiberglass duct is an acceptable substitute for rigid round duct. Sections of this duct may be fabricated from standard duct board using a groov-

Riveting Ideas

Mastic and Glass Fabric Tape over Sheet Metal

There are places in some duct systems where fiberglass duct must be connected to sheet metal duct, such as a connection to the flanges of central air equipment. In these cases, sheet metal screws and washers are used as fasteners to carry the mechanical load. Mastic and glass fabric tape is applied to seal the connections at these points. Usually two widths of glass fabric tape are required.

ing machine. The circumferential joints may be shiplapped or butt-edged. The male and female ends should be fabricated while the material is in flat board form. Internal metal sleeves must be used on all return air ducts and on supply ducts without shiplap joints. Ten-sided fiberglass duct can be used to fabricate standard fittings such as elbows, tees, offsets, and reducers (*Figure 28*). *Table 2* indicates the stretchout dimensions for 10-sided duct.

9.0.0 ◆ CONNECTING DUCT BOARD TO SHEET METAL

In a duct system, there are areas where duct board must be connected to sheet metal. Study *Figure 29* to see how to connect duct board to sheet metal flanges on air distribution equipment. The four possible connections are as follows:

- Pressure-sensitive tape over sheet metal screws on a 26-gauge sheet metal U-channel (*Figure 29A*)
- Pressure-sensitive tape over sheet metal screws on a 26-gauge pocket lock (*Figure 29B*)
- Pressure-sensitive tape over sheet metal screws on a 22-gauge sheet metal sleeve (*Figure 29C*)
- Pressure-sensitive tape over sheet metal screws on a 22-gauge sheet metal flange (*Figure 29D*)

Riveting Ideas

Handling Glass Fabric Tape

Once you have applied a thin coat of mastic, unroll a small section of tape and press it into place. Hold the roll of tape in one hand and the mastic brush in the other. Unroll the tape with a smooth, steady motion while smoothing it into place with the brush. Work with steady, even movements. When you reach the end of the area to be taped, set the brush across the top of the mastic container and use your utility knife to cut the tape. By using this method, you won't have to measure the amount of tape you need, and you have better control over the tape.

STEP 1

STEP 2

STEP 3

STEP 4

209F26.EPS

Figure 26 ◆ Fabricating a duct module.

You must carefully clean the sheet metal and seal the connections properly with pressure-sensitive tape. The mechanical connections (screws or screws and washers) must be placed at a maximum of 12 inches on center. Washers must be 0.028 inches thick with turned edges to prevent them from cutting into the duct board or the facing.

Apply the pressure-sensitive tape using the same procedures outlined earlier in this module. Remember to avoid putting too much pressure on the sealing tool when rubbing the tape over mechanical connections.

Riveting   **Handling Tape**

Tape is a convenient and easy way to seal joints and seams and to make repairs. However, the tape can fold back on itself or wrinkle as you apply it. To better control the tape, pull out a small section and firmly press it into place while holding the roll with your other hand. Hold the adhered section in place briefly with one hand as you steadily pull the tape roll with the other. Smooth the tape into place, moving your free hand down the tape as it comes off the roll.

STEP 1

STEP 2

STEP 4

209F27.EPS

Figure 27 ◆ Joining two duct modules.

10.0.0 ◆ FABRICATING RIGID DUCT BOARD

In this section, you will learn how to fabricate fittings using fiberglass duct board. You will also learn how to install tap-ins, **end caps**, dampers, and registers or grilles. Some of the instructions will refer you to the tape tab schedule shown in *Table 3*. Some of the instructions will refer you to the fastener schedule shown in *Table 4*. Review the approved closure systems covered earlier in this module.

10.1.0 Elbow 45 Degrees or Less from Duct Board Section

In this section, you will learn how to cut into a duct board section to form elbows with angles up to 45 degrees. Using elbows of 30 degrees or less will produce the best air flow results. Standard **turning vanes** may not be used for angles other than 90 degrees. To fabricate this fitting, follow these steps (*Figure 30*):

Step 1 Using an insulation knife, cut into the duct section as shown by lines A, B, C, and D in *Figure 30A*. Maintain the cheek panel angle when making the side panel cuts.

Step 2 Separate the two sections (*Figure 30B*). Rotate one section 180 degrees (*Figure 30C*). Push the sections together.

Step 3 Tape all joints, starting with tape tabs on the heel panels according to the tape tab schedule in *Table 3* (*Figure 30D*).

Step 4 As an option for the cheek panels only, make cuts with the shiplap tool on lines AB and CD. Maintain the shiplap orientation of the original duct section (*Figure 30E*).

10.2.0 One-Way 30-Degree Offset from Duct Board Section

An offset is used to run the ductwork around an obstruction such as a plumbing stack, electrical chase, or framing member. Note that because the angle is less than 90 degrees, turning vanes cannot be used. To fabricate this fitting, follow these steps (*Figure 31*):

Step 1 Using an insulation knife, cut along lines A, B, C, and D and lines A', B', C', and D'. Maintain a 15-degree cheek panel angle when making the side panel cuts (*Figure 31A*). The heel dimensions equal 0.27 times the outer diameter (OD).

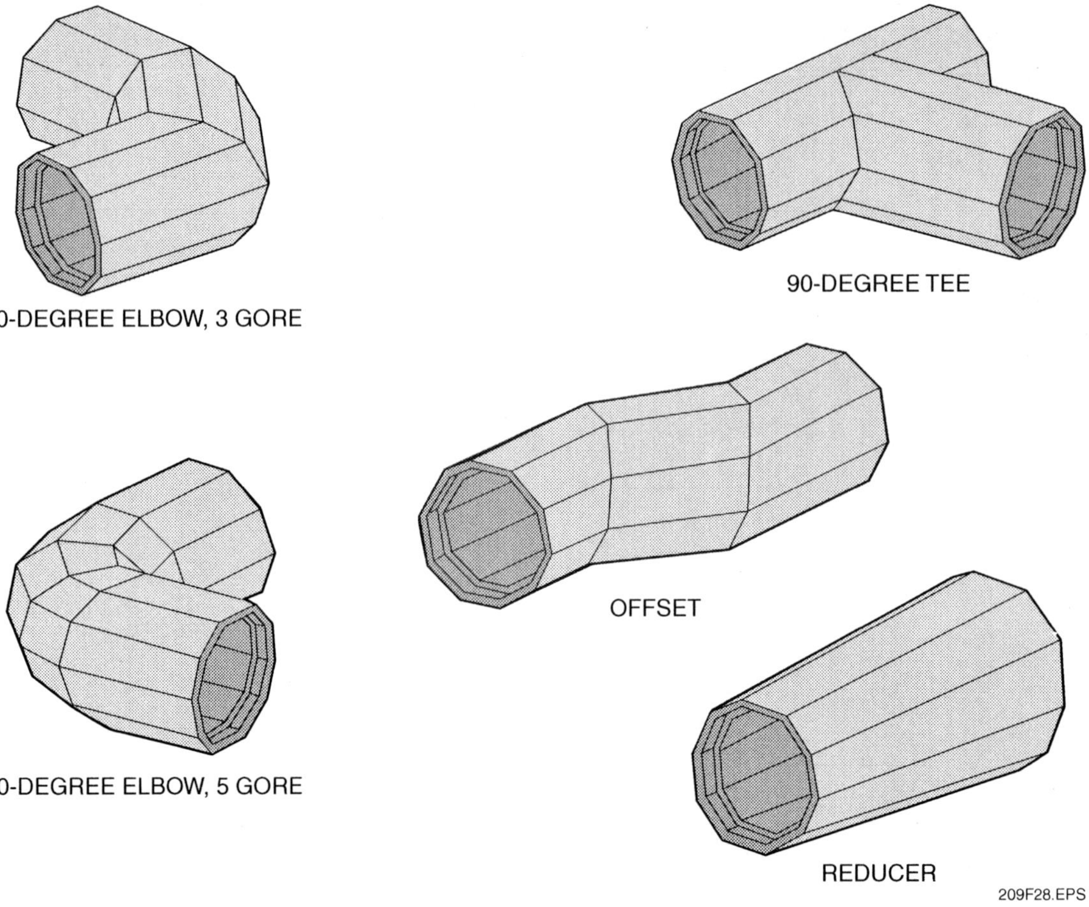
90-DEGREE ELBOW, 3 GORE
90-DEGREE TEE
OFFSET
90-DEGREE ELBOW, 5 GORE
REDUCER
209F28.EPS

Figure 28 ◈ Typical 10-sided fiberglass duct fittings.

Table 2 Ten-Sided Duct Board Stretchout

Nominal Inside Diameter		1" (25mm) Duct Board				1½" (38mm) Duct Board			
in.	**(mm)**	**Full Section 10 sides**		**Half Section 5 sides**		**Full Section 10 sides**		**Half Section 5 sides**	
5	(127)	23½	(597)	12⅜	(314)	26⅝	(676)	14	(356)
6	(152)	26⅝	(676)	14	(356)	28⅜	(721)	15⅝	(397)
7	(178)	29¾	(756)	15½	(394)	32⅞	(835)	17⅛	(435)
8	(203)	32⅞	(835)	17⅛	(435)	36	(914)	18¾	(476)
9	(229)	36	(914)	18⅝	(473)	39¼	(997)	20¼	(514)
10	(254)	39⅛	(994)	20¼	(514)	42⅜	(1076)	21⅞	(556)
12	(305)	45⅜	(1153)	23⅜	(594)	48⅝	(1235)	25	(635)
14	(356)	51⅝	(1311)	26½	(673)	55	(1397)	28⅛	(714)
16	(406)	57⅞	(1470)	29⅝	(752)	61¼	(1556)	31⅜	(797)
18	(457)	64	(1626)	32¾	(832)	67⅝	(1718)	34½	(876)
20	(508)	70½	(1791)	36	(914)	72⅝	(1845)	37⅝	(956)

Step 2 Separate the three sections (*Figure 31B*). Rotate the center section 180 degrees (*Figure 31C*). Push the three sections together.

Step 3 Tape all joints, starting with tape tabs on the heel panels according to the tape tab schedule in *Table 3*. Tab all around (*Figure 31D*).

Step 4 As an option for the cheek panels only, make cuts with the shiplap tool on lines A'B', AB, C'D', and CD. Maintain the shiplap orientation of the original duct section (*Figure 31E*).

Table 3 Tape Tab Schedule

8 inches nominal length
12 inches nominal centers
At least one piece of tape per side

Table 4 Fastener Schedule

No. 10 plated sheet metal screws, board thickness + ¼ inch with 2½-inch square or 3-inch round plated steel washers
12 inch nominal on centers

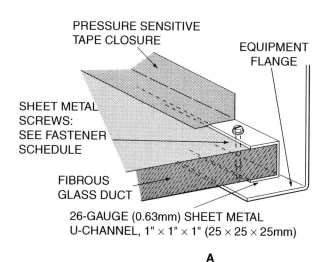

A

PRESSURE SENSITIVE TAPE CLOSURE

EQUIPMENT FLANGE

SHEET METAL SCREWS: SEE FASTENER SCHEDULE

FIBROUS GLASS DUCT

26-GAUGE (0.63mm) SHEET METAL U-CHANNEL, 1" × 1" × 1" (25 × 25 × 25mm)

C

SHEET METAL SCREWS AND WASHERS: SEE FASTENER SCHEDULE

PRESSURE SENSITIVE TAPE CLOSURE

EQUIPMENT

SCREW POP RIVET OR SPOT WELD TO EQUIPMENT FLANGE

FIBROUS GLASS DUCT

22-GAUGE (0.75mm) SHEET METAL SLEEVE, MIN. 3" (76mm) WIDE

B

PRESSURE SENSITIVE TAPE CLOSURE

EQUIPMENT FLANGE

SHEET METAL SCREWS: SEE FASTENER SCHEDULE

FIBROUS GLASS DUCT

26-GAUGE (0.63mm) SHEET METAL POCKET LOCK, 1" × 1" × 1" (25 × 25 × 25mm)

D

SHEET METAL SCREWS AND WASHERS: SEE FASTENER SCHEDULE

PRESSURE SENSITIVE TAPE CLOSURE

EQUIPMENT FLANGE

FIBROUS GLASS DUCT

22-GAUGE (0.75mm) SHEET METAL 1" × 3" (25mm × 76mm)

SCREW, POP RIVET OR WELD TO FLANGE

209F29.EPS

Figure 29 ◆ Duct board to sheet metal connections.

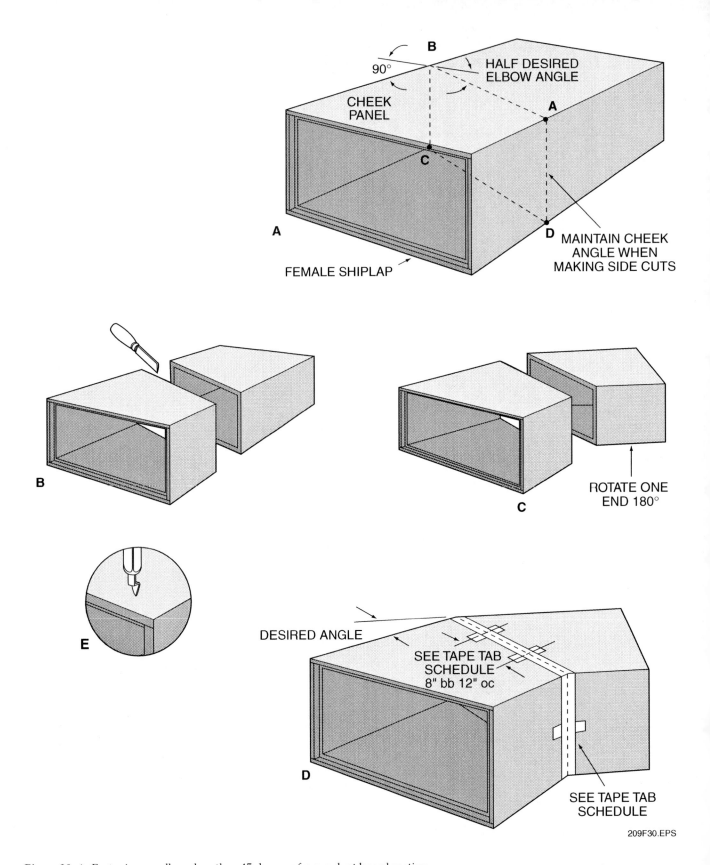

90°

B

HALF DESIRED ELBOW ANGLE

CHEEK PANEL

A

C

A

D MAINTAIN CHEEK ANGLE WHEN MAKING SIDE CUTS

FEMALE SHIPLAP

B

C ROTATE ONE END 180°

E

D DESIRED ANGLE

SEE TAPE TAB SCHEDULE 8" bb 12" oc

SEE TAPE TAB SCHEDULE

209F30.EPS

Figure 30 ◆ Fastening an elbow less than 45 degrees from a duct board section.

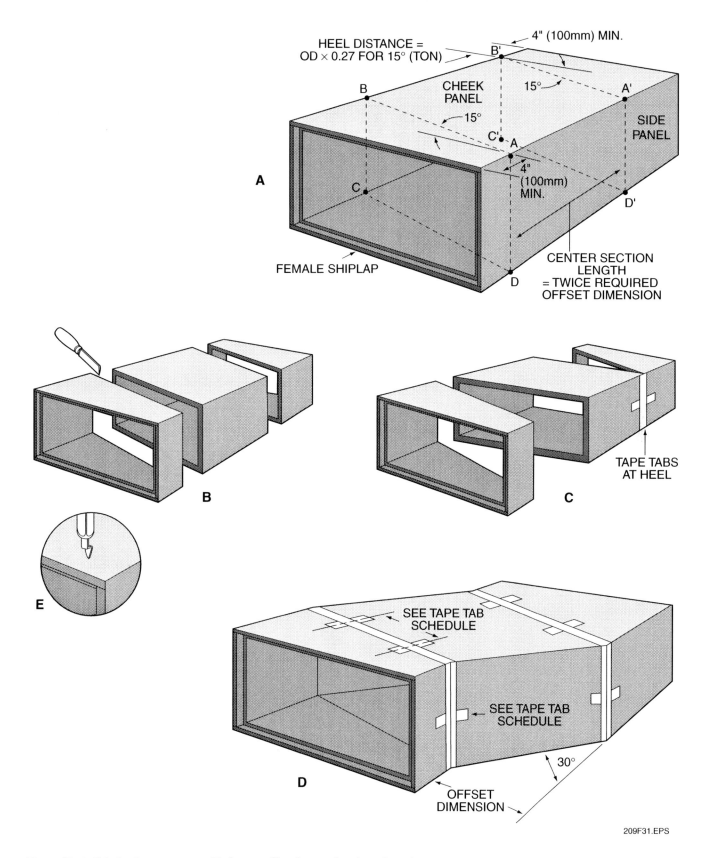

HEEL DISTANCE = OD × 0.27 FOR 15° (TON)

4" (100mm) MIN.

B'

CHEEK PANEL

15°

A'

SIDE PANEL

B

15°

C' A

A

4" (100mm) MIN.

C

D'

FEMALE SHIPLAP

D

CENTER SECTION LENGTH = TWICE REQUIRED OFFSET DIMENSION

B

C

TAPE TABS AT HEEL

E

SEE TAPE TAB SCHEDULE

SEE TAPE TAB SCHEDULE

30°

D

OFFSET DIMENSION

209F31.EPS

Figure 31 ◆ Fabricating a one-way 30-degree offset from a duct board section.

10.3.0 One-Way Offset of Any Angle from Duct Board Section

In this section, you will learn how to fabricate a one-way offset of any angle from a duct board section. In this example, you will use some of the geometry principles you learned in earlier modules. This is a useful technique to use in the field when shop machines and equipment are not available. To fabricate this fitting, follow these steps (*Figure 32*):

Step 1 Determine starting point A, offset length L, and offset width O (*Figure 32A*).

Step 2 Draw line OO' parallel to the duct edge to establish the offset dimension from the edge.

Step 3 Locate point B by measuring distance L from point A. Strike an arc using B as the center and L as the radius. The intersection of the arc with line OO' establishes point A'.

Step 4 Draw line AA', extending it across the cheek panel to point C. Mark point C. Line AC establishes the offset angle.

Step 5 Measure distance L from point C to locate and mark point D. Draw line BD, extending the lines around the duct section.

Step 6 Using an insulation knife, make two cuts around the duct section. Extend the layout lines square with the corner. Maintain the cheek panel angle when making the side panel cuts.

Step 7 Separate the three sections (*Figure 32B*). Rotate the center section 180 degrees (*Figure 32C*). Push the sections together.

Step 8 Tape all joints, starting with tape tabs on the heel panels according to the tape tab schedule in *Table 3* (*Figure 32D*).

Step 9 As an option for the cheek panels only, make cuts with the shiplap tool on lines AC, BD, and on the diagonal lines on the opposite cheek panel. Maintain the shiplap orientation of the original duct section (*Figure 32E*).

10.4.0 Three-Piece 90-Degree Elbow from Duct Board Section

In this example, you will fabricate a fitting from a formed duct board section. You can use the following method on elbows from 30 degrees to 90 degrees. To fabricate this fitting, follow these steps (*Figure 33*):

Step 1 Using an insulation knife, cut along lines A, B, C, D and A', B', C', and D' (*Figure 33A*). Maintain the cheek panel angle when making the side panel cuts.

Step 2 Separate the three sections (*Figure 33B*). Rotate the center section 180 degrees (*Figure 33C*). Push the sections together and apply tape tabs to heel panels according to the tape tab schedule in *Table 3* (*Figure 33D*).

Step 3 Tape all joints (*Figure 33F*).

Step 4 As an option for the cheek panels only, make cuts with the shiplap tool on lines AB, A'B', CD, and C'D'. Maintain the shiplap orientation of the original duct section (*Figure 33E*).

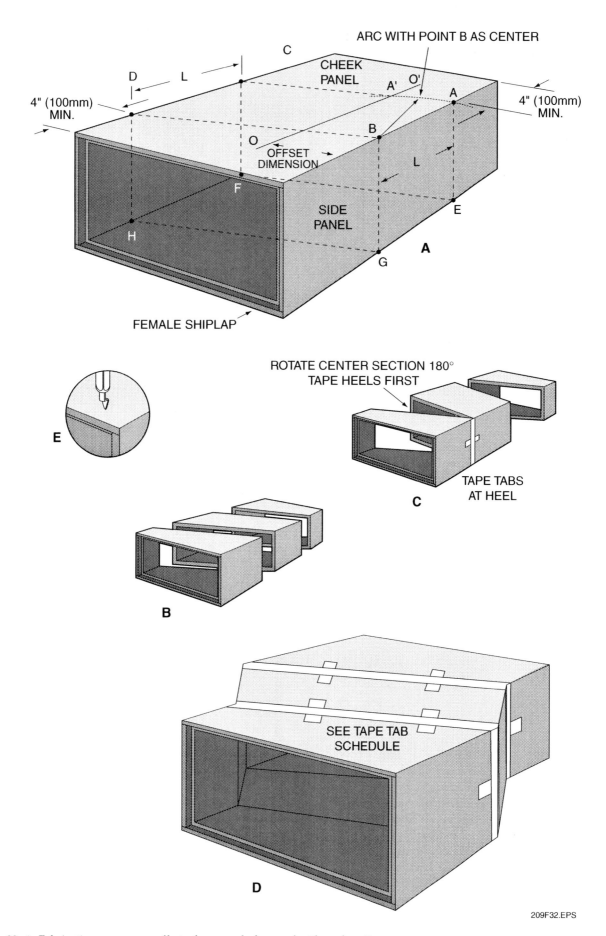

Figure 32 ◆ Fabricating a one-way offset of any angle from a duct board section.

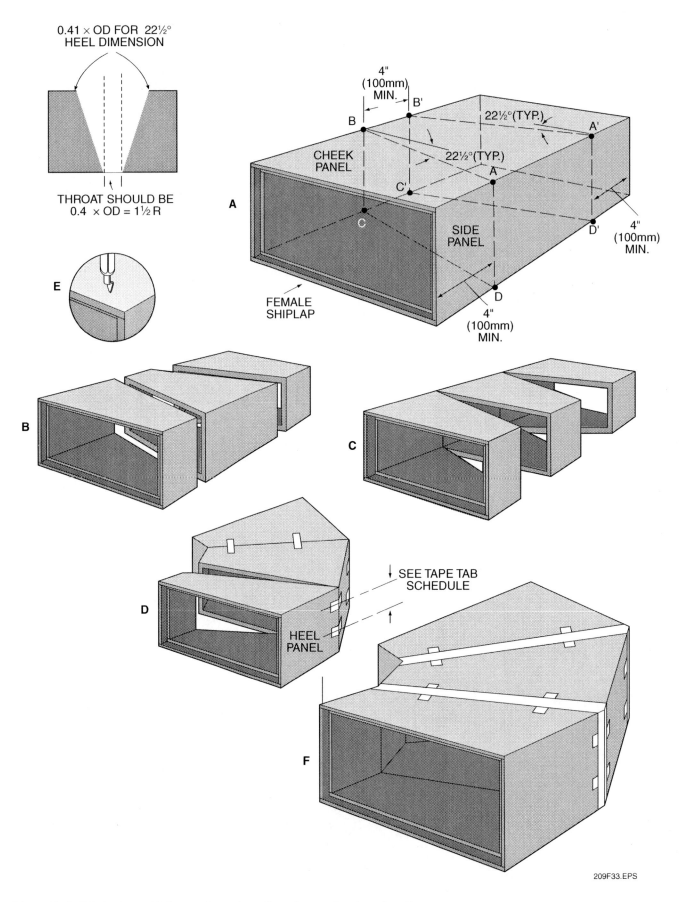

0.41 × OD FOR 22½°
HEEL DIMENSION

THROAT SHOULD BE
0.4 × OD = 1½ R

4"
(100mm)
MIN.

22½°(TYP.)

B'

B

22½°(TYP.)

A'

CHEEK
PANEL

C'

A

A

C

SIDE
PANEL

D'

4"
(100mm)
MIN.

FEMALE
SHIPLAP

D

4"
(100mm)
MIN.

E

B

C

D

SEE TAPE TAB
SCHEDULE

HEEL
PANEL

F

209F33.EPS

Figure 33 ◈ Fabricating a 90-degree three-piece elbow from a duct board section.

10.5.0 Elbow with Turning Vanes from Flat Duct Board

In this section, you will learn how to fabricate a 90-degree elbow from flat grooved duct board. Because elbows significantly change the direction of a duct's airflow, they create a considerable amount of turbulence inside the duct. This turbulence creates a noticeable drop in static pressure between the inlet of the elbow and its outlet. To help with this, turning vanes are installed in elbows to smooth the airflow through the fitting which makes the pressure drop less severe.

You will install turning vanes inside the duct section at the point where the duct changes direction. Each vane is curved to help create a more uniform airflow. The vanes may be made of rigid fiberglass or sheet metal. Note that sheet metal vanes work best when placed at a 45-degree angle. The openings on both sides must be equal. To fabricate a 90-degree elbow with turning vanes, follow these steps (*Figure 34*):

Step 1 On flat grooved duct board, draw diagonal lines to locate the placement of the turning vanes. Notice that the lines are drawn on the top and bottom duct panels (*Figure 34A*).

Step 2 Using a vane hole cutter, cut the vane holes. Remove and discard loose insulation.

Step 3 Using an insulation knife, cut the turning vanes to the correct length and insert them in the holes in the bottom of the panel (*Figure 34B* and *Figure 34C*). For 1-inch-thick duct board, cut vanes 1½ inches to 1¾ inches long. For 1½-inch-thick duct board, cut vanes 2½ inches to 2¾ inches long.

Step 4 Using an insulation knife, cut tap-out panel A from the duct board (refer to *Figure 34B*).

Step 5 Rotate the tap-out panel 180 degrees. Place it in the end of the duct section so that the female shiplap is at the downstream, open end of the elbow (*Figure 34D*).

Step 6 Staple the flap of the tap-out panel on the bottom to secure it to the duct section.

Step 7 Tape all joints, starting with tape tabs according to the tape tab schedule in *Table 3*.

Riveting Ideas — Turning Vanes

Two types of turning vanes are commonly used in fiberglass duct systems: fiberglass vanes and sheet metal vanes. Fiberglass vanes are inserted individually into holes in the duct wall. Each vane is inserted into a hole that has been cut to fit that particular vane. Sheet metal vanes come assembled with top and bottom rails that can be bolted to the duct wall. The project engineer will specify the type of vanes to use on each project.

The straight duct section that will complete this elbow will branch off from the fitting you have just fabricated (refer to *Figure 34D* to see the dotted lines that show the duct extension). The length of the straight duct section must be no less than one-sixth of the branch width or at least four inches long.

10.6.0 Elbow with Turning Vanes from Duct Board Section

In this section, you will learn how to cut into a duct board section to install turning vanes in a 90-degree elbow. In this example, the turning vanes are made of sheet metal, rail mounted, and fastened in place with sheet metal screws and washers. Some of the steps in this example are similar to the steps in the previous example. To cut into a duct board section and install rail-mounted turning vanes, follow these steps (*Figure 35*):

Step 1 Using an insulation knife, cut tap-out panel A. To preserve the shiplap, cut through only the facing along the duct corners. Cut through the board and facing across tap-out panel A (*Figure 35A*).

Step 2 Install the rail-mounted turning vanes and fasten top and bottom of the rails according to the fastener schedule in *Table 4* (*Figure 35B*).

Step 3 Rotate tap-out panel A 180 degrees. Place the tap-out panel in the end of the duct panel so that the female shiplap is at the downstream, open end of the panel.

Step 4 Tape all joints, starting with tape tabs, according to the tape tab schedule in *Table 3* (*Figure 35C*).

The straight duct section that will complete this elbow will branch off from the fitting you have just fabricated (refer to *Figure 35C* to see the dotted lines that show the duct extension). The length of the straight section of duct must be no less than one-sixth of the branch width or at least four inches long.

10.7.0 Tee from Duct Board Section

In this section, you will learn how to form a tee with turning vanes from a duct board section. The vanes must be adjustable to the angle required to achieve the 90-degree turn on both sides of the tee. To form this tee, follow these steps (*Figure 36*):

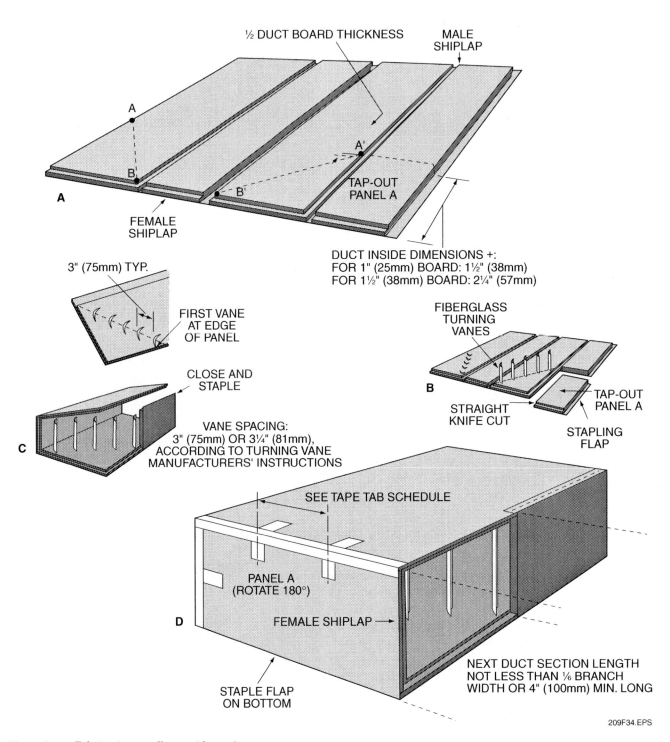

Figure 34 ◆ Fabricating an elbow with turning vanes.

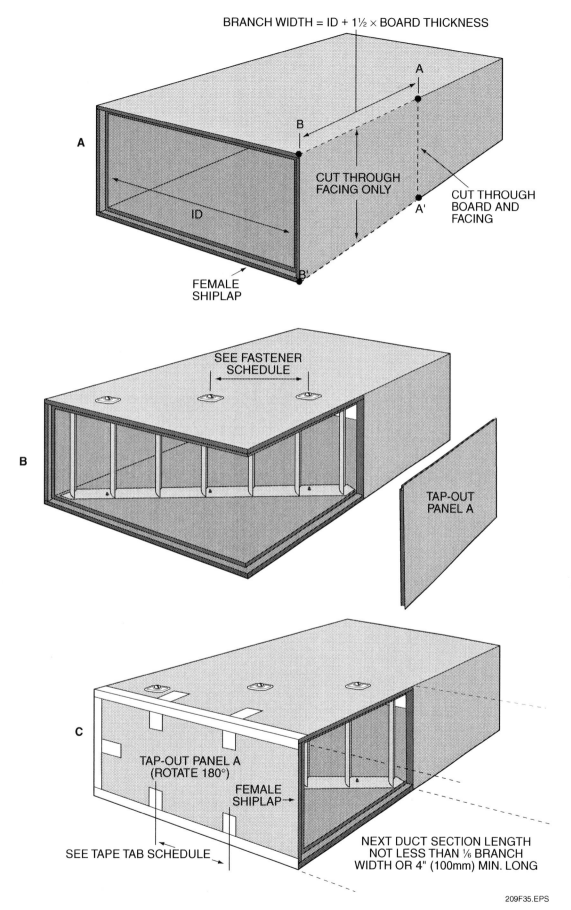

BRANCH WIDTH = ID + 1½ × BOARD THICKNESS

A

ID

FEMALE
SHIPLAP

CUT THROUGH
FACING ONLY

CUT THROUGH
BOARD AND
FACING

SEE FASTENER
SCHEDULE

B

TAP-OUT
PANEL A

C

TAP-OUT PANEL A
(ROTATE 180°)

FEMALE
SHIPLAP

SEE TAPE TAB SCHEDULE

NEXT DUCT SECTION LENGTH
NOT LESS THAN ⅙ BRANCH
WIDTH OR 4" (100mm) MIN. LONG

209F35.EPS

Figure 35 ◆ Fabricating a 90-degree elbow with turning vanes from a duct board section.

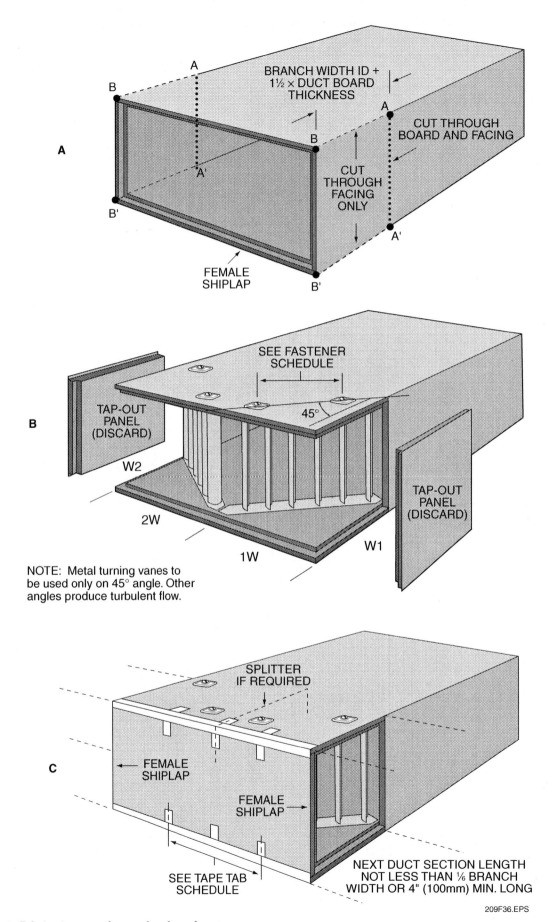

A

A

B

BRANCH WIDTH ID + 1½ × DUCT BOARD THICKNESS

A

CUT THROUGH BOARD AND FACING

B

CUT THROUGH FACING ONLY

A'

B'

A'

FEMALE SHIPLAP

B'

B

SEE FASTENER SCHEDULE

TAP-OUT PANEL (DISCARD)

45°

W2

TAP-OUT PANEL (DISCARD)

2W

1W

W1

NOTE: Metal turning vanes to be used only on 45° angle. Other angles produce turbulent flow.

C

SPLITTER IF REQUIRED

FEMALE SHIPLAP

FEMALE SHIPLAP

SEE TAPE TAB SCHEDULE

NEXT DUCT SECTION LENGTH NOT LESS THAN ⅙ BRANCH WIDTH OR 4" (100mm) MIN. LONG

209F36.EPS

Figure 36 ◈ Fabricating a tee from a duct board system.

Step 1 Using an insulation knife, cut out tap-out panels A and B as shown in *Figure 36A*. To preserve the shiplap, cut through only the facing along the duct corners. Cut through the board and facing across tap-out panels A and B. Discard the tap-out panels.

Step 2 Install the rail-mounted turning vanes according to the fastener schedule in *Table 4* (*Figure 36B*).

Step 3 Fabricate an end cap with two female shiplaps to close the end of the duct section (*Figure 36C*). An end cap is a piece of duct board sized to close the open end of the duct. If a splitter section is required, place it where shown.

Step 4 Tape all joints, starting with tape tabs according to the tape tab schedule in *Table 3*.

The straight duct sections that will complete the tee will branch off from the fitting you have just fabricated. Refer to *Figure 36C* to see the dotted lines that show the duct extension. The length of each straight duct section must be no less than one-sixth of the branch width or at least four inches long.

10.8.0 Tee from Flat Grooved Duct Board

In this section, you will learn how to fabricate a tee with turning vanes from flat grooved duct board. Some of the steps are similar to those in the previous example. To fabricate this tee, follow these steps (*Figure 37*):

Step 1 Measure and draw lines to locate the position of the turning vanes and to outline the tee openings (*Figure 37A*).

Step 2 Using an insulation knife, cut along the lines for the tee openings. Discard the cutout sections as shown in *Figure 37B*.

Step 3 Using a vane hole cutter, cut vane holes 3 inches apart along the lines you drew in Step 1. Remove and discard loose insulation. (Note: You may also use metal rail-mounted vanes.)

Step 4 Using an insulation knife, cut the turning vanes to the correct length and insert them in the holes in the bottom of the panel. For 1-inch-thick duct board, cut vanes 1½ inches to 1¾ inches long. For 1½-inch-thick duct board, cut vanes 2½ inches to 2¾ inches long (*Figure 37C*).

Step 5 Fabricate an end cap with two female shiplaps to close the end of the duct section.

Step 6 Tape all joints, starting with tape tabs, according to the tape tab schedule in *Table 3* (*Figure 37D*).

The straight duct sections that will complete the tee will branch off from the fitting you have just fabricated. Refer to *Figure 37D* to see the dotted lines that show the duct extension. The length of each straight duct section must be no less than one-sixth of the branch width or at least four inches long.

10.9.0 One-Way Transition from Flat Duct Board with Shiplap Panel

In this section, you will fabricate a one-way transition. This is the type of task you will most likely do in the shop, working from flat duct board to the finished duct panel. To fabricate this fitting, follow these steps (*Figure 38*):

Step 1 Draw diagonal lines to establish the transition angle (*Figure 38A*).

Step 2 Using an insulation knife, cut through the insulation and facing along the diagonal lines. Trim the facing for staple flaps about 2 inches wide (*Figure 38B*). Discard the triangular pieces.

Step 3 Using the female shiplap tool, make shiplap cuts along the diagonal lines (*Figure 38C*).

Step 4 Using an insulation knife, make a relief cut if needed to allow the panel to conform to the transition angle.

Step 5 Staple the flaps and tape joints with approved tape (*Figure 38D*).

Step 6 If the sloping panel is short of the section end by more than ⅜-inch, insert a shiplapped filler panel to fill the gap.

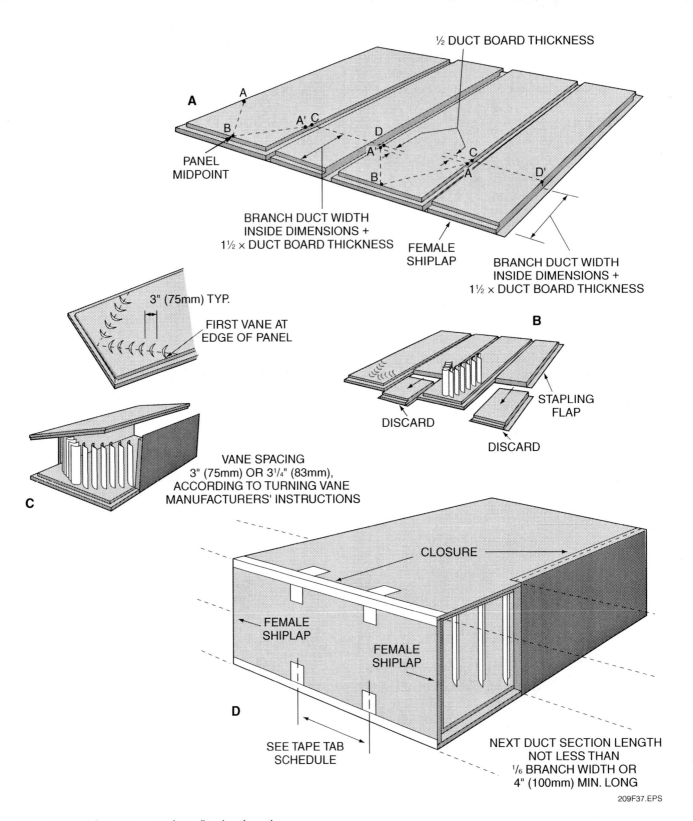

A

½ DUCT BOARD THICKNESS

A

A' C

D

B

C

A'

A'

D'

A

B'

PANEL MIDPOINT

BRANCH DUCT WIDTH INSIDE DIMENSIONS + 1½ × DUCT BOARD THICKNESS

FEMALE SHIPLAP

BRANCH DUCT WIDTH INSIDE DIMENSIONS + 1½ × DUCT BOARD THICKNESS

3" (75mm) TYP.

FIRST VANE AT EDGE OF PANEL

B

DISCARD

STAPLING FLAP

DISCARD

VANE SPACING 3" (75mm) OR 3¼" (83mm), ACCORDING TO TURNING VANE MANUFACTURERS' INSTRUCTIONS

C

CLOSURE

FEMALE SHIPLAP

FEMALE SHIPLAP

D

SEE TAPE TAB SCHEDULE

NEXT DUCT SECTION LENGTH NOT LESS THAN ⅙ BRANCH WIDTH OR 4" (100mm) MIN. LONG

209F37.EPS

Figure 37 ◆ Fabricating a tee from flat duct board.

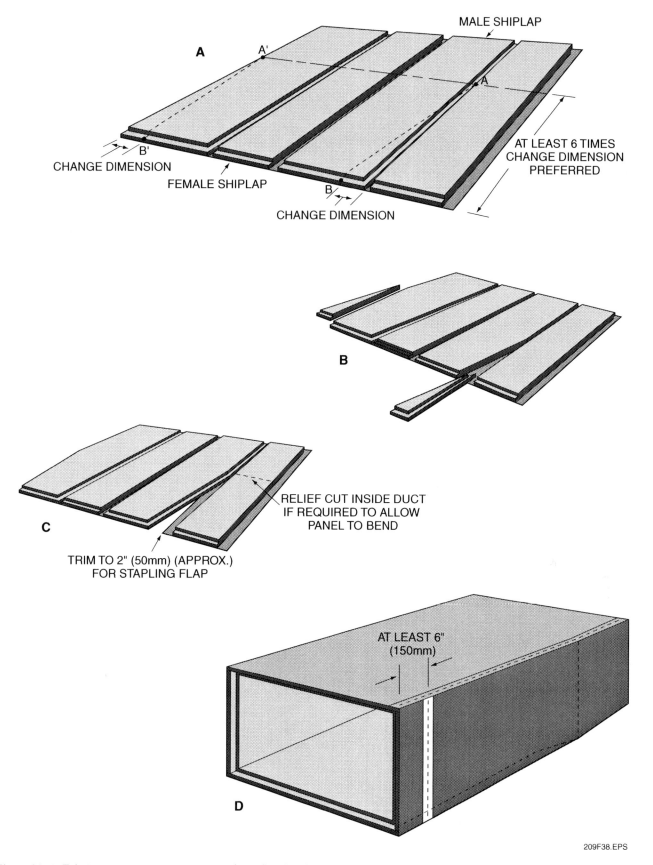

MALE SHIPLAP

A

A'

AT LEAST 6 TIMES
CHANGE DIMENSION
PREFERRED

B'

CHANGE DIMENSION

FEMALE SHIPLAP

B

CHANGE DIMENSION

A

B

C

RELIEF CUT INSIDE DUCT
IF REQUIRED TO ALLOW
PANEL TO BEND

TRIM TO 2" (50mm) (APPROX.)
FOR STAPLING FLAP

AT LEAST 6"
(150mm)

D

209F38.EPS

Figure 38 ◆ Fabricating a one-way transition from flat duct board with shiplap panel.

10.10.0 One-Way Transition from Duct Board Section with Shiplap

In this section, you will learn how to modify a one-way transition using a hand shiplap tool. This is the type of task you would most likely do in the field when you have to modify a fabricated duct section. The process is similar to that of the previous example; however, because you are cutting along the long dimension of the duct, you must maintain the shiplap edge. To modify this fitting, follow these steps (*Figure 39*):

Step 1   Using a hand shiplap tool, cut through the insulation and facing along the dotted lines as shown in *Figure 39A*. Be sure to maintain the shiplap orientation of the duct section.

Step 2   Peel back the facing and trim both sides, leaving staple flaps about 2 inches wide. Remove and discard the triangular pieces (*Figure 39B*).

Step 3   Using an insulation knife, make a relief cut inside the duct if required to allow the panel to conform to the transition angle.

Step 4   Staple the flaps and tape the joints with approved tape (*Figure 39C*).

Step 5   If the sloping panel is short of the section end by more than ⅜-inch, insert a shiplapped filler panel to fill the gap.

10.11.0 One-Way Transition from Duct Board Section

A transition is used to bridge the gap between ducts of different dimensions. This is the type of task that you would most likely do in the field when you must modify a fitting to create the transition. The change in dimension is equal to the difference between the height of the upstream and downstream ends of the duct. To modify this fitting, follow these steps (*Figure 40*):

Step 1   Using an insulation knife, cut through the insulation and facing as shown in *Figure 40A*.

Step 2   Trim both sides, leaving staple flaps about 2 inches wide. Remove and discard the triangular pieces (*Figure 40B*).

Step 3   Using an insulation knife, make a relief cut if needed to allow the panel to conform to the transition angle (*Figure 40C*).

Step 4   Staple flaps and seal joints with approved tape (*Figure 40D*).

Step 5   If the sloping panel is short of the section end by more than ¾-inch, insert a shiplapped filler panel to fill the gap.

10.12.0 Square and Round Tap-Ins

Square and round tap-ins allow you to attach runouts to the main trunk of the duct system. They are called tap-ins, or taps, because they tap into the main trunk duct. After being connected to the main trunk, these fittings are secured with staples and sealed with tape. These fittings may be made using rigid fiberglass duct for square tap-ins or round flexible duct for round tap-ins.

For square tap-ins, the preferred method is to use a full-height fitting (*Figure 41*). This method provides the greatest strength. Note that if the width of the hole exceeds 16 inches or the hole occurs at the connection joint, you may need to install a tie rod with washers to restore the strength that was lost when the panel was removed. An alternative method in which the tap-in is less than the full height of the main trunk is acceptable (*Figure 42*). However, this method often requires that the trunk duct be reinforced.

Cutting a hole in the duct may reduce the strength of the side of the duct panel. If the hole is wider than 16 inches or if it occurs at a connection joint, a tie rod and washers may be needed to restore the lost strength of the side panel.

You can use either a dovetail or spin-in collar to make round tap-in connections to the trunk duct (*Figure 43*). It is best to use a round hole cutter instead of an insulation knife to cut the hole in the main trunk. A hole cutter is quicker and easier to use than a knife and will give you an accurately sized opening. Refer to the general installation instructions for these collars covered earlier in this module. For a specific installation, follow the manufacturer's instructions. The thickness of a conical collar must be the same as the thickness of the fiberglass duct. Collars are most commonly available in 1-inch and 1½-inch thicknesses.

10.13.0 Wide Mouth Tap-In

A wide mouth tap-in, as its name implies, is a tap-in with one end that is wider than the other. It is also called a shoe fitting. The wider part of the tap-in connects to the main trunk. To fabricate this fitting, use a shiplap tool and a straight knife (*Figure 44*). Although *Figure 44* illustrates a four-piece tap-in, some prefer to make tap-ins as a mitered C-shape duct and a side panel.

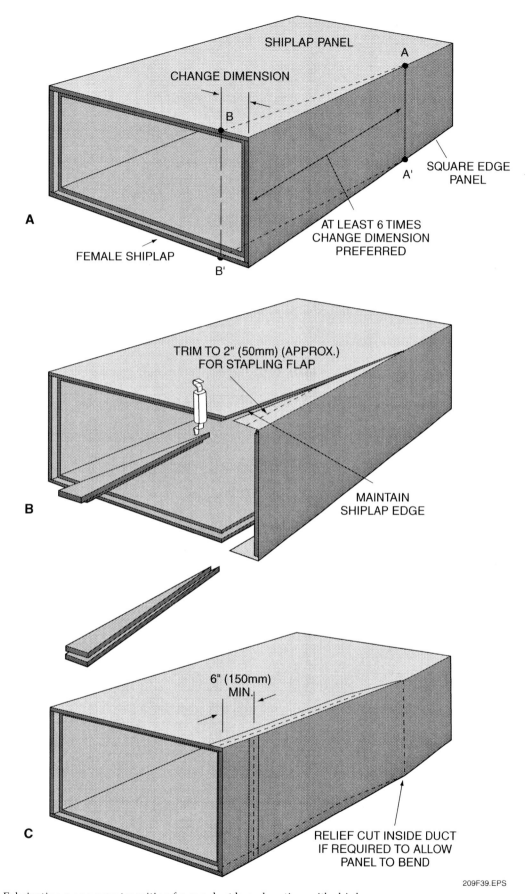

SHIPLAP PANEL

CHANGE DIMENSION

A

A'

SQUARE EDGE PANEL

AT LEAST 6 TIMES CHANGE DIMENSION PREFERRED

B

B'

FEMALE SHIPLAP

A

TRIM TO 2" (50mm) (APPROX.) FOR STAPLING FLAP

MAINTAIN SHIPLAP EDGE

B

6" (150mm) MIN.

RELIEF CUT INSIDE DUCT IF REQUIRED TO ALLOW PANEL TO BEND

C

209F39.EPS

Figure 39 ◆ Fabricating a one-way transition from a duct board section with shiplap.

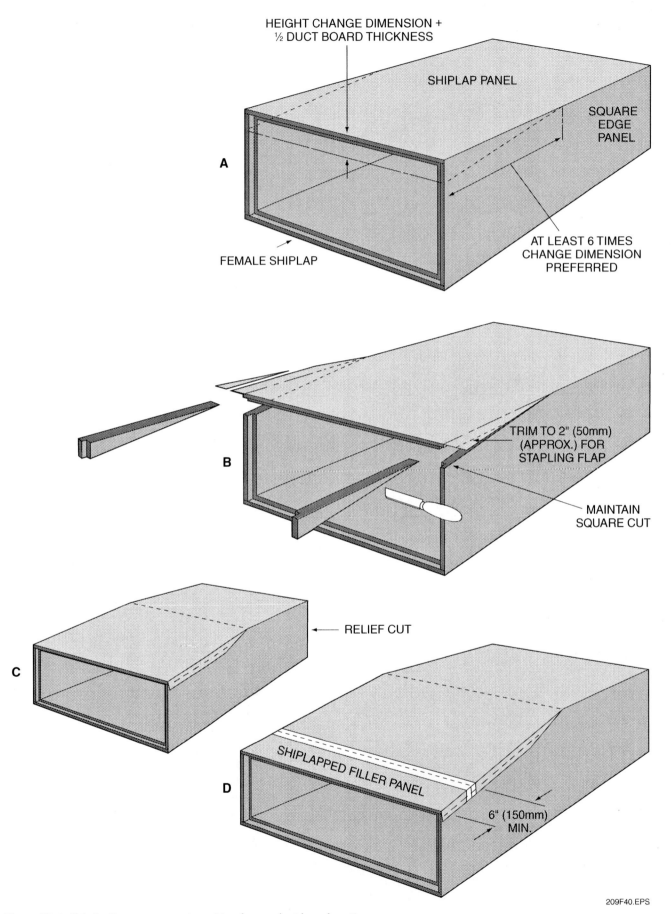

HEIGHT CHANGE DIMENSION +
½ DUCT BOARD THICKNESS

SHIPLAP PANEL

SQUARE
EDGE
PANEL

A

FEMALE SHIPLAP

AT LEAST 6 TIMES
CHANGE DIMENSION
PREFERRED

B

TRIM TO 2" (50mm)
(APPROX.) FOR
STAPLING FLAP

MAINTAIN
SQUARE CUT

RELIEF CUT

C

D

SHIPLAPPED FILLER PANEL

6" (150mm)
MIN.

209F40.EPS

Figure 40 ◆ Fabricating a one-way transition from a duct board section.

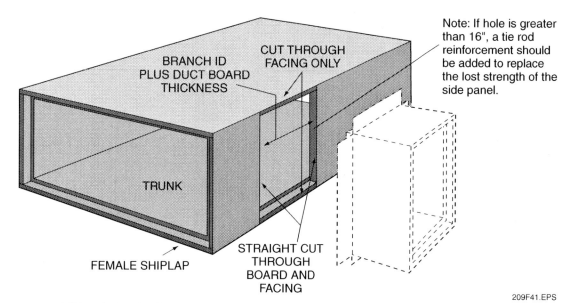

Note: If hole is greater than 16", a tie rod reinforcement should be added to replace the lost strength of the side panel.

BRANCH ID PLUS DUCT BOARD THICKNESS

CUT THROUGH FACING ONLY

TRUNK

FEMALE SHIPLAP

STRAIGHT CUT THROUGH BOARD AND FACING

209F41.EPS

Figure 41 ◆ Square tap-in, full height (or depth) method.

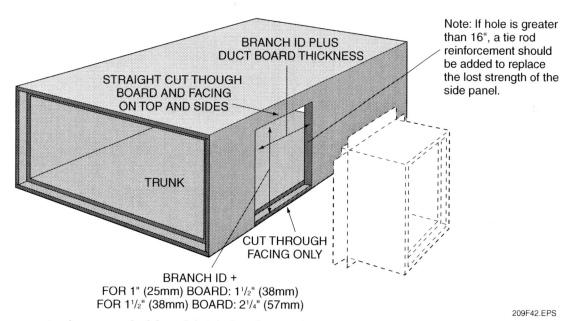

Note: If hole is greater than 16", a tie rod reinforcement should be added to replace the lost strength of the side panel.

BRANCH ID PLUS DUCT BOARD THICKNESS

STRAIGHT CUT THOUGH BOARD AND FACING ON TOP AND SIDES

TRUNK

CUT THROUGH FACING ONLY

BRANCH ID +
FOR 1" (25mm) BOARD: 1$\frac{1}{2}$" (38mm)
FOR 1$\frac{1}{2}$" (38mm) BOARD: 2$\frac{1}{4}$" (57mm)

209F42.EPS

Figure 42 ◆ Square tap-in, alternate method, branch height less than trunk.

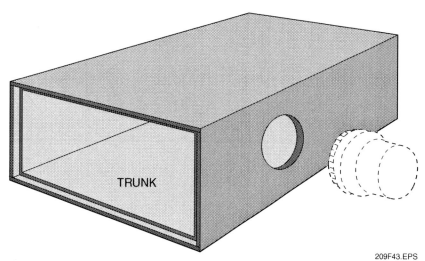

TRUNK

209F43.EPS

Figure 43 ◆ Round tap-in.

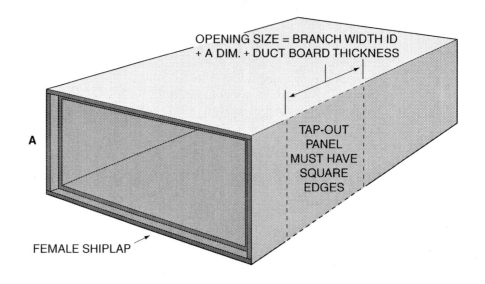

OPENING SIZE = BRANCH WIDTH ID + A DIM. + DUCT BOARD THICKNESS

TAP-OUT PANEL MUST HAVE SQUARE EDGES

FEMALE SHIPLAP

A

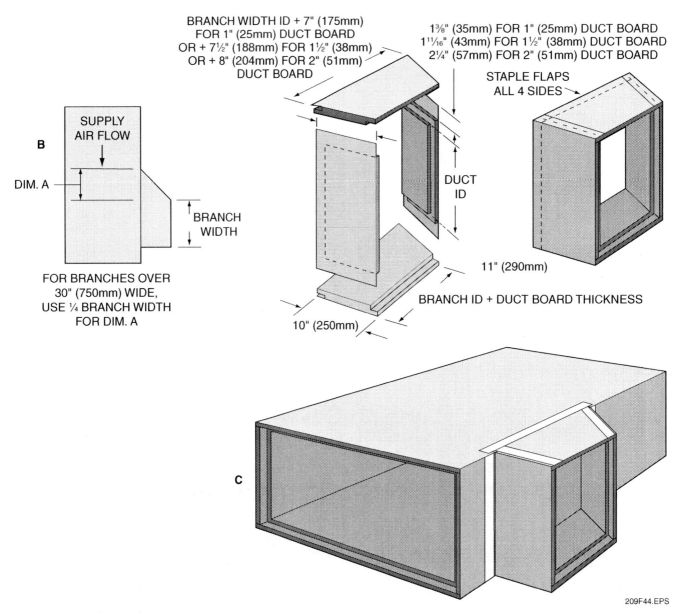

BRANCH WIDTH ID + 7" (175mm) FOR 1" (25mm) DUCT BOARD OR + 7½" (188mm) FOR 1½" (38mm) OR + 8" (204mm) FOR 2" (51mm) DUCT BOARD

1⅜" (35mm) FOR 1" (25mm) DUCT BOARD
1¹¹⁄₁₆" (43mm) FOR 1½" (38mm) DUCT BOARD
2¼" (57mm) FOR 2" (51mm) DUCT BOARD

STAPLE FLAPS ALL 4 SIDES

DUCT ID

B

SUPPLY AIR FLOW

DIM. A

BRANCH WIDTH

FOR BRANCHES OVER 30" (750mm) WIDE, USE ¼ BRANCH WIDTH FOR DIM. A

10" (250mm)

11" (290mm)

BRANCH ID + DUCT BOARD THICKNESS

C

209F44.EPS

Figure 44 ◆ Wide mouth tap-in.

10.14.0 End Caps

End caps are simply pieces of rigid duct board that are sized and cut to fit or cap the open end of a duct section. An end cap may be attached with tape or with staple flaps. Three types of end cap fabrications are shown in this section. In end cap Type A, the seams are sealed with tape and tape tabs. To fabricate end cap Type A, follow these steps (*Figure 45*):

Step 1 Measure and cut a piece of rigid duct board to fit the duct end opening (*Figure 45A*).

Step 2 Insert the end cap into the female ship-lapped grooves of the duct opening.

Step 3 Tape all joints, starting with tape tabs according to the tape tab schedule in *Table 3*.

The staple flaps on end cap Type B are folded, not cut, at the corners. In this version of end-cap placement, the staple flaps are on all four sides of the duct board section. To fabricate this end cap, follow these steps (*Figure 45B*):

Step 1 Measure and cut a piece of rigid duct board to fit the duct end opening.

Step 2 Insert the end cap into the female ship-lapped grooves of the duct opening and pull the staple flaps over the end cap.

Step 3 Fold, do not cut, the staple flaps at the corners. Staple the end cap in place.

The staple flaps on end cap Type C also are folded and not cut. In this version of the end-cap placement, the staple flaps are fabricated on all four sides of the end cap. Type C is the usual and preferred method for fabricating end caps. To fabricate this end cap, follow these steps (*Figure 45C*):

Step 1 Measure and cut a piece of rigid duct board to fit the duct end opening. Leave staple flaps on all four sides of the end cap.

Step 2 Insert the end cap into the female ship-lapped grooves of the duct opening and pull the staple flaps over the duct section.

Step 3 Fold, do not cut, the staple flaps at the corners. Staple the end cap in place.

10.15.0 Single Blade Volume Damper

Dampers allow adjustment of the airflow in the duct. When a single blade volume damper is mounted in ducts less than two feet square, sheet metal sleeves and extra hangers for support are not required. Dampers in ducts larger than 24 inches by 12 inches should be in a separate sheet metal sleeve. An example of a single blade volume damper mounted inside a duct section is shown in *Figure 46*.

Installation is fairly simple and can be accomplished with commercially available hardware. Once the damper is installed, test it to ensure that it swings easily within the space and fits snugly inside the duct when closed. It is easy to install the endplates incorrectly so that the screws interfere with the operation of the damper.

10.16.0 Fire Damper

The installation of a fire damper is covered in detail in applicable building codes. Most codes require that a sheet metal sleeve be installed through the firewall. These codes give the proper sheet metal gauge and attachment angle. The fiberglass duct board may be sealed to the sheet metal sleeve with *UL 181A/P* pressure-sensitive tape. The metal surfaces must first be carefully cleaned. An example of a fire damper assembly is shown in *Figure 47*. Notice the access door, which allows access to service the fire damper.

10.17.0 Sleeved Accessory Mounting

Sleeved accessory mountings include such equipment as electrostatic air cleaners, electric heaters, and humidifiers. Hangers must be installed at the sheet metal sleeves to support each accessory separately if it weighs more than 50 pounds. An example of how a sleeved accessory is installed and supported is shown in *Figure 48*. Mechanical fasteners are used to attach the sheet metal sleeve to the duct board. As an option, you can use a full length of separately supported straight metal duct.

10.18.0 Registers or Grilles

Registers or grilles allow air to flow into occupied spaces. They also allow return air to flow back into the duct. These accessories may be installed in the duct wall or at the end of the duct. If they are installed in the duct wall, a U-shaped channel is cut in the duct board to allow insertion of the register or grille. Approved tape is placed on all four sides of the opening, and the register or grille is fastened in place with sheet metal screws.

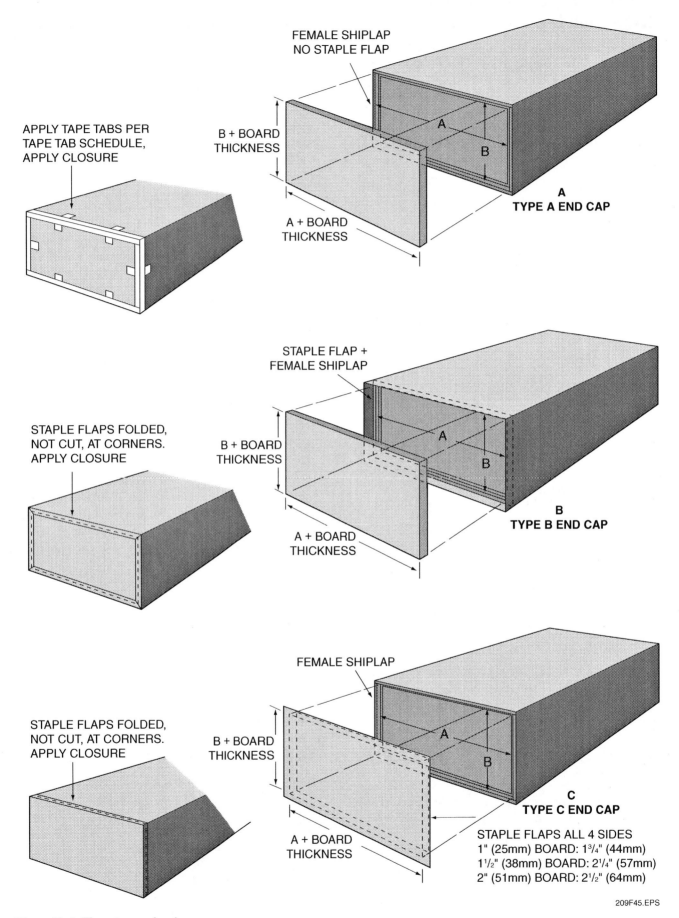

APPLY TAPE TABS PER
TAPE TAB SCHEDULE,
APPLY CLOSURE

FEMALE SHIPLAP
NO STAPLE FLAP

B + BOARD
THICKNESS

A

B

A + BOARD
THICKNESS

A
TYPE A END CAP

STAPLE FLAPS FOLDED,
NOT CUT, AT CORNERS.
APPLY CLOSURE

STAPLE FLAP +
FEMALE SHIPLAP

B + BOARD
THICKNESS

A

B

A + BOARD
THICKNESS

B
TYPE B END CAP

STAPLE FLAPS FOLDED,
NOT CUT, AT CORNERS.
APPLY CLOSURE

FEMALE SHIPLAP

B + BOARD
THICKNESS

A

B

A + BOARD
THICKNESS

C
TYPE C END CAP

STAPLE FLAPS ALL 4 SIDES
1" (25mm) BOARD: 1³/₄" (44mm)
1¹/₂" (38mm) BOARD: 2¹/₄" (57mm)
2" (51mm) BOARD: 2¹/₂" (64mm)

209F45.EPS

Figure 45 ◈ Three types of end cap.

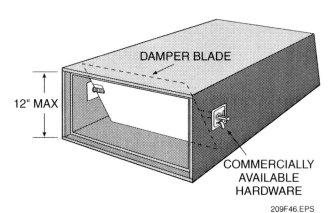

Figure 46 ◆ Single blade volume damper.

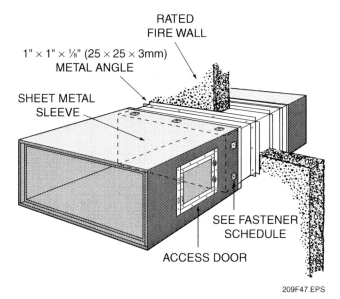

Figure 47 ◆ Fire damper.

If the register or grille is installed in a duct end, a metal frame is nailed to the wall studs. The register or grille is then attached to both the frame and the wall studs with sheet metal screws. Both the duct wall and duct end installation, including a detail of the duct end installation, are shown in *Figure 49*.

11.0.0 ◆ FLEXIBLE ROUND DUCT CONNECTIONS

Flexible round duct is used for runouts from low-pressure ducts, mixing boxes, diffusers, or other low-air velocity units. Flexible duct is not recommended for use on the return air side of an air distribution system. In this section, you will learn how to connect flexible duct to rigid duct board using two methods: the closure strap method and the insulated collar method. In both methods you must pull back the insulation that covers the inner core of the flexible duct to properly make the connection. Once the connection is made, you will pull the insulation back into place so that the entire assembly butts tightly against the duct board.

Panduit® straps are a commonly used brand of closure straps. To use the closure strap method, follow these steps (*Figure 50*):

Step 1 Cut a hole in the duct board using a hole cutter sized to accept a sheet metal collar. You may use either a spin-in collar or a dovetail collar. If you use a spin-in collar, also cut a 1-inch slit radial to the hole (*Figure 50A*).

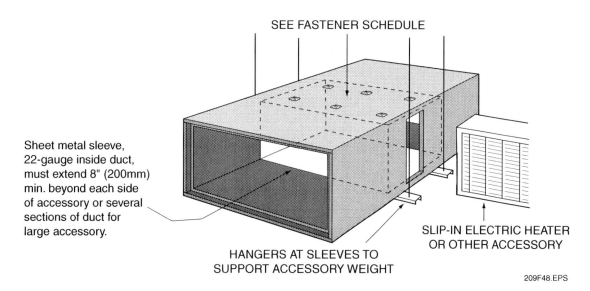

Figure 48 ◆ Sleeved accessory mounting.

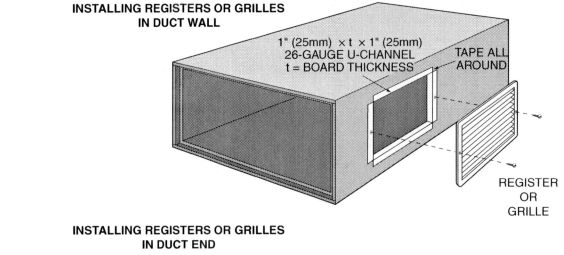

INSTALLING REGISTERS OR GRILLES IN DUCT WALL

1" (25mm) × t × 1" (25mm)
26-GAUGE U-CHANNEL
t = BOARD THICKNESS

TAPE ALL AROUND

REGISTER OR GRILLE

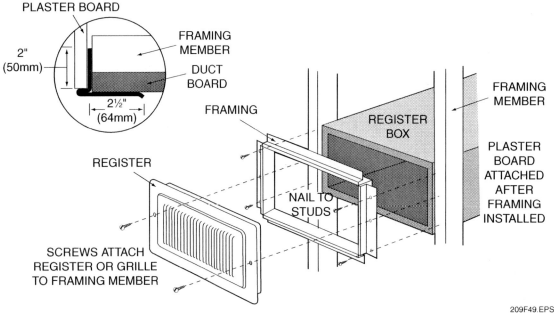

INSTALLING REGISTERS OR GRILLES IN DUCT END

PLASTER BOARD

FRAMING MEMBER

2" (50mm)

DUCT BOARD

2½" (64mm)

FRAMING

REGISTER

SCREWS ATTACH REGISTER OR GRILLE TO FRAMING MEMBER

NAIL TO STUDS

REGISTER BOX

FRAMING MEMBER

PLASTER BOARD ATTACHED AFTER FRAMING INSTALLED

209F49.EPS

Figure 49 ◈ Register or grille installations.

Step 2 To reduce air leakage, coat the collar flange with mastic to seal the collar to the duct board.

Step 3 Insert either a spin-in collar (*Figure 50B*) or a dovetail collar (*Figure 50C*) made for fiberglass duct. If you use a spin-in collar, bend the leading edge of the inner flange down at an angle and slip it through the slit in the duct board. Screw the collar into the hole in the duct board until the flange is snug against the duct board facing. If you use a dovetail collar, push it into the hole in the duct board until the flange is snug against the duct board facing. Bend all tabs 90 degrees to lock the collar into place.

Step 4 Push the insulation back and slide 1 inch of the duct core over the collar. Seal the core to the collar if required by the manufacturer or by the job specifications (*Figure 50D*).

Step 5 Pull the insulation back over the core so that it butts firmly against the duct wall (*Figure 50E*). Complete the installation in accordance with the manufacturer's instructions.

In the insulated collar method, you will fabricate a ring from fiberglass duct board to insulate a sheet metal collar. Depending on the job requirements, you may use either a spin-in collar or a dovetail collar. The insulated collar method maintains full insulation thickness across the flexible duct connection and helps prevent condensation. To use the insulated collar method, follow these steps (*Figure 51*):

Step 1 Cut a duct board ring to the following specifications:

– An inside diameter equal to the diameter of the sheet metal collar being installed.

– An outside diameter with a thickness corresponding to the insulation's R-value (*Table 5*).

Step 2 Slide the ring onto the sheet metal collar with the foil facing of the duct board away from the flange. Tape the insulation to the back of the flange with pressure-sensitive tape (*Figure 51A*).

Step 3 Cut a hole in the duct board sized to accept the sheet metal collar. You may use either a spin-in collar or a dovetail collar. If you use a spin-in collar, cut a 1-inch slit radial to the hole. Bend the leading edge of the inner flange down at an angle and slip it through the slit in the duct board. Screw the collar into place with the outer flange snug against the foil facing and the inner ring fully visible inside the duct. To reduce air leakage, coat the collar of the flange with approved mastic. If you use a dovetail collar, push the collar into the hole until the outer flange is snug against the facing. Bend the dovetails 90 degrees outward to lock them into place.

Step 4 Slide the insulation back from the core of the flexible duct and pull the duct core over the collar. Apply sealant if required by the manufacturer or the job specifications. Secure with a closure strap placed between the bead on the collar and the duct board (*Figure 51B*).

CAUTION

When using foam insulating sealant, remember to wear gloves and safety goggles. Apply the foam sealant in small amounts. The sealant is messy and expands when it is released from its can.

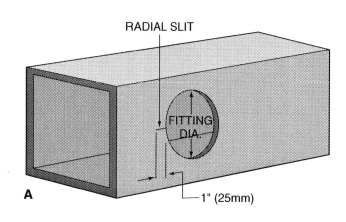

A

RADIAL SLIT

FITTING DIA.

1" (25mm)

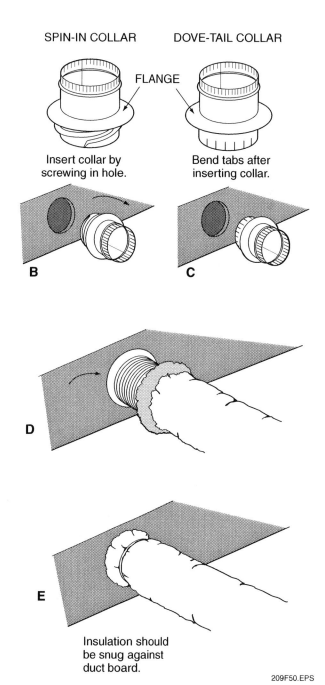

SPIN-IN COLLAR   DOVE-TAIL COLLAR

FLANGE

Insert collar by screwing in hole.

Bend tabs after inserting collar.

B   C

D

E

Insulation should be snug against duct board.

209F50.EPS

Figure 50 ◆ Closure strap method.

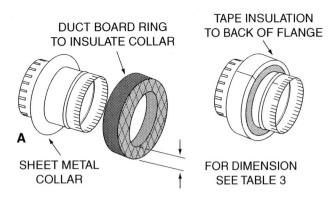

DUCT BOARD RING
TO INSULATE COLLAR

TAPE INSULATION
TO BACK OF FLANGE

A

SHEET METAL
COLLAR

FOR DIMENSION
SEE TABLE 3

FLEXIBLE DUCT CORE

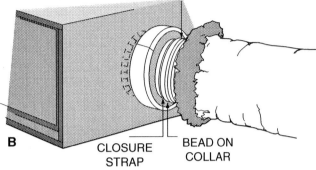

B

CLOSURE
STRAP

BEAD ON
COLLAR

BUTT FLEXIBLE DUCT
TIGHTLY TO COLLAR

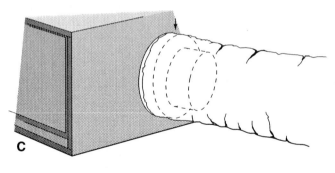

C

TAPE VAPOR BARRIER

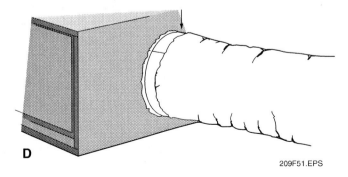

D

209F51.EPS

Figure 51 ◆ Insulated collar method.

Step 5 Butt the flexible duct and insulation firmly against the duct board ring (*Figure 51C*). Pull the vapor barrier over the ring so that it covers about one-half of the width of the ring.

Step 6 Tape the vapor barrier to the ring using tape that is compatible with the jacket of the flexible duct (*Figure 51D*). Complete the installation in accordance with the manufacturer's instructions.

12.0.0 ◆ HANGERS AND SUPPORTS

Fiberglass duct is lightweight and, for the most part, self-supporting. A minimum of carefully placed hangers can be used for support. Ensure that the hanging method you use has adequate load-bearing capability for the installation, without placing too much stress on either the hanger or the fiberglass duct system.

Remember to install enough hangers to support additional accessories, including heaters and dampers. Hanger treatment and spacing requirements depend on duct dimensions. Channel gauge and profile vary with duct size. Refer to your local code for specifications regarding the method and materials. Recommendations for hanging and supporting 10-sided fiberglass duct can be found in NAIMA's *Fibrous Glass Duct Construction Standards, Section II*.

12.1.0 Hanging and Supporting Rectangular Fiberglass Duct

Fiberglass duct up to 48 inches wide can be suspended using either 12-gauge (minimum) or 22-gauge metal straps that are a minimum of one inch wide (*Figure 52*). Use of a 12-gauge wire is the preferred method. When using channel reinforcement members, consider the maximum unreinforced duct dimension to determine spacing. The gauge and profile of the channel will vary with the size of the duct. Sheet metal straps may be bolted to channel reinforcement.

Table 5 Inside Dimension and R-Value

Inside Diameter in Inches	R-Value
2	4.2
3	6.0
4	8.0

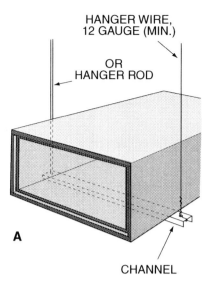

A

HANGER WIRE,
12 GAUGE (MIN.)

OR
HANGER ROD

CHANNEL

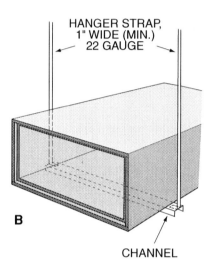

B

HANGER STRAP,
1" WIDE (MIN.)
22 GAUGE

CHANNEL

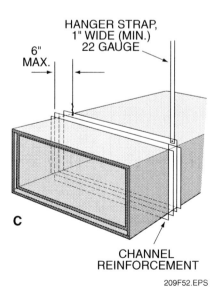

C

HANGER STRAP,
1" WIDE (MIN.)
22 GAUGE

6"
MAX.

CHANNEL
REINFORCEMENT

209F52.EPS

Figure 52 ◆ Suspending fiberglass duct.

12.1.1 Determining Hanger Extension

To determine hanger extension, add the distance between the hanging wires and the duct walls (*Figure 53*). For rectangular fiberglass duct, the supporting channel should never be less than two inches wide. If the hanger extension is less than 6 inches, use 24-gauge (minimum) wire and 3-inch by 1½-inch-wide channels. If the hanger extension is less than 18 inches, use 22-gauge (minimum) wire and 3-inch by 2-inch-wide channels. If the hanger extension is less than 30 inches, use 18-gauge (minimum) wire and 3-inch by 2-inch-wide channels.

For ducts less than 48 inches wide and 24 inches high, 2-inch-wide hangers can be used instead of 3-inch-wide hangers. Use 22-gauge wire and space the 2-inch hangers at maximum 4 feet apart (*Figure 54*). Refer to *Figure 55* to determine the maximum hanger spacing for straight duct using 3-inch channels.

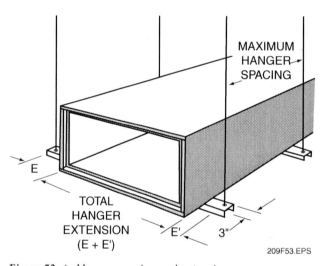

MAXIMUM
HANGER
SPACING

E

TOTAL
HANGER
EXTENSION
(E + E')

E'

3"

209F53.EPS

Figure 53 ◆ Hanger spacing and extension.

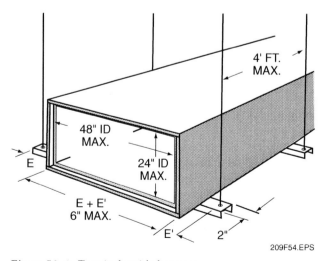

4' FT.
MAX.

48" ID
MAX.

24" ID
MAX.

E

E + E'
6" MAX.

E'

2"

209F54.EPS

Figure 54 ◆ Two-inch-wide hangers.

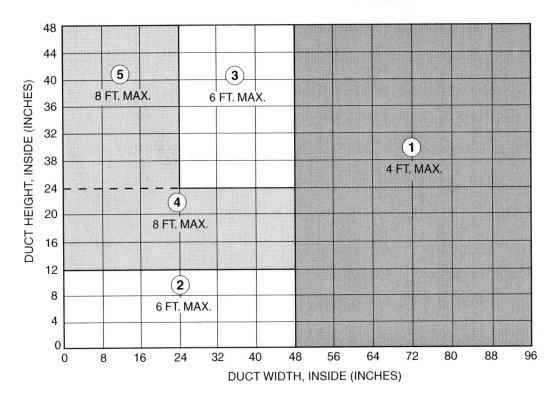

**MAXIMUM HANGER
SPACING BY DUCT SIZE, ID**

	DUCT SIZE, INCHES	MAXIMUM HANGER SPACING
1	48" wide or greater	4 ft.
2	Less than 48" wide and less than 12" high	6 ft.
3	Width between 24" and 48" and greater than 24" high	6 ft.
4	Less than 48" wide and height between 12" and 24"	8 ft.
5	Width 24" or less and height greater than 12"	8 ft.

209F55.EPS

Figure 55 ◆ Maximum hanger spacing.

12.1.2 Fiberglass Duct Fittings

Fiberglass duct fittings up to 48 inches wide will require additional support. If duct is wider than 18 inches, support elbows with a channel reinforcement located two-thirds of the diagonal distance from the throat to the heel (*Figure 56*). If a trunk duct hanger should be located where a branch duct connects with the trunk, install hangers on either side of the branch duct (*Figure 57*). If a tee run-out hanger should be located where a branch duct connects with the trunk duct, install run-out hangers on either side of the trunk duct (*Figure 58*). Do not exceed maximum hanger spacing when adding branch and tee supports.

Offset supports are needed when the angled portion of offset is longer than 48 inches (*Figure 59*). For transition support, locate hangers as you would for straight duct (*Figure 60*). If there is an inclined bottom surface on a duct with a width greater than 48 inches, support the offsets and transitions (*Figure 61*). When supporting offsets and transitions, hangers must comply with hanger spacing requirements.

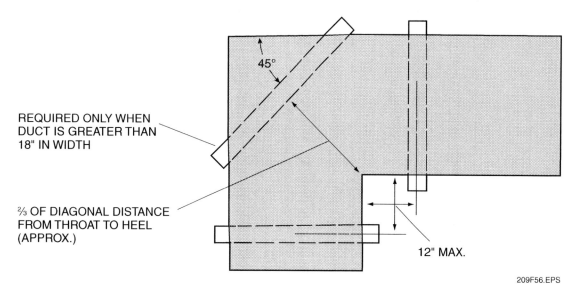

REQUIRED ONLY WHEN
DUCT IS GREATER THAN
18" IN WIDTH

⅔ OF DIAGONAL DISTANCE
FROM THROAT TO HEEL
(APPROX.)

45°

12" MAX.

209F56.EPS

Figure 56 ◆ Elbow support.

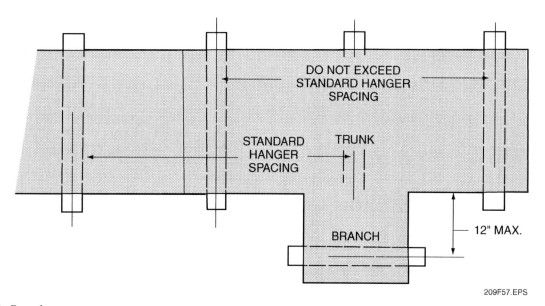

DO NOT EXCEED
STANDARD HANGER
SPACING

STANDARD
HANGER
SPACING

TRUNK

BRANCH

12" MAX.

209F57.EPS

Figure 57 ◆ Branch support.

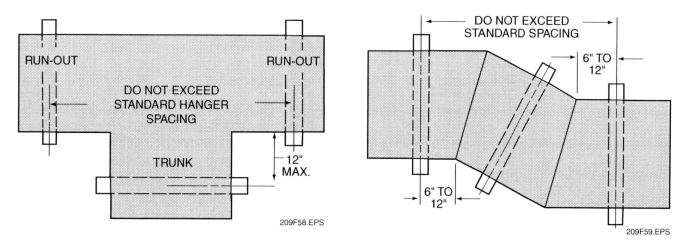

RUN-OUT

RUN-OUT

DO NOT EXCEED
STANDARD HANGER
SPACING

TRUNK

12"
MAX.

209F58.EPS

Figure 58 ◆ Tee support.

DO NOT EXCEED
STANDARD SPACING

6" TO
12"

6" TO
12"

209F59.EPS

Figure 59 ◆ Offset support, flat bottom surface.

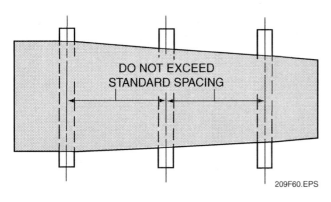

Figure 60 ◆ Transition support, flat bottom surface.

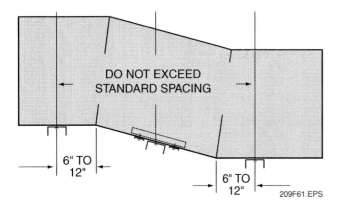

Figure 61 ◆ Offset and transition support, inclined bottom surfaces.

12.2.0 Hanging and Supporting Flexible Duct

Flexible duct is round duct that may or may not be insulated. A wire coil that runs around the inside of the duct gives flexible duct some stiffness and makes it easier to handle. Flexible duct is used as a connection between the rigid duct and heating/cooling equipment and ceiling diffusers. Standard practice permits the installation of flexible duct and connectors only for indoor comfort HVAC systems.

12.2.1 Installing Flexible Duct

Before installing flexible duct, review the project plans for proper placement. To install flexible duct, follow these steps:

Step 1 Measure to determine the length of duct required. Be sure to allow the correct amount for any turns.

Step 2 Cut completely around and through the duct with a sharp utility knife.

Step 3 Trim the interior wire with lineman pliers (side cutters).

Step 4 Pull back the insulation (if any) from the core and slide at least one inch of the core over the collar, pipe, or fitting (*Figure 62*). Note that sheet metal collars must be at least 2 inches long.

Step 5 Tape with at least two wraps of approved duct tape. The tape must be plenum-rated if the ceiling space is to be used as a return air plenum. You may use an approved clamp or strap in place of or with the tape (*Figure 63*).

Step 6 Fully extend the duct for installation. Do not compress the duct. Radius bends at the duct centerline should not be less than the diameter of the duct (*Figure 64*).

Step 7 Support the duct at the manufacturer's recommended intervals, but at least every 4 feet. The maximum permissible sag is ½-inch per foot of spacing between supports. For long horizontal runs that have sharp bends, use additional supports before and after the bend.

Step 8 Splice flexible duct together with a sheet metal sleeve at least 4 inches long (*Figure 65A*). Clamp or strap (*Figure 65B*) the splices and tape with at least two wraps of approved duct tape.

Step 9 Place a hanger at connections to rigid ducting, splice connections, or equipment supports. Hangers or saddles used to support flexible duct must be at least 1½ inches wide (*Figure 66*). Other types of approved supports—for horizontal duct on ceiling joists, for angled duct, and for vertical duct—are shown in *Figure 66*.

Step 10 Repair any tears or damage to the vapor barrier jacket with approved duct tape or according to the manufacturer's instructions. If the internal core is penetrated, replace flexible duct or splice and seal it as in Step 8.

12.2.2 Testing

To ensure that flexible duct is properly installed, use the following checklist:

- Do the hangers support the duct without cutting into it, especially at the bends?

- Have tears or damage to the vapor barrier jacket been repaired with approved duct tape or according to the manufacturer's instructions?

- Have you checked to ensure that the duct does not sag more than ½-inch between supports?

209F62.EPS

Figure 62 ◆ Pull back vapor barrier jacket and insulation.

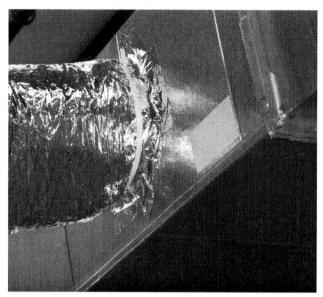

209F63.EPS

Figure 63 ◆ Duct clamps.

CORRECT

INCORRECT

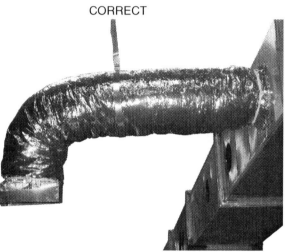

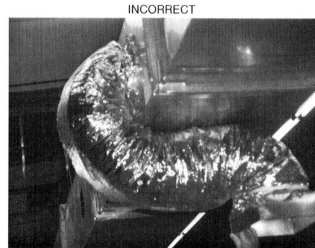

209F64A.EPS

209F64B.EPS

Figure 64 ◆ Extend duct fully, do not bend sharply.

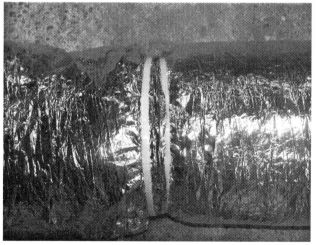

209F65A.EPS

209F65B.EPS

Figure 65 ◆ Metal sleeve inserted and clamped, finished with tape.

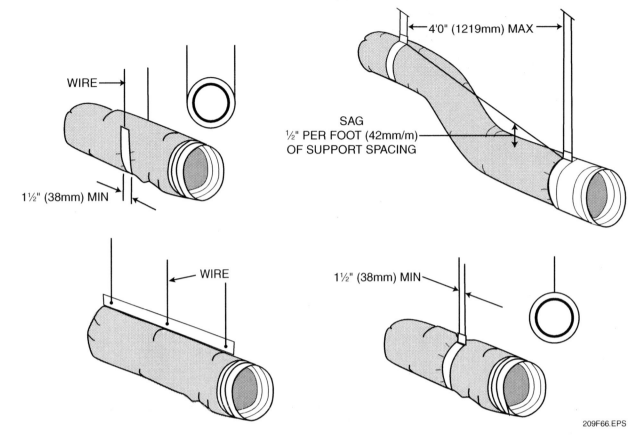

Figure 66 ◆ Flexible duct supports.

12.3.0 Supporting Vertical Risers

Risers of eight feet or more in fiberglass duct systems require additional support. Use No.10 plated sheet metal screws and 2½-inch square washers. Install wall support angles to brace the channel reinforcement as shown in *Figure 67*. Install vertical riser supports at 12-foot (maximum) intervals.

13.0.0 ◆ REINFORCING PRESSURE CLASSES GREATER THAN ½-INCH

This section focuses on fiberglass duct used in pressure applications greater than ½-inch. Straight sections and fittings of fiberglass duct systems can be reinforced using two methods: tie rod reinforcement and channel reinforcement. Tie rods are used wherever possible, and channel reinforcement is used where tie rods cannot be used.

Some duct dimensions may not require reinforcement in straight sections. However, fittings with these same dimensions may require additional reinforcement. Use the same basic principles to reinforce fiberglass duct fittings that you would use to reinforce straight duct sections.

13.1.0 Tie Rod Reinforcement

Tie rod reinforcement is used in positive pressure duct systems (*Figure 68*). Refer to NAIMA's *Fibrous Glass Duct Construction Standards* for a complete tie rod system reinforcement schedule. Tie rod reinforcement is used with duct sections or fittings that have shiplapped joints and butt joints.

13.1.1 Shiplapped Joints

If the duct section or fitting has shiplapped joints, locate the reinforcement within four inches of the shiplap on the female side. Install 2½-inch square washers and face the turned edges away from the duct board facing (*Figure 69*). The weight of the duct board when the system is not pressurized may cause top panels to sag. Install sag supports in ducts wider than 48 inches. Sag supports should be installed in addition to tie rod assemblies. Place the hangers within 12 inches of the sag supports (*Figure 70*).

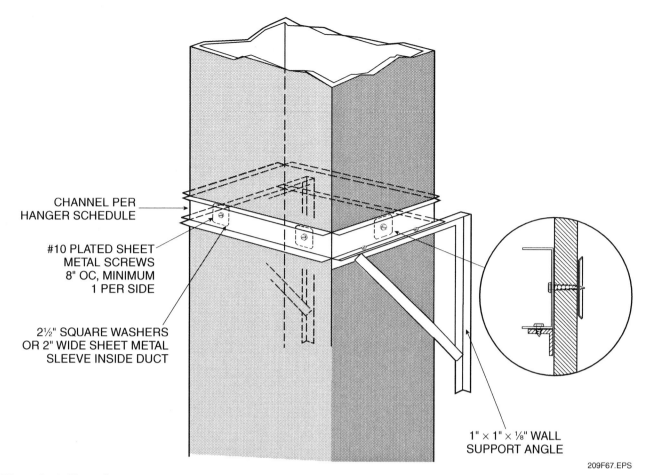

CHANNEL PER
HANGER SCHEDULE

#10 PLATED SHEET
METAL SCREWS
8" OC, MINIMUM
1 PER SIDE

2½" SQUARE WASHERS
OR 2" WIDE SHEET METAL
SLEEVE INSIDE DUCT

1" × 1" × ⅛" WALL
SUPPORT ANGLE

209F67.EPS

Figure 67 ◆ Vertical riser support.

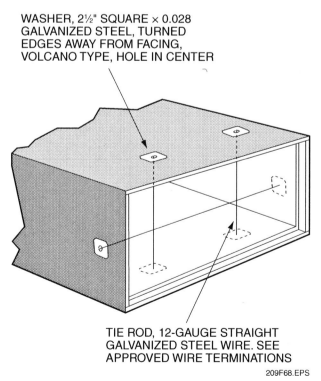

WASHER, 2½" SQUARE × 0.028
GALVANIZED STEEL, TURNED
EDGES AWAY FROM FACING,
VOLCANO TYPE, HOLE IN CENTER

TIE ROD, 12-GAUGE STRAIGHT
GALVANIZED STEEL WIRE. SEE
APPROVED WIRE TERMINATIONS

209F68.EPS

Figure 68 ◆ Tie rod reinforcement.

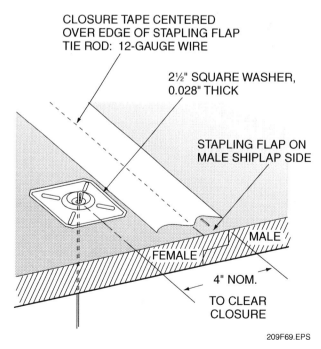

CLOSURE TAPE CENTERED
OVER EDGE OF STAPLING FLAP
TIE ROD: 12-GAUGE WIRE

2½" SQUARE WASHER,
0.028" THICK

STAPLING FLAP ON
MALE SHIPLAP SIDE

MALE

FEMALE

4" NOM.

TO CLEAR
CLOSURE

209F69.EPS

Figure 69 ◆ Tie rod reinforcement, shiplapped joint.

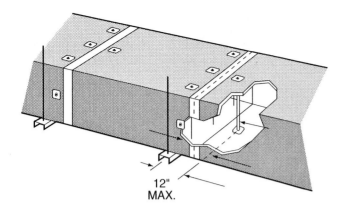

12"
MAX.

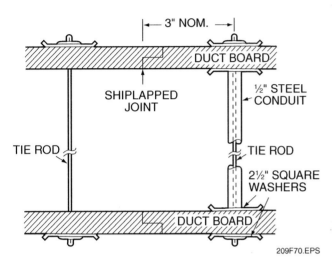

Figure 70 ◆ Sag control, tie rod reinforcement.

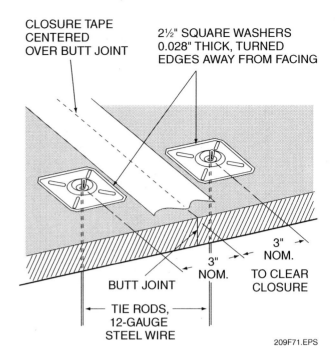

Figure 71 ◆ Butt joint reinforcement.

13.1.2 Butt Joints

If a duct section or fitting that must be reinforced has butt joints, install 2½-inch square washers and face the turned edges away from the facing. Place closure tape centered over the butt joint (*Figure 71*). Refer to the duct board manufacturer's specifications for an appropriate adhesive system. Alternately, you can use a single tie rod reinforcement if you glue the butt joint using an approved adhesive.

To control sag, install supports on both sides of the butt joint. Use 2½-inch square washers and ½-inch steel conduit added to the standard tie rods inside the duct (*Figure 72*).

13.1.3 Tie Rod Termination

There are three accepted methods of tie rod termination—the FasLoop™ method, the pop rivet sleeve method, and the locking cap method. Tie rod termination methods should only be used for low-pressure duct systems, 0 to 2 inch water gauge. Be aware of any limitations for using cer-

tain methods with specific fittings. For example, you cannot use the locking cap method on sloped panels of fittings.

For the FasLoop™ method (*Figure 73*), you will need a FasLoop™ bending tool or an equivalent tool, wire cutters, and a tape measure. Use 12-gauge galvanized steel wire. The wire should be 1¾ inches longer than the outside duct dimension. Use two 2½-inch-square by 0.028-inch-thick galvanized steel washers. The volcano-type washers should have beveled edges and a 0.150-inch diameter hole in the center. Other manufactured flat washers are not suitable for this method.

For the pop rivet sleeve method (*Figure 74*), you will need 14-inch horseshoe nipper or carpenter's pincers, wire cutters, and a tape measure. If you use pincers, ensure that they are modified to close to a gap that equals 18-gauge sheet metal. Use 12-gauge galvanized steel wire. The wire should be cut exactly to the outside duct dimension. Use two 2½-inch-square by 0.020-inch-thick galvanized steel washers. The washers should have beveled edges and a ⁷⁄₃₂-inch diameter hole in the center. The two ³⁄₁₆-inch steel pop rivet sleeves should be ³⁄₈-inch long.

For the locking cap method (*Figure 75*), you will need wire cutters and a tape measure. Use 12-gauge galvanized steel wire. The wire should be cut ⁷⁄₁₆-inches longer than the outside duct dimension. Use two 2½-inch-square by 0.020-inch-thick galvanized steel washers. The washers should have beveled edges and a 0.150-inch diameter hole in the center. The two ⅞-inch diameter locking caps should have spring steel or stainless steel locking inserts. The wire must freely move within the 2½-inch washer.

> **CAUTION**
> An ordinary insulation locking washer does not have sufficient holding power to use with fiberglass duct. Never re-use locking caps. Do not use the locking cap method on sloped panels of fittings.

13.2.0 Channel Reinforcement

Channel reinforcement can be used in either positive or negative pressure duct systems. This method uses formed sheet metal channels wrapped around the perimeter of the duct. When attaching channels to the duct, use No.10 plated sheet metal screws and 2½-inch-thick sheet metal washers. Ensure that the washers have turned edges to prevent cutting into the duct board. Consult NAIMA's *Fibrous Glass Duct Construction Standards* for information on sizing, spacing, and dimensions.

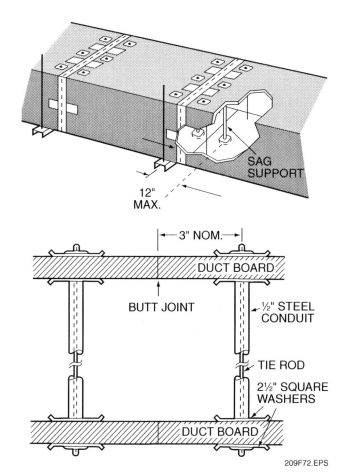

Figure 72 ◆ Sag control, butt joint reinforcement.

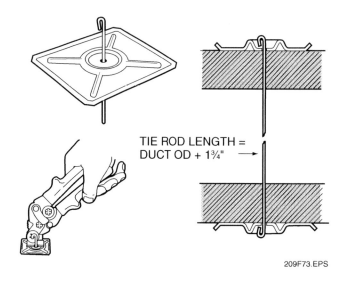

Figure 73 ◆ Tie rod termination, FasLoop™ method.

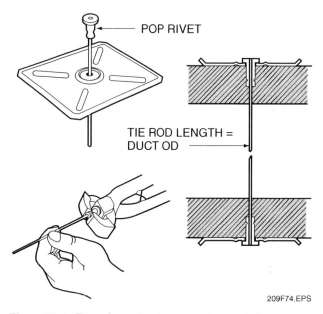

Figure 74 ◆ Tie rod termination, pop rivet and sleeve method.

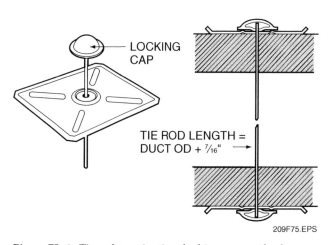

Figure 75 ◆ Tie rod termination, locking cap method.

13.2.1 Positive Pressure

In positive pressure applications, wrap-around channels do not have to be attached to the duct board unless it is required for sag control. Reinforcements are fabricated from a continuous length of channel (*Figure 76*). Each length of channel has three 90-degree bends and a 90-degree corner. The corner is fastened with bolts, screws, rivets, spot welds, or staples. Reinforcements may also be fabricated with two, three, or four fastened corners.

For duct wider than 48 inches, secure channel reinforcements to the top of the duct to provide sag support. Use No.10 plated sheet metal screws and 2½-inch square washers. If wrap-around channels with sag support are used, slip the channel over the closure tape after the tap has been applied. Center the channel over the female shiplap end for maximum support.

To determine the length of a single-piece reinforcing member, use the following formula. Add the outside duct (OD) perimeter to twice the height of the channel (*Figure 77*):

single length = outside duct perimeter + (2 × height of the channel)

To determine the length of a two-piece reinforcing member, use the following formula. The length of each reinforcing member must equal the length of the outside duct width plus the outside duct height plus twice the height of the channel (refer to *Figure 78*):

two-piece length = outside duct width + outside duct height + (2 × height of the channel)

Offset the channel to clear the closure (*Figure 78*). If sag support is not required, center the channel over the shiplap joint.

13.2.2 Negative Pressure

In negative pressure applications, apply channel reinforcements over the male shiplap (*Figure 79*). Attach additional channel reinforcements to the duct with No.10 plated sheet metal screws and 2½-inch square washers. To fasten channel reinforcements to the duct board, follow these steps:

Step 1 Thread and strip screws into the channel.

Step 2 Place the channel in the proper location on the duct.

Step 3 Push the screws through the duct board.

Step 4 Use a powered driver to thread the screws into the washers or clips. Do not space the clips more than 16 inches apart and not more than 16 inches from the longitudinal edge of a duct side (*Figure 80*).

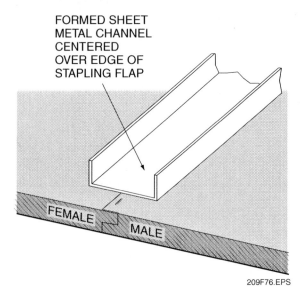

Figure 76 ◆ Channel reinforcement, positive pressure system.

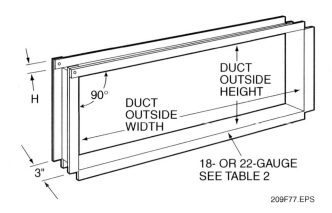

Figure 77 ◆ Detail of channel reinforcement.

Step 5 Fasten at the corners.

Channel reinforcements for negative pressure applications are fabricated the same way as those for positive pressure applications. Use the same steps to determine the length of the reinforcing member.

CAUTION

Always follow the specific arrangements for negative pressure channel reinforcement. Otherwise, the system may fail. Consult NAIMA's *Fibrous Glass Duct Construction Standards* to determine the number of channels required and for fastener requirements in negative pressure applications.

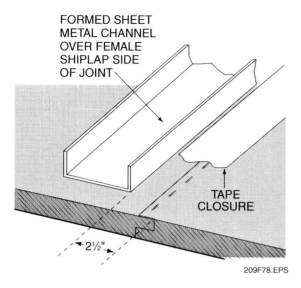

Figure 78 ◆ Channel reinforcement offset to clear closure.

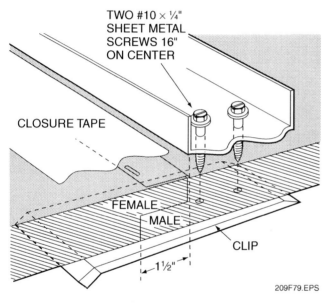

Figure 79 ◆ Channel reinforcement, negative pressure system.

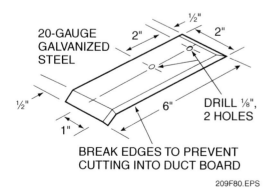

Figure 80 ◆ Clip detail, negative pressure system.

13.2.3 Partial Wrap-Around Reinforcement

If reinforcement is required, but it cannot be fastened to the opposite sides of the duct section or fitting, install formed sheet metal channels (*Figure 81*). These channels partially wrap around the fiberglass duct system fitting at the specified location. Refer to NAIMA's *Fibrous Glass Duct Construction Standards* for appropriate gauge and profile. Use No.10 by 1¼-inch plated sheet metal screws and 2½-inch square washers to attach the ends of the channels to the duct board.

13.3.0 Reinforcing Fittings

Some fittings—such as 90-degree elbows, branch connections, tees, offsets, transitions, access doors, and end caps—may require reinforcement even though schedules for straight duct of the same dimension may show reinforcement is not required.

13.3.1 90-Degree Elbows, Shiplap Construction

When reinforcing 90-degree elbows with shiplap construction, adhere to the following guidelines. Refer to NAIMA's tie rod reinforcement schedule for the allowable dimensions. To reinforce mitered elbows, refer to the reinforcement standards for offsets.

For cheek panels in positive pressure applications, use the following standards. If either of the inside dimensions shown are less than the maximum unreinforced duct dimension, but the diagonal line labeled XY is greater than the maximum unreinforced duct dimension, install a tie rod reinforcement at the mid-span of the diagonal. Refer to *Figure 82*. If either of the inside dimensions shown are greater than the maximum unreinforced duct dimension, reinforce the fitting four inches up from the female shiplap joints. Also reinforce the fitting where the centerlines intersect. Refer to *Figure 83*.

Reinforce heel and throat panels in positive pressure applications according to the following guidelines. If the duct does not require reinforcement, but the dimensions are greater than 24 inches, install a sheet metal angle at the throat (*Figure 84*). Use 2½-inch square washers, No.10 plated sheet metal screws, and 3" × 3" × 4" 20-gauge angles. The angles may also be installed on the inside of the throat. If the duct dimensions require reinforcement, install sheet metal angles at the throat as shown in *Figure 85*. Use 2½-inch square washers and tie rod termination, No.10

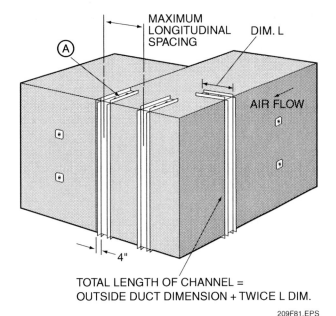

Figure 81 ◆ Partial wrap-around channel reinforcement.

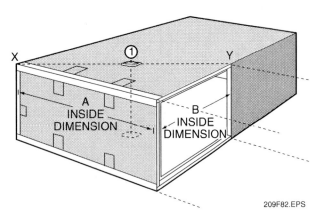

Figure 82 ◆ Tie rod reinforcement at diagonal XY, mid-span, 90-degree elbow.

plated sheet metal screws, and 3" × 3" 20-gauge angles. Install tie rods through the angle on the upstream side of the fitting, 16 inches on center, according to the length in *Table 6*. Install three-inch channel reinforcements on the heel panels.

13.3.2 Branch Connections

For connection reinforcement in positive pressure applications, adhere to the following guidelines. When possible, the branch height should equal the trunk height. This eliminates the need to reinforce the top of the branch. Use a transition to reduce the branch to the desired size.

Use the following standards to reinforce angled branches and positive take-offs:

- If the height is less than 16 inches and the width is greater than half the maximum unreinforced duct dimension, reinforce the top of the branch (*Figure 86*).
- If the height is greater than 16 inches and the width is less than the maximum unreinforced duct dimension, reinforce the sides of the branch (*Figure 87*).

Table 6 Number of Tie Rods According to Angle Length

Number of Tie Rods	Angle Length L (in.)
1	4
2	20
3	36
4	52
5	68

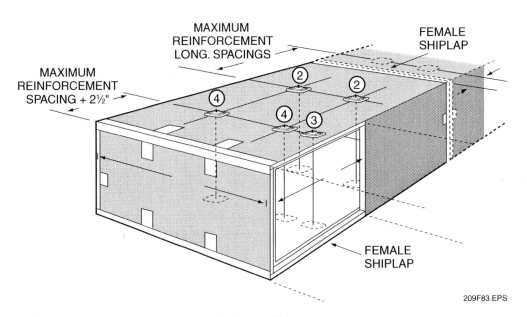

Figure 83 ◆ Tie rod reinforcement, cheek panels, large 90-degree elbow.

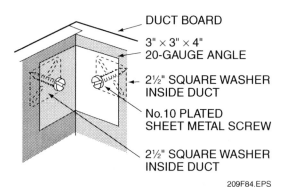

Figure 84 ◆ Angle reinforcement not required at throat, greater than 24 inches, 90-degree elbow.

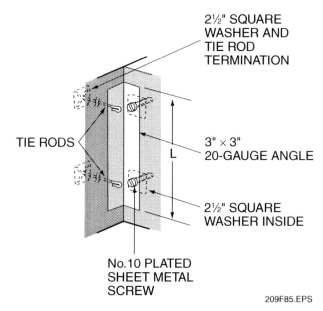

Figure 85 ◆ Angle reinforcement required at throat, 90-degree elbow.

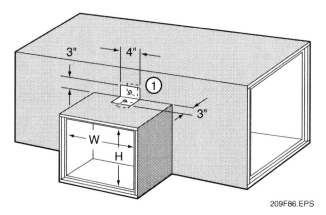

Figure 86 ◆ Sheet metal angle reinforcement, top of branch.

• If the height is greater than 16 inches and the width is greater than the maximum unreinforced duct dimension, reinforce the top and sides of the branch (*Figure 88*). For angle length, refer to *Table 6*.

> **NOTE**
>
> It is best to attach the angles after the closure is complete. Remember that the sections should be short enough to allow the technician to reach inside to install the washers.

Use the following standards to reinforce trunk duct reinforcement in positive pressure applications:

• If the width is greater than one half the maximum unreinforced duct dimension, but is less than the maximum unreinforced duct dimension, reinforce four inches off of the female shiplap (*Figure 89*).
• If the branch interferes with standard tie rod locations, install the tie rods four inches from both sides of the branch opening (*Figure 90*).
• If the width is greater than the maximum longitudinal reinforcement spacing of the trunk duct, and/or the height is greater than 16 inches, reinforce with 3-inch channels (*Figure 91*).

13.3.3 Tees

Use the following guidelines when reinforcing cheek panels in tees in positive pressure applications. If dimension A is less than the maximum unreinforced duct dimension, but diagonals XY and YZ exceed the maximum unreinforced duct dimension, install tied rods at the female shiplap joints. Refer to *Figure 92*. If dimension A is greater than the maximum unreinforced duct dimension, but dimension B is greater than the maximum unreinforced duct dimension, install tie rods four inches from the female shiplap joints. Also install tie rods along branch and trunk center lines according to the spacing shown in *Figure 93*.

> **NOTE**
>
> Attach the angles to the trunk and branch ducts after closure is complete. Use a short branch so that the technician can reach inside to install the washers.

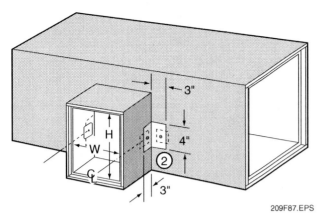

Figure 87 ◆ Sheet metal angle reinforcement, sides of branch.

Use the following guidelines when reinforcing heel and throat panels in tees in positive pressure applications:

- Refer to *Figure 94*. If dimension H is less than the maximum unreinforced duct dimension, but is greater than 24 inches, reinforce with a 3" × 3" × 4" sheet metal angle at the throat. Reinforcement is not required at the heel.
- Refer to *Figure 95*. If dimension H is greater than the maximum unreinforced duct dimension, install 3" × 3" 20-gauge angle. Refer to *Table 6* to determine the appropriate number of tie rods. Install 3-inch channels on the heel at the female shiplaps and as required to maintain the specified spacing.

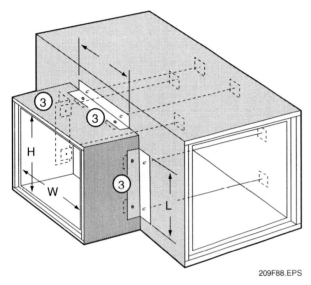

Figure 88 ◆ Sheet metal angle reinforcement, top and sides of branch.

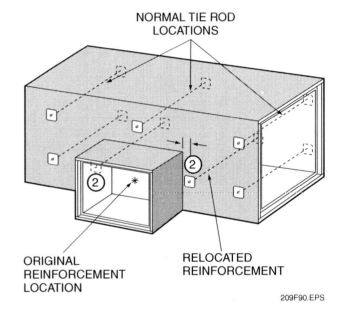

Figure 90 ◆ Trunk duct reinforcement with relocated tie rods.

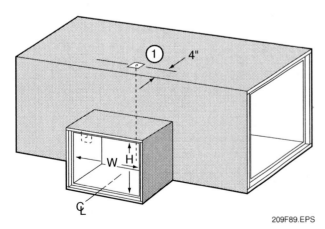

Figure 89 ◆ Trunk duct reinforcement four inches off female shiplap.

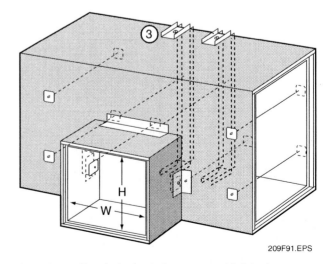

Figure 91 ◆ Trunk duct reinforcement with 3-inch channels.

- Refer to *Figure 96*. If splitter dampers interfere with tie rod reinforcements, use wrap-around channels in place of the tie rods. After closure, attach angles at the throats inside or outside of the fitting.

NOTE

Attach the channels to trunk ducts after the closure is complete. Use a short branch so that the technician can reach inside to install the washers. To simplify attachment, wrap channels around the near corners of the trunk.

NOTE

If a splitter damper interferes with tie rod reinforcement, use wrap-around channels in their place. Remember that turning vanes are not a substitute for reinforcement.

13.3.4 Offsets

Reinforce cheek panels in offsets in positive pressure applications according to the following guidelines:

- Refer to *Figure 97*. If dimension B is greater than the maximum unreinforced duct dimension, and if cheek panels have shiplap joints, reinforce the fitting four inches from the female shiplap.
- Refer to *Figure 98*. If dimension B is greater than the maximum unreinforced duct dimension, and if cheek panels have butt joints, install reinforcements at all butt joints.
- If the dimension labeled B is greater than the maximum unreinforced duct dimension, and if the cheek panels are shiplapped, install additional tie rods. Install the tie rods along the lines that are parallel to the panel edges (*Figure 99*). The tie rods must be located four inches off the female shiplap joints.

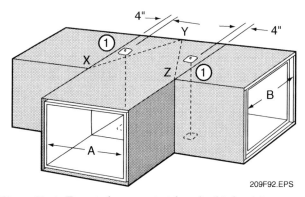

209F92.EPS

Figure 92 ◆ Tee reinforcement at female shiplap joint.

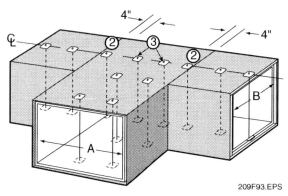

209F93.EPS

Figure 93 ◆ Tee reinforcement, normal tie rod location.

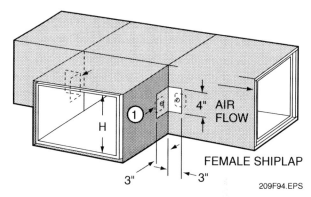

209F94.EPS

Figure 94 ◆ Tee reinforcement with 3"× 3"× 4" sheet metal angle at throat.

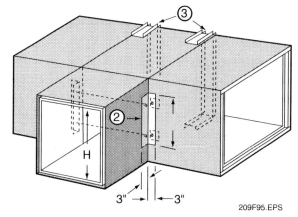

209F95.EPS

Figure 95 ◆ Tee reinforcement with throat angles and heel channels.

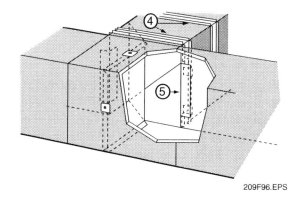

209F96.EPS

Figure 96 ◆ Tee reinforcement with wrap-around channels.

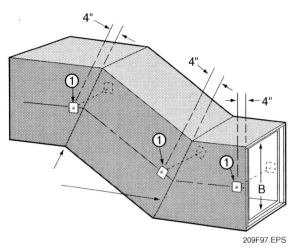

Figure 97 ◆ Offset reinforcement, cheek panels with shiplap joints.

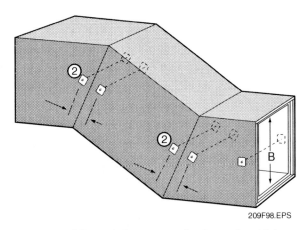

Figure 98 ◆ Offset reinforcement, cheek panels with butt joints.

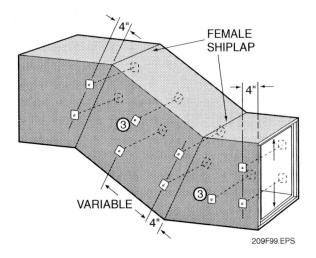

Figure 99 ◆ Offset reinforcement, tie rods along lines parallel to cheek panel edges.

Reinforce heel and throat panels in offsets in positive pressure applications according to the following guidelines:

- Refer to *Figure 100*. If dimension A is greater than the maximum unreinforced duct dimension, reinforce the fitting with sheet metal plates and tie rods.
- Refer to *Figure 101*. If dimension XY is greater than the maximum longitudinal reinforcement spacing, install additional tie rods. If the dimension labeled YZ is greater than the maximum longitudinal reinforcement spacing, install additional tie rods. Remember to maintain the required spacing.

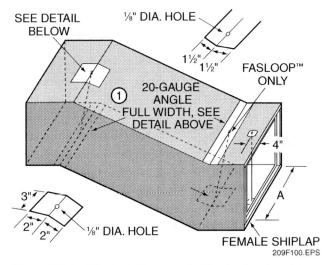

Figure 100 ◆ Offset reinforcement with sheet metal plates and tie rods.

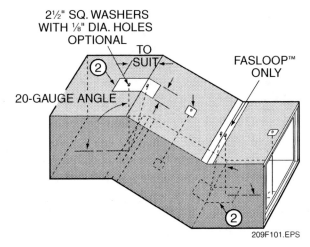

Figure 101 ◆ Offset reinforcement with tie rods through throat.

- If the spacing between any two tie rods is greater than the maximum longitudinal reinforcement spacing, install intermediate tie rods through the throat (*Figure 102*).

13.3.5 Transitions

Reinforce cheek panels in transitions in positive pressure applications according to the following guidelines. Refer to *Figure 103*. If dimension B is greater than the maximum unreinforced duct dimension, reinforce the fitting as shown. Determine the appropriate tie rod spacing from the larger duct dimension. Maintain the spacing and number of tie rods throughout the length of the transition.

Reinforce sloped panels in transitions in positive pressure applications according to the following guidelines (refer to *Figure 104*):

- If the dimension labeled A is greater than the maximum unreinforced duct dimension, reinforce the fitting four inches from the female shiplap, identified as 2 in *Figure 104*.
- If the facing is cut, use a 3" × 4" 20-gauge sheet metal plate. Bend the plate to match the transition angle, identified as 4 in *Figure 104*. Tie rod terminations must be made with the FasLoop™ method or pop rivet terminations. Tie rod terminations may not be made using the locking cap method.
- If the facing is not cut, use a 2½-inch square washer to secure the tie rods. The washer should be pre-bent to match the slope angle.

NOTE

If tie rods interfere with damper access or use, install channel reinforcement instead of tie rods between the access door and the fire damper.

13.3.6 Access Doors

To reinforce access doors in positive pressure applications, adhere to the following guidelines. If the access door is narrower than the maximum longitudinal reinforcement spacing but interferes with the location of reinforcements, install tie rods four inches from both sides of the door opening (*Figure 105*). Refer to NAIMA standards for maximum reinforcement spacing. If the height of the access door is greater than 16 inches and its width is greater than the maximum longitudinal reinforcement spacing, frame the inside. Install tie rods near the vertical sides of the door frame (*Fig-*

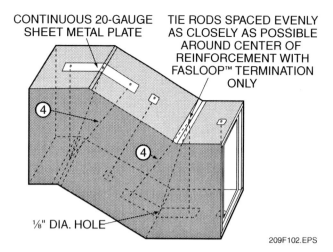

CONTINUOUS 20-GAUGE SHEET METAL PLATE TIE RODS SPACED EVENLY AS CLOSELY AS POSSIBLE AROUND CENTER OF REINFORCEMENT WITH FASLOOP™ TERMINATION ONLY

⅛" DIA. HOLE

209F102.EPS

Figure 102 ◆ Offset reinforcement with intermediate tie rods and extended sheet metal plates.

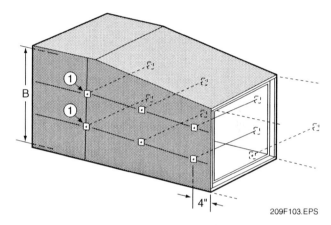

209F103.EPS

Figure 103 ◆ Transition reinforcement, cheek panels.

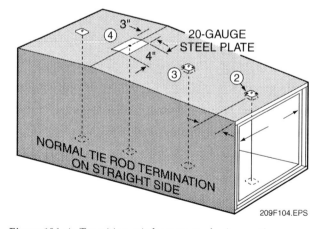

3" 20-GAUGE STEEL PLATE

4"

NORMAL TIE ROD TERMINATION ON STRAIGHT SIDE

209F104.EPS

Figure 104 ◆ Transition reinforcement, sloping section.

ure 106). Use 2½-inch square washers outside of the duct. Measure upstream from the vertical tie rod locations and install tie rods near the horizontal sides of the door frame.

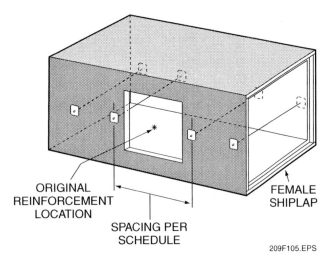

Figure 105 ◆ Access door location interfering with original reinforcement location.

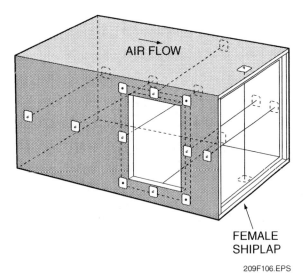

Figure 106 ◆ Reinforcement of access door framing.

13.3.7 End Caps, Shiplapped Construction

Install channel reinforcements on the inside of the duct so that the end cap can withstand static and velocity pressures to which it will be subjected (*Figures 107* and *108*). Refer to the appropriate NAIMA standards for longitudinal spacing and the number of necessary screws for applicable duct spans and static pressure. Staple and tape the end cap in place.

NOTE

End-cap reinforcement can be applied in two ways. It can be installed parallel to the longest inside dimension or parallel to the shortest inside dimension. The method you use depends on the requirements for sheet metal and fastener use.

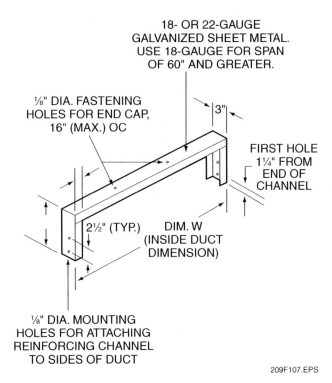

Figure 107 ◆ End cap channel reinforcement.

14.0.0 ◆ REPAIRING DAMAGE TO FIBERGLASS DUCT

Fiberglass duct can be damaged during installation. In addition, workers from other trades on the job site can damage it as they do their own installations. In this section, you will learn how to identify and repair different types of damage.

14.1.0 Repairing Minor Damage

Minor damage includes small, straight slits in the facing material that have not damaged the insulation. Use the appropriate closure material (pressure-sensitive tape, heat-activated tape, or mastic and glass fabric tape). Clean and dry the surface before applying the closure material to ensure a good bond.

If the facing damage is more than just a straight slit but is not more than ½-inch wide, make the repair as shown in *Figure 109*. The closure material must extend at least one inch beyond all sides of the tear.

If the facing damage is wider than ½-inch but is less than the width of the closure material, smooth the facing and make the repair as shown in *Figure 110*. You will apply two layers of closure material. The first layer is a single piece of closure material centered over the tear. The second layer is two pieces of closure material butted side by side so that the second layer covers the first layer with a one-inch minimum overlap.

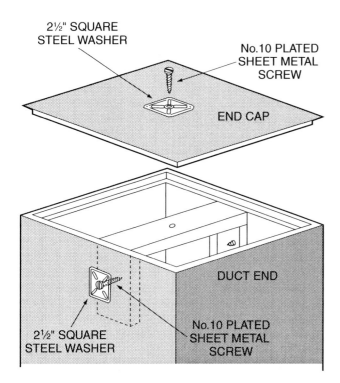

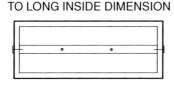

End cap reinforcement may be applied either parallel to the longest inside dimension or parallel to the shortest, depending on sheet metal and fastener usage required.

REINFORCEMENT PARALLEL
TO LONG INSIDE DIMENSION

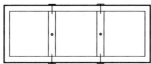

REINFORCEMENT PARALLEL
TO SHORT INSIDE DIMENSION

STAPLE AND TAPE END CAP IN PLACE.
SEE SECTION III, CLOSURES.

209F108.EPS

Figure 108 ◆ End cap channel reinforcement installed in duct section.

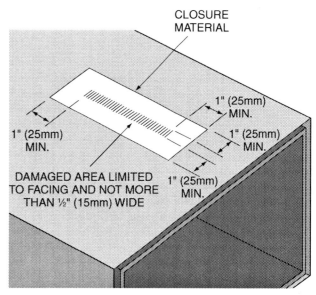

209F109.EPS

Figure 109 ◆ Repair of minor facing damage less than ½-inch wide.

209F110.EPS

Figure 110 ◆ Repair of minor facing damage ½-inch wide or wider.

14.2.0 Repairing Major Damage

Major damage occurs when the fiberglass insulation is damaged or displaced. When this happens, remove the damaged section and fabricate a replacement plug (*Figure 111*). To replace a damaged panel, follow these steps:

Step 1 Using a shiplap tool, cut out the damaged area and discard it. Cut a square or rectangle that is slightly larger than the damaged area. Doing this will allow you to fabricate the replacement plug more easily.

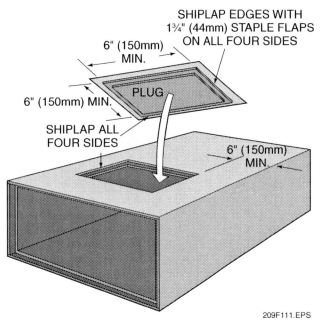

Figure 111 ◆ Repairing major damage to one panel.

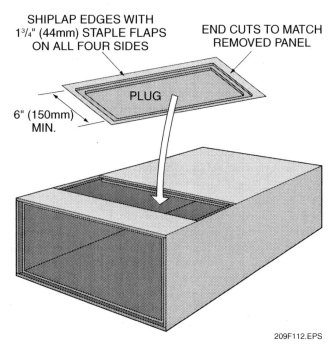

Figure 112 ◆ Repairing an entire panel.

Step 2 Measure and cut a replacement plug from fiberglass duct board. Shiplap the edges and leave a 1¾-inch staple flap on all four sides. Be sure the board thickness matches the thickness of the duct panel being repaired.

Step 3 Insert the plug and test the fit. The plug should fit smoothly and snugly.

Step 4 Staple the flaps and seal with the approved closure system.

CAUTION

To ensure a strong bond, use only approved closure materials and methods. Always read and follow the manufacturer's instructions. Do not use heat-activated tape over either pressure-sensitive tape or mastic and glass fabric closure systems. The heat may cause these other closure systems to fail.

If the damaged area extends to within six inches of the edge of the panel, cut out and repair the entire width of the damaged panel, using the previous steps as shown in *Figure 112*. Make end cuts on the plug to match the ends of the damaged panel.

14.2.1 Replacing an Entire Shiplapped Panel

Sometimes the damage to a duct panel is so great that you will have to remove the entire panel. This usually happens when something falls on the panel, causing it to crack or crease. To make this repair, follow these steps (*Figure 113*):

Step 1 Using a shiplap tool, cut away and discard the damaged panel.

Step 2 Measure and fabricate a new panel. Shiplap the edges of the new panel.

Step 3 Remove insulation from the new panel to form a staple flap on two sides as shown. See the detail drawings in the figure for 1⅜-inch and 11/16-inch minimum staple flaps.

Step 4 Clean out any debris from inside the duct panel and insert the replacement panel.

Step 5 Test the fit. It should be smooth and snug.

Step 6 Staple the flaps and seal with the approved closure system.

14.2.2 Replacing an Entire Square-Edged Panel

The process for replacing an entire square-edged panel is similar to that for replacing an entire shiplapped panel. Cut the replacement panel with square, instead of shiplapped, edges, using the following steps (*Figure 114*):

Step 1 Using a shiplap tool, cut away and discard the damaged panel.

Step 2 Measure and fabricate a new panel with square edges.

Step 3 Remove insulation from the new panel to form a staple flap on two sides as shown.

Step 4 Clean out any debris from inside the duct panel and insert the replacement panel.

Step 5 Test the fit. It should be smooth and snug.

Step 6 Staple the flaps and seal with the approved closure system.

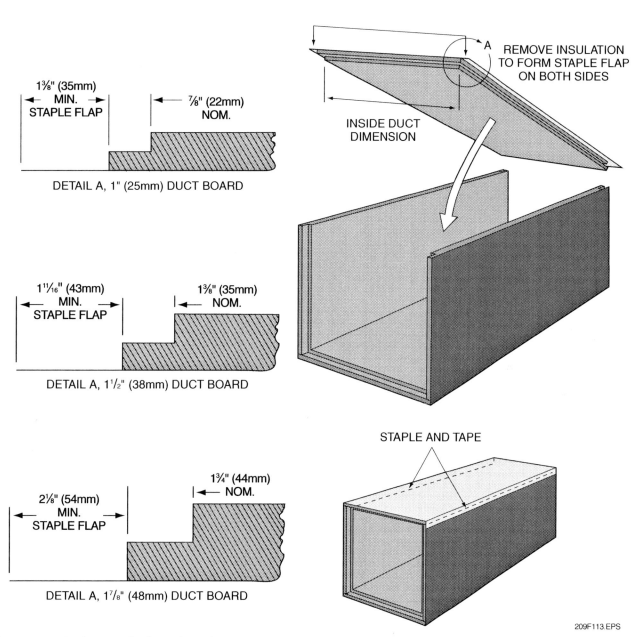

DETAIL A, 1" (25mm) DUCT BOARD

1⅜" (35mm) MIN. STAPLE FLAP

⅞" (22mm) NOM.

DETAIL A, 1½" (38mm) DUCT BOARD

1¹¹⁄₁₆" (43mm) MIN. STAPLE FLAP

1⅜" (35mm) NOM.

DETAIL A, 1⅞" (48mm) DUCT BOARD

2⅛" (54mm) MIN. STAPLE FLAP

1¾" (44mm) NOM.

REMOVE INSULATION TO FORM STAPLE FLAP ON BOTH SIDES

A

INSIDE DUCT DIMENSION

STAPLE AND TAPE

209F113.EPS

Figure 113 ◆ Replacing a shiplapped panel.

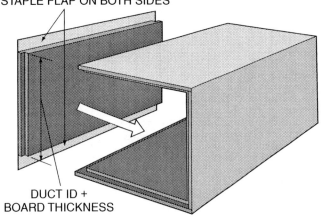

REMOVE INSULATION TO FORM
STAPLE FLAP ON BOTH SIDES

DUCT ID +
BOARD THICKNESS

STAPLE AND TAPE

209F114.EPS

Figure 114 ◆ Replacing an entire square-edge panel.

Review Questions

1. The strength and stiffness of rigid duct board is identified by its _____.
   a. IE rating
   b. EI rating
   c. IS rating
   d. SI rating

2. The two types of fiberglass duct are _____.
   a. 475 and 800
   b. 375 and 700
   c. shiplapped and square-edged
   d. rigid duct board and flexible round

3. ASHRAE standards recommend _____ for duct board based on climate zones.
   a. EI rating
   b. UL 181
   c. R-values
   d. leak class

4. The extended plenum supply system is ideal for _____ installations because it is sturdy and efficient and provides excellent acoustical and thermal performance.
   a. basement
   b. attic
   c. outdoor
   d. rooftop

5. The two methods for setting up the cutting tools in a grooving machine are called _____.
   a. stage one tooling and stage two tooling
   b. standard tooling and reverse tooling
   c. modified shiplap setting and standard shiplap setting
   d. forward groove setting and reverse groove setting

6. You can lay out fiberglass duct with either of two methods, _____.
   a. the V-groove method or the modified shiplap method
   b. the centerline method or the guide edge method
   c. the reverse tooling method or the standard tooling method
   d. the direct layout method or the indirect layout method

7. The formula for calculating stretchout is _____.
   a. SO = 2W + 2H + 2A
   b. SO = W + H + A
   c. SO = 3W + 3H + A
   d. SO = 2W + 2H + A

8. Tape tabs _____.
   a. may be placed either over or under pressure-sensitive or heat-activated tape
   b. may be placed under pressure-sensitive tape or over heat-activated tape
   c. may be placed over pressure-sensitive tape or under heat-activated tape
   d. must not be used with pressure-sensitive tape or heat-activated tape

9. When preparing the surface for a closure system, you can use _____ to clean the area.
   a. a paper towel
   b. a piece of sandpaper
   c. a cloth
   d. a brush

10. When fabricating a duct module, you should staple the long flap with _____.
    a. 1-inch staples approximately 2.5 inches on center
    b. 1-inch outward clinching staples approximately 2 inches on center
    c. 0.5-inch outward clinching staples approximately 2 inches on center
    d. 0.5-inch stainless steel staples approximately 2.5 inches on center

11. Turning vanes in fiberglass duct systems _____ pressure drop.
    a. increase
    b. eliminate
    c. reduce
    d. redirect

12. For square tap-ins, the preferred method is to use a _____.
    a. full-height fitting
    b. half-height fitting
    c. square-edged fitting
    d. shiplap-edged fitting

13. According to accepted practice, flexible round duct is used _____.
    a. as vertical risers in buildings of three or more stories
    b. for runouts from low-air velocity units
    c. to replace entire runs of trunk duct
    d. to make repairs in rigid round duct sections

14. Two methods for supporting rectangular fiberglass duct are _____.
    a. channels suspended from 12-gauge (minimum) wire or 22-gauge 1-inch wide (minimum) metal straps
    b. channels suspended from 12-gauge (minimum) wire or 1-inch wide (minimum) flexible fiberglass straps
    c. channels suspended from 10-gauge (minimum) wire or 1½-inch wide (minimum) metal straps
    d. channels suspended from 10-gauge (minimum) wire or ½-inch wide (minimum) metal straps

15. An entire duct panel should be removed if a damaged area extends to within _____ of the duct panel edges.
    a. 2 inches
    b. 4 inches
    c. 6 inches
    d. 8 inches

Summary

In this module, you were introduced to fiberglass duct which, like sheet metal duct, is used to fabricate the straight duct sections and fittings that make up an air distribution system. You learned how to work with fiberglass materials efficiently and safely. Included in this module are general guidelines for safely handling and disposing of fiberglass materials as well as a reference to a safe work practices guide published by NAIMA.

Like sheet metal, fiberglass must be properly measured, laid out, cut, and assembled before installation. Special tools, called grooving tools, are used to form the correctly sized shiplapped edges. You learned about both the hand and machine versions of these tools and practiced grooving the duct boards, cutting the shiplap edges, and assembling the duct modules and fittings. Assembling fiberglass duct sections is somewhat easier than assembling sheet metal duct sections. Instead of rivets and soldering and welding techniques, you use staples and closure systems made up of tape or tape and mastic.

Finally, you learned how to properly hang and support this material and how to make repairs. Though strong and lightweight, fiberglass duct can easily be damaged during and after installation. Fortunately, repairs are also easy, and you learned several quick and effective repair methods.

Notes

Closure system: Any one of the three types of sealing products used with fiberglass duct: pressure-sensitive tape, heat-activated tape, and mastic and glass fabric tape.

EI rating: A strength and rigidity rating assigned to rigid duct board that is based on a calculation called Young's modulus of elasticity.

End cap: A piece of rigid duct board with square edges, shiplaps, or a staple flap used to close the open end of a duct panel.

Fatigue test: A test in which a material is subjected to normal wear and tear to gauge how long it will remain intact and effective before beginning to break down.

FSK (foil-scrim-kraft): A vapor barrier made from layers of aluminum foil, fiberglass yarn or scrim, and kraft paper. FSK is also flame retardant.

Plenum: In an air distribution system, an enclosed volume of air that is at a slightly higher pressure than the atmosphere and is connected to several branch supply ducts. In a return air system, the air in the plenum is at a slightly lower pressure than the atmosphere and is connected to several return air grilles or registers.

Reverse tooling: On grooving machines, a tool setup in which the duct section is fabricated with a closure flap on the left-hand side of the duct board. See also *standard tooling.*

Runout: A term describing the connection between the main trunk line of duct and the diffuser, grille, or register.

R-value: A measure of the ability of a material (such as insulation) to slow heat transfer, expressed as a numerical rating. Also known as thermal-resistance value. The higher the rating number, the better the insulating properties.

Standard tooling: On grooving machines, a tool setup in which the duct section is fabricated with a closure flap on the right-hand side of the duct board. See also *reverse tooling.*

Staple flap: A piece of duct board facing from which the insulation has been removed that is pulled over a joint or seam and then stapled into place.

Stretchout: Describes a section of duct board or a fitting when it is laid out flat before fabrication; the amount of material required to form a duct panel or fitting.

Tap-ins: Square or round fittings made of fiberglass duct board, flexible duct, or sheet metal that tap into the main trunk duct to connect the main duct to a runout.

Torsion: The twisting of a building's structural member along its length by two equal and opposite forceful rotations at each end.

Turning vanes: Curved pieces of rigid duct board or sheet metal installed inside duct that turns sharply (as in a 90-degree turn) to provide consistent airflow and lessen pressure drop.

Vertical riser: In an air distribution system, a duct that extends vertically one full story or more to deliver air to branch ducts on different floors of a building.

Additional Resources

This module is intended to present thorough resources for task training. The following reference works are suggested for further study. These are optional materials for continued education rather than for task training.

Environmental Protection Agency Web site, www.epa.gov, "Biocontaminant Control," www.epa.gov/appcdwww/iemb/biocontam.htm, reviewed March 2008.

Environmental Protection Agency Web site, www.epa.gov, "Should You Have the Air Ducts in Your Home Cleaned," Indoor Environments Division, EPA-402-K-97-002, October 1997, www.epa.gov/iaq/pubs/airduct.html, reviewed March 2008.

Fibrous Glass Duct Construction Standard, Fourth Edition, 2001. Alexandria, VA: North American Insulation Manufacturers Association.

Fibrous Glass Residential Duct Construction Standard, Third Edition, 2002. Alexandria, VA: North American Insulation Manufacturers Association. Available at: www.naima.org/pages/resources/library/pdf/AH119.PDF.

NFPA 90A: Standard for the Installation of Air Conditioning and Ventilating Systems, 2002. Quincy, MA: National Fire Protection Association.

Standard for Closure Systems for Use with Rigid Air Ducts and Air Connectors, UL 181A, Third Edition, 2005. Northbrook, IL: Underwriters Laboratories, Inc.

Standard for Closure Systems for Use with Flexible Air Ducts and Air Connectors, UL 181B, Second Edition, 2005. Northbrook, IL: Underwriters Laboratories, Inc.

Working With Fiber Glass, Rock Wool, and Slag Wool Products, 2001. Alexandria, VA: North American Insulation Manufacturers Association.

Acknowledgments

The material and figures in this module have been adapted from *Fibrous Glass Duct Construction Standard* (fourth edition), *Fibrous Glass Duct Construction Standard* (third edition), and *Fibrous Glass Residential Duct Construction Standard* (second edition) with the permission of the North American Insulation Manufacturers Association (NAIMA). For more information about NAIMA and its publications, contact NAIMA, 44 Canal Center Plaza, Suite 310, Alexandria, VA 22314.

NAIMA assumes no responsibility and accepts no liability for the application of principles or techniques covered in this module. NAIMA makes no warranty of any kind, express or implied, or regarding merchantability or fitness for any particular purpose in connection with the information supplied herein. Authorities considering adoption of NAIMA standards should review all federal, state, local, and contractual regulations applicable to specific installations. These materials are not intended to address issues relating to thermal or acoustical insulation within and furnished as integral parts of HVAC equipment such as air handling units, coils, air cleaners, silencers, humidifiers, and terminal devices. Manufacturers of such equipment are responsible for design, specification, and installation of appropriate insulation components in their products so that thermal, acoustical, and indoor air quality requirements are met.

Resources & Acknowledgments

Figure Credits

North American Insulation Manufacturer's
Association (NAIMA), 209F02, 209F04, 209F05,
209F09, 209F10, 209F14–209F61, 209F67–209F114

Sketch by Michael Bergen, 209F03

NAIMA, plus sketch by Michael Bergen, 209F06,
209F07, 209F08, 209F11, 209F12, 209F13

NCCER makes every effort to keep these textbooks up-to-date and free of technical errors. We appreciate your help in this process. If you have an idea for improving this textbook, or if you find an error, a typographical mistake, or an inaccuracy in NCCER's Contren® textbooks, please write us, using this form or a photocopy. Be sure to include the exact module number, page number, a detailed description, and the correction, if applicable. Your input will be brought to the attention of the Technical Review Committee. Thank you for your assistance.

Instructors – If you found that additional materials were necessary in order to teach this module effectively, please let us know so that we may include them in the Equipment/Materials list in the Annotated Instructor's Guide.

Write: Product Development and Revision
National Center for Construction Education and Research
3600 NW 43rd St, Bldg G, Gainesville, FL 32606

Fax: 352-334-0932

E-mail: curriculum@nccer.org

Craft _____ Module Name _____

Copyright Date _____ Module Number _____ Page Number(s) _____

Description _____

_____

_____

_____

(Optional) Correction _____

_____

_____

(Optional) Your Name and Address _____

_____

_____

_____

Glossary of Trade Terms

ABS (acrylonitrile-butadiene-styrene): A plastic used to make pipes for drainage systems, storm sewers, and underground electrical conduit.

Absolute pressure: The sum of gauge pressure plus atmospheric pressure.

Air bend: Any punch and die that is pneumatically controlled.

Air-handling unit: In an air distribution system, a device that consists of the cooling and heating coils, fan or blower, filters, air mixing section, and dampers enclosed in a sheet metal casing.

Apex: In a tapered fitting, the point where lines drawn from the corners cross each other. The apex is located over the center of the base.

As-built drawing: A drawing of a completed installation that is used as a reference to make a drawing showing actual changes to the project.

Atmospheric pressure: The pressure exerted on the earth by the weight of the atmosphere.

Bearing plate: A piece of steel placed between a support and a pipe to protect the surface of the pipe from corrosion.

Black iron pipe: A noncoated pipe made of carbon steel.

British thermal unit (Btu): The amount of heat required to raise the temperature of 1 pound of water by 1°F.

Change order: A written order signed by the architect and the owner and given to the contractor after the execution of the contract that authorizes a change in the work.

Chord: A line drawn that goes from one part of a circle's circumference to another without passing through the circle's center.

Closure system: Any one of the three types of sealing products used with fiberglass duct: pressure-sensitive tape, heat-activated tape, and mastic and glass fabric tape.

Coefficient: A number or mathematical sign (plus or minus) that appears directly in front of a variable or grouping symbol.

Complement: In a right triangle, a non-right angle that, when added to the other non-right angle, equals 90 degrees.

Constant air volume (CAV) system: An air distribution system that provides a constant airflow while varying the temperature to meet demand for heated or cooled air.

Cosine: In a right triangle, the ratio between the side adjacent to an acute angle and the hypotenuse.

CPVC (chlorinated polyvinyl chloride): A plastic used in hot and cold water piping systems and in drainage systems.

Cubic feet per minute (cfm): A measure of the amount or volume of air in cubic feet flowing past a point in one minute.

Dew point: The temperature at which water vapor in the air becomes saturated and starts to condense into water droplets.

Downhand welding: A welding process that is done with both the weld axis and the base metal in a nearly horizontal or flat position.

Dry-bulb thermometer: A standard thermometer used to measure the amount of sensible heat in the air.

EI rating: A strength and rigidity rating assigned to rigid duct board that is based on a calculation called Young's modulus of elasticity.

Elevation: A drawing that shows the vertical elements of a building (exterior or interior) as a projection on a vertical plane. Elevation may also be the height of a point on land.

Empirical method: A method that relies on observation supplemented by experiments to develop a working practice, procedure, or formula.

End cap: A piece of rigid duct board with square edges, shiplaps, or a staple flap used to close the open end of a duct panel.

Enthalpy: The total heat content of the mixture of air and water vapor measured from a predetermined base.

Evaporator: The part of a refrigeration system in which the refrigerant is evaporated, thereby taking up outside heat and producing cool air.

Fatigue test: A test in which a material is subjected to normal wear and tear to gauge how long it will remain intact and effective before beginning to break down.

Flame cutting: The process of cutting metal using an oxyacetylene torch.

Flux: A fusible substance used in welding and soldering that helps fuse the metals and that prevents surface oxidation.

Footing: The concrete base on which a building's foundation walls are built.

Forming punch: Any punch and die set used for forming air bends.

Fractional percentage: A percentage that is less than one percent.

Free air delivery: The condition that exists when there are no effective restrictions to airflow (no static pressure) at the inlet or outlet of an air-moving device.

Free area: The total minimum area of the openings in an air inlet or outlet, such as an air diffuser, grille, or register, through which air can pass. It is usually expressed as a percentage of the total area.

FSK (foil-scrim-kraft): A vapor barrier made from layers of aluminum foil, fiberglass yarn or scrim, and kraft paper. FSK is also flame retardant.

Function: For an angle in a right triangle, the ratio of the sides and hypotenuse.

Galvanized pipe: A steel or wrought iron pipe that has been coated with a thin layer of zinc to protect it from corrosion.

Gauge pressure: The pressure measured on a gauge, expressed as pounds per square inch gauge or as inches in mercury vacuum.

Geometric method: A method of calculating bend allowance where the length of metal in the bend is represented as if it were part of a circle. The bend allowance is found by calculating the circumference of a circle with a radius equal to the radius of the neutral axis.

Gore: A tapering wedge-shaped unit on a sheet metal fitting.

Gouging: The process of cutting a groove in a metal surface.

Hypotenuse: The side of a right-angled triangle that is opposite the right angle.

Inches of water gauge: The American system unit of measurement that measures pressure.

Joist hanger: A metal hanger or strap used to attach a joist to a beam.

Joist: One of a series of parallel beams of timber, reinforced concrete, or steel used to support floor and ceiling loads.

KEVLAR®: A silky, man-made fiber that is stronger than steel and is used in a wide variety of products including tires, sporting goods, and protective gloves.

Laminate construction: A building process in which a product is made by bonding together two or more layers of material—for example, wood or plastic.

Lap joint: A joint formed by overlapping the edges of metal sheets.

Latent heat: The heat content of moisture vapor in the air at a constant temperature. Latent heat added to sensible heat equals total heat.

Lateral area: In a cylinder, the area of the surface that is perpendicular to the two circular sides.

Legend: A list that defines the pictorial symbols on a construction drawing.

Liquidus: The temperature at which a metal or an alloy becomes completely molten or liquid. Compare with *solidus*.

Load capacity: The amount of load that a structural member, support, pipe, or an electric circuit can safely withstand without failing.

Machining: The process of grinding metal by hand or by machine to shape the edge or grind it to a desired size.

Member: In structural engineering, any part of a structure that is complete in itself.

MIG (metal inert gas): A type of welding that uses an inert gas such as argon or carbon dioxide to shield the weld.

Mild steel: Ductile (flexible) steel that is nearly pure iron with very low carbon content.

Muriatic acid: The commercial name for a diluted solution of hydrochloric acid.

Neutral axis: In a bend, the point where the tensile and compression stresses disappear. The position of the neutral axis depends on the radius of the bend and the thickness of the metal. The neutral axis is also called the zero-stress line.

Oxidation: The reaction of metal to a chemical compound, such as oxygen, that can weaken a metal weld.

Pascal: The metric system unit of measurement that measures pressure.

Pitch: A slope (usually of a roof) expressed as a ratio of vertical rise to horizontal run.

Pitot tube: A device used with a manometer or other pressure-reading instrument to measure the velocity of air in a duct or water in a pipe.

Glossary of Trade Terms

Plain end (PE): Pipe manufactured without threads.

Plenum: In an air conditioning system, a sealed chamber at the inlet or outlet of an air-handling unit to which the duct is attached. In construction, the space between a ceiling and the structure above it.

Plenum: In an air distribution system, an enclosed volume of air that is at a slightly higher pressure than the atmosphere and is connected to several branch supply ducts. In a return air system, the air in the plenum is at a slightly lower pressure than the atmosphere and is connected to several return air grilles or registers.

Pointing up: The process of using the tip of a soldering iron to distribute molten solder across a seam. Most often used with vertical seams.

Polyethylene (PE): A plastic used to make flexible plastic pipe.

Polypropylene (PP): A tough plastic with excellent resistance to heat and chemicals.

Propeller fan: A fan that uses curved blades to direct the airflow in a parallel path to the propeller shaft. Propeller fans are often mounted in venturis.

Proprietary connector: Any connector or connection system developed by a company and registered with a trademark.

Psychrometric chart: A chart used to determine how air properties vary based on changes in any of the air properties included on the chart.

Psychrometry: The study of air and its properties.

PVC (polyvinyl chloride): A thermoplastic resin that resists chemicals and corrosion and is used for pipe fittings and piping in cold water systems and in sewage and waste lines.

Radial: An arc defined by the sweep of the radius of a circle.

Radiant heating system: A space heating system that consists of a network of hot water pipes installed in the floor or ceiling.

Ratio: The relationship in quantity, amount, or size between two or more things.

Reinforcement: Used at transverse connector joints or at intermediate points on certain sizes and pressure classes of ductrun sections to provide strength and reduce deflection of the duct sheet metal. In some cases, it enables the use of lighter-gauge sheet metal.

Relative humidity: The ratio of the amount of moisture present in a given sample of air to the amount it can hold at saturation, expressed as a percentage.

Reverse tooling: On grooving machines, a tool setup in which the duct section is fabricated with a closure flap on the left-hand side of the duct board. See also *standard tooling*.

Riser diagram: A diagram that shows the vertical path of plumbing, electrical wiring, or duct from one story of a building to another.

Roof pitch: The pitch of a roof generally expressed as the angle of pitch in degrees or as a ratio of the vertical rise to the horizontal run.

Runout: A term describing the connection between the main trunk line of duct and the diffuser, grille, or register.

R-value: A measure of the ability of a material (such as insulation) to slow heat transfer, expressed as a numerical rating. Also known as thermal-resistance value. The higher the rating number, the better the insulating properties.

Sal ammoniac block: A material used to clean and tin soldering iron tips.

Schedule: A detailed list of components to be installed in a building such as doors, windows, flooring, and wall treatments.

Scupper: An opening in a wall or parapet (the part of a wall that extends above the roof) that allows water to drain from a roof.

Seal class: Any one of three types of sealing methods used to control duct leakage on certain pressure classes of ductruns.

Section: A drawing that shows how a building or part of a building would look if it were cut open to show the interior.

Sensible heat: An amount of heat that, when added to the air, causes a change in temperature with no change in the amount of moisture.

Setback: The amount of metal that is saved when metal is bent around a radius (as opposed to bending it around a square).

Sheet metal blank: The measured piece of sheet metal from which a fitting is made.

Shielded metal-arc welding: A type of welding that uses the heat produced by an arc between a covered metal electrode and the items to be welded.

Sight line: A reference line established to enable the operator of a bending machine to position the metal accurately so that it may be bent on the tangent line, which is hidden from direct view by machine parts.

Sine: In a right triangle, the ratio of the side opposite an acute angle and the hypotenuse.

Slag: A grayish aggregate resulting from heat applied to metal—for example, from a blast furnace or from a welding process.

Solidus: The temperature at which a metal or an alloy first begins to melt. Compare with *liquidus*.

Solvent weld: A pipe-joining method in which a liquid solvent adhesive is brushed onto the pipe to form a strong bond between pipe sections or between a pipe and a fitting.

Specific density of air: The weight of 1 pound of air. At sea level at 70°F, 1 pound of dry air weighs 0.075 pounds per cubic foot.

Specific heat of air: The amount of heat required to raise 1 pound of air 1°F. Expressed as Btu/lb/°F. At sea level, air has a specific heat of 0.24 Btus per pound, per degree Fahrenheit.

Specific volume of air: The space 1 pound of dry air occupies. At sea level at 70°F, 1 pound of dry air occupies a volume of 13.33 cubic feet.

Splitter cone: The inverted cone in an exhaust weather cap that prevents gases from accumulating in the cone's peak.

Springback: The slight reshaping of metal toward its original shape after the release of stress.

Standard tooling: On grooving machines, a tool setup in which the duct section is fabricated with a closure flap on the right-hand side of the duct board. See also *reverse tooling*.

Staple flap: A piece of duct board facing from which the insulation has been removed that is pulled over a joint or seam and then stapled into place.

Static pressure tip: A probe used with a manometer or other pressure-reading instrument to measure air pressure in a duct system.

Static pressure: The pressure that is exerted uniformly in all directions within a duct system.

Stays: The metal bands that attach the weather cap to the curb. Stays are usually located in the corners, but sometimes in the center, of the weather cap.

Stretchout: Describes a section of duct board or a fitting when it is laid out flat before fabrication; the amount of material required to form a duct panel or fitting.

Submittal: A detailed drawing and description of an accessory or piece of equipment to be installed in a building.

Sweat solder: The process of coating metal pieces with solder, pressing them together, and then reheating the metal to melt the solder and form a bond.

Swinging: The movement of a metal blank, either up or down, as it is bent in a forming punch.

Tacking: A soldering or welding method used to hold metal pieces in place temporarily until a solid seam can be made.

Tangent line: On a sheet metal blank, the line that establishes where a bend is to begin and end.

Tangent: In a right triangle, the ratio of the side opposite an acute angle to the side adjacent to the angle.

Taper: To narrow from the base to the apex.

Tap-ins: Square or round fittings made of fiberglass duct board, flexible duct, or sheet metal that tap into the main trunk duct to connect the main duct to a runout.

Thermistor: An electrical resistor that varies sharply in a predictable manner with the temperature.

Thermocouple: A device used for measuring temperatures; it uses the voltage generated by the junction of two dissimilar metals at different temperatures to produce a reading.

Thermoplastic: A plastic that, when heated, becomes soft and pliable without losing any of its other properties and becomes hard and rigid when cooled.

Thermoset: A material such as a synthetic resin that hardens when heated and does not soften when reheated.

Thread engagement: On a length of threaded pipe, the area of threaded pipe that is screwed into the fitting.

Threaded and coupled (T&C): Pipe manufactured with either threads or coupling grooves.

Tie rod: Used with transverse joints and with intermediate reinforcement for certain sizes and pressure classes of ductrun sections to provide strength and reduce deflection of the sheet metal. In some cases, it enables the use of lighter-gauge reinforcement members.

TIG (tungsten inert gas): A welding method that uses a nonconsumable tungsten electrode to create an arc and weld metal together.

Tinning: The process of applying a thin coat of solder to metal parts to be joined. Also called wetting.

Torsion: The twisting of a building's structural member along its length by two equal and opposite forceful rotations at each end.

Total heat: The sum of sensible heat plus latent heat.

Total pressure: The sum of static pressure plus velocity pressure in an air duct.

Transite: A heat-resistant board made from Portland cement and nonasbestos fibers.

Transverse connector: A type of joint used to connect sections of rectangular ductruns.

Turning vanes: Curved pieces of rigid duct board or sheet metal installed inside duct that turns sharply (as in a 90-degree turn) to provide consistent airflow and lessen pressure drop.

Variable air volume (VAV) system: A system in which the temperature of each conditioned space is controlled by the temperature and/or volume of air supplied to that space.

Velocity pressure: The difference between total pressure and static pressure in an air duct.

Venturi: A ring or panel surrounding the blades on a propeller fan that is used to improve fan performance.

Vertical riser: In an air distribution system, a duct that extends vertically one full story or more to deliver air to branch ducts on different floors of a building.

Volume: The amount of air in cubic feet flowing past a point in one minute.

Wet-bulb thermometer: A standard thermometer with a saturated wick wrapped around a sensing bulb that measures the moisture content of the air.

Wetting: The process of applying a thin coat of solder to metal parts to be joined. Also called tinning.

Yield strength: The point at which metal no longer maintains its proportion of stress to strain.

Index

Index

shoe (boot), intersecting a taper on center, 3.23–3.26
    support for, 9.54, 9.55
        10-sided fiberglass, 9.28
    pipe, 8.19
Teflon®, 8.7
Temperature
    ABS pipe, 8.17
    comfort zone, 5.6, 5.14
    common HVAC value, 5.2
    CPVC pipe, 8.16
    dry-bulb, 5.10, 5.11
    duct joint tape, 9.22
    effects on relative humidity, 5.13
    fiberglass duct, 9.4
    heat-activated tape, 9.22
    pressure and volume relationships of gases, 5.3–5.5
    PVC pipe, 8.16
    for soldering, 7.2, 7.3, 7.4
    solvent cement, 8.8
    specific heat of air, 5.5, 5.6
    wet-bulb, 5.10, 5.11
Tension (stress), 6.2, 6.3, 6.5, 6.7
Thermal Insulation Manufacturers Association (TIMA), 4.3
Thermistor, 5.24, 5.32
Thermite, 8.7
Thermocouple, 5.24, 5.32
Thermometers
    air duct, 2.45
    digital and analog displays, 5.25
    dry-bulb, 5.7, 5.24, 5.25, 5.32
    electronic, 5.24, 5.25
    wet-bulb, 5.7, 5.24, 5.25, 5.32
Thermoplastic, 8.9, 8.27
Thermoset, 8.18, 8.27
Thickness of metal. *See* Gauge (metal thickness)
Thread, pipe, 8.2, 8.7, 8.27
Threaded and coupled end (T&C), 8.3, 8.27
Threading machine, 8.4
Threading process for pipe, 8.3–8.4
Ticket, 2.14
Tie rod
    attachments, 4.12, 4.13
    definition, 4.36
    and duct gauge, 4.6
    for fiberglass duct, 9.58–9.61, 9.67–9.69
    number per angle length, 9.64
    with tap-in fitting, 9.42
    termination, 9.60–9.61
TIG. *See* Welding, pipe, tungsten inert gas
TIMA. *See* Thermal Insulation Manufacturers Association
Tin, 7.2, 7.3, 7.16
Tinning, 7.2, 7.4, 7.9, 7.10, 7.19
Tip (point), on soldering iron, 7.4, 7.5, 7.9–7.10
Titanium, 8.6
Title block, 2.2
Tobacco, blowpipe system for, 3.18
Tolerance, 1.22, 8.8
Tooling, standard and reverse, 9.10, 9.13, 9.77
Tools
    to bend mating hat channel, 6.12
    end cut-off, 9.9
    fiberglass duct fabrication, 9.7–9.11
    iron

heating, 9.22, 9.24
        heat sealing, 9.11
        soldering, 7.4–7.11
    muscle memory working with, 3.4
    radial line development, 3.4
    reamer, 6.12, 8.3, 8.4
    safety, 3.3
    saws, 3.4, 8.2, 8.3
    shiplap, 9.9, 9.39, 9.42
        for soldering, 7.9, 7.10, 7.12–7.13
    stakes, 6.2
    staple gun, 9.11
    for tie rod termination, 9.60
    tube cutter, 8.2, 8.3
    wrenches, 3.4, 8.8
Torches, 7.5, 8.6, 8.7, 8.27, 9.24
Torricelli, Evangelista, 5.2
Torsion, 9.77
Training, 4.2, 4.3
Trammel point, 3.4
Transite, 7.11, 7.12, 7.13, 7.15, 7.19
Transition fittings
    one-way
        from duct board section, 9.42, 9.44
        from duct board section with shiplap, 9.42, 9.43
        from flat duct board with shiplap panel, 9.39, 9.41
    reinforcement, 9.69
    support for, 9.54, 9.56
Transverse section, 2.7
Trapezoid, 1.33
Trenching operations, 8.18
Triangle, right, 1.8, 1.9, 1.15–1.18
Triangular solid, 1.33
Trigonometric functions table, 1.17
Trigonometry, 1.15–1.18
Trough section, 6.2
Trunkline, 3.22, 5.15, 5.16. *See also* Tap-in
Tubing, copper. *See* Piping, copper
Turbulence, 5.16, 5.19

U
UL. *See* Underwriters Laboratories
Ultraviolet rays, 8.6, 8.17
UMC. *See* Uniform Mechanical Code
Underwriters Laboratories (UL), 4.3, 9.3, 9.4, 9.5, 9.23, 9.47
Uniform Mechanical Code (UMC), 4.2
Union Carbide Corporation, 8.15
Union fittings, pipe, 8.19–8.20
Unistrut®, 8.22
Utilities, 2.6, 2.14

V
Vacuum, 5.2, 5.3, 5.14
Vanes
    centrifugal blower, 5.18, 5.20
    turning, in duct system, 9.35–9.36, 9.37, 9.39, 9.40, 9.77
    velometer, 5.27, 5.28
Vapor barrier, 5.14, 9.2, 9.52
Variable air volume systems (VAV), 4.5, 4.36
Variables, 1.2
VAV. *See* Variable air volume systems
Velocity, of air in ductwork, 5.15–5.16, 5.27–5.28
Velometer, 5.27, 5.28
Vent, exhaust, 4.4

Ventilator, 3.12–3.14, 5.23, 8.5
Venturi, 5.21, 5.32
Vertex, 1.8, 1.20
Vibration, 8.23, 9.2
Vise, 7.9, 7.10
Vise grip, 7.11
Volume
    of air in ductwork, 5.15–5.16
    definition, 5.32
    formulas, 1.33
    pressure and temperature relationships of gases, 5.3–5.5
    specific volume of air, 5.5, 5.6, 5.11, 5.32

W